VALLEY OF GENIUS

VALLEY OF GENIUS

The Uncensored History of Silicon Valley,
as Told by the Hackers, Founders, and
Freaks Who Made It Boom

ADAM FISHER

GLEN COVE PUBLIC LIBRARY
4 GLEN COVE AVENUE
GLEN COVE, NEW YORK 11542-2885

TWELVE

NEW YORK   BOSTON

Copyright © 2018 by Adam Fisher

Cover design and illustration by Yves Béhar / Fuse Projects.
Cover copyright © 2018 by Hachette Book Group, Inc.

Hachette Book Group supports the right to free expression and the value of copyright. The purpose of copyright is to encourage writers and artists to produce the creative works that enrich our culture.

The scanning, uploading, and distribution of this book without permission is a theft of the author's intellectual property. If you would like permission to use material from the book (other than for review purposes), please contact permissions@hbgusa.com. Thank you for your support of the author's rights.

Twelve
Hachette Book Group
1290 Avenue of the Americas, New York, NY 10104
twelvebooks.com
twitter.com/twelvebooks

First Edition: July 2018

Twelve is an imprint of Grand Central Publishing. The Twelve name and logo are trademarks of Hachette Book Group, Inc.

The publisher is not responsible for websites (or their content) that are not owned by the publisher.

The Hachette Speakers Bureau provides a wide range of authors for speaking events. To find out more, go to www.hachettespeakersbureau.com or call (866) 376-6591.

Library of Congress Cataloging-in-Publication Data has been applied for.

ISBNs: 978-1-4555-5902-2 (hardcover), 978-1-4555-5901-5 (ebook), 978-1-5387-1449-2 (international trade)

Printed in the United States of America

LSC-C

10 9 8 7 6 5 4 3 2 1

Contents

BOOK THREE
Network Effects

EPILOGUE

for Kiri, forever and always

Stay hungry. Stay foolish.

—STEVE JOBS, QUOTING STEWART BRAND

Preface

I grew up in what is now known as Silicon Valley. Only in retrospect does it seem like an unusual place. As a kid, it seemed mostly suburban and safe and even dull, except for the fact that there were also a lot of nerdy hacker types around, and they kept things interesting in their own way.

The woman who lived next door to us ran the computer center at the local community college. In the late seventies my mom would drop me off there so I could play *Colossal Cave Adventure*—a text-only choose-your-own-adventure type game: YOU ARE STANDING AT THE END OF A ROAD BEFORE A SMALL BRICK BUILDING. AROUND YOU IS A FOREST. A SMALL STREAM FLOWS OUT OF THE BUILDING AND DOWN A GULLY. WHAT'S NEXT?

I never even touched the actual computer, as it was a mainframe and kept safely behind glass. I played *Adventure* by poking away at a so-called dumb terminal: a keyboard and teletype machine at one end of a long cord. *Adventure* was primitive but fun, and still the best babysitter I've ever had:

KILL DRAGON, I typed.

WITH WHAT? YOUR BARE HANDS? rattled the printer.

YES, I pecked.

CONGRATULATIONS! YOU HAVE JUST VANQUISHED A DRAGON WITH YOUR BARE HANDS! (UNBELIEVABLE, ISN'T IT?)

A few years later, in 1979, the family next to our next-door neighbors bought their own computer: an Apple II. It was astonishing. You could touch it, take it apart, modify it. It used an ordinary color TV as a screen. I vividly remember helping to insert a chip into the motherboard that enabled lowercase.

On the Apple II, the *Adventure* was conducted in both upper- and lowercase. *Wow!* But it wasn't just text-based adventure games. The

Apple II could also play video games. *Little Brick Out* was the classic: a copy of Atari's *Breakout* arcade game—and in one crucial dimension even better than the original. *Little Brick Out* was written in BASIC, and thus the source code could be examined and even vaguely understood. Take line 130, for example:

130 PRINT "CONGRATULATIONS, YOU WIN."

That line could be rewritten to take advantage of the lowercase chip. Like so:

130 PRINT "Congratulations, you win."

Or even:

PRINT "Congratulations, Adam!!! You have just vanquished *Little Brick Out* with your bare hands!"

I learned that with a little hacking, one could make the computer say—and do—*anything*.

And I was hooked.

The book you are holding in your bare hands is a compendium of the most told, retold, and talked-about stories in the Valley. They're all true, of course, but structurally speaking, most of the stories have the logic of myth. The oldest of them have acquired the sheen of legend. Doug Engelbart's 1968 demonstration of his new computer system is known as the Mother of All Demos. Steve Jobs and Steve Wozniak have become archetypes: the Genius Entrepreneur and the Genius Engineer. Collectively, these tales serve as the Valley's distinctive folklore. They are the stories that Silicon Valley tells itself.

To capture them, I went back to the source. I tracked down and interviewed the real people who were there at these magic moments: the heroes and heroines, the players on the stage and the witnesses who saw the stories unfold. Almost everyone is still alive—many are, in fact, still young. And I had them tell me their stories: What happened? What did you see? What did you feel? What does it all mean?

I interviewed more than two hundred people, most of them for many hours. Along the way, I learned a lot of things. The first surprise was the range of types of people whom I encountered. Silicon Valley grew from a few suburban towns to encompass the cities around it, and that growth was fueled by a rather remarkable diversity. There's no Silicon Valley ethnic type, per se. Silicon Valley is racially diverse—it is the proverbial melting pot—although it's also true that black people are still far and few between. Women are also underrepresented, although there are many more than one might imagine. And there's no typical age, either. Silicon Valley focuses on its young—that's where the new ideas usually come from—but it's also been around for a long time.

However, there were some commonalities. Almost to a person, their childhoods sounded like mine. There was an early exposure and then fascination with computers—usually because of computer games—which ultimately led to a fascination with hacking, computer science, or even electrical engineering. The names of the games change, but the pattern remains the same.

After the interviewing came the transcribing. I had hundreds of hours of transcripts, millions of words of memories. The printouts filled an entire bookcase in my office. Then came the real work: I cut the transcripts together as one might so many reels of film.

The editing process was often literally done with printouts, a pair of scissors, and a roll of tape. I'm a big fan of computers: I can't write without one. But the old-fashioned way of editing still works best, I've found. Artificial intelligence is nowhere near good enough to transcribe an interview with any accuracy, either. And of course there is no automating the process of reporting: tracking people down and convincing them to talk to me.

Shoe leather, sharp scissors, and Scotch tape made this book possible—an irony that didn't always seem funny to me. The process took four solid years of full-time work and constant focus. By another measure, it has taken even longer. I've been gathering string—banking interviews—for the past decade.

It would have been much easier to simply write a history of Silicon Valley instead of to painstakingly construct one, but I felt that would have missed the point. These aren't my stories: They're the collective property of the Valley itself. I wanted to disappear because I want you, the reader, to hear the stories as I had heard them: unfiltered and uncensored, straight from the horse's mouth. The stories are told collaboratively, as a chorus, in the words of the people who were actually there.

There's not much of me in this book, but of course all journalists have their biases, their point of view. Here's mine: I think that what's most interesting, most important, about Silicon Valley is the culture, that aspect of the Valley that gets under people's skin and starts making them think—and act—different.

I was indoctrinated at computer camp (at the time, 1982, there was only one). There I met a counselor who signed his postcards home with a squiggle, like this: ⋀⋁⋁⋁. The squiggle is the symbol for a resistor—the electrical kind that one might see on a circuit diagram. Silicon Valley is a sheltered place, but the intense vibrations emanating from Berkeley and San Francisco did penetrate.

I found Mr. Resistor fascinating. He was working on a secret project for something he called "a start-up company," and it was going to "change the world." The next summer Mr. Resistor hitchhiked to my house and gave me and my father the full vision: "One day," he predicted, "everyone will have a mobile telephone—in their car." He even had a color brochure that showed a man happily chatting away on a handset plugged into the dashboard of a Honda Accord. "Because soon car phones won't just be for rich people," he explained. Then he said his good-byes and left, via the public bus system. I was skeptical: Would the future be invented by a hippie who didn't even own a car? My father was less so: "Maybe," he mused.

I left Silicon Valley for college and then returned a dozen years later as an editor for *Wired* magazine. There I started to notice something odd: The stories about Silicon Valley emanating from the New York media world were vastly different from those stories that I had heard at sleepaway camp and in computer rooms, and then later in barrooms and at Burning Man. There was a cognitive dissonance there. *New York just doesn't get it*, I told myself.

Eventually I came to understand that it all came down to perspective. The mainstream media sees Silicon Valley as a business beat, a money story: Who's up and who's down in the new economy? Who's the latest billionaire? Those are valid questions, maybe even interesting ones—but not to me.

In the Silicon Valley where I'm from, the stories were almost never about money. They were tales about resistance, heroism, and struggle, yarns about the creation of something out of nothing—and the derring-do required to pull such a feat off. In short, they were about dragon slaying. That's still true, at least in the Silicon Valley I know. Those were the stories that got me excited. And they still do.

I'm not saying there isn't an economic story to be told. In fact, I think that we are witnessing the greatest transition since the industrial revolution. A new economy—the information economy—is being created, and the center of that new economic order will be Silicon Valley. And if that's not the business story of the century, what is?

Still, the bigger question, in my humble opinion, is how that transformation will transform us. We begin to see the answer in the culture that's being created in Silicon Valley, now. It's future obsessed and forward thinking. It's technical and quantitative. It's market oriented. It's simultaneously practical and utopian. It's brainy, even in its humor. In short, it's a nerd culture. And of course there have been nerds since time immemorial. Leonardo da Vinci was a nerd. Ben Franklin was a nerd. Albert Einstein was the quintessential nerd. But the new thing is that the nerd culture is becoming the popular culture.

Evidence for that idea, once grokked, is everywhere. Exhibit A: *The Big Bang Theory*—a show by, for, and about nerds—is one of the highest-rated and longest-running television sitcoms ever. Exhibit B: *The Martian*'s unlikely journey from self-published NASA fan fiction to blockbuster hit. Exhibit C: The fact that *xkcd*, a web-based comic devoted to "romance, sarcasm, math, and language," has any audience at all.

Even more astonishing, at least to me, is that this new popular culture is a youth culture. The kids who are searching for an exciting life no longer want to be rock stars, or rap stars, but rather Silicon Valley–style tech stars. They want to be Steve Jobs, or Mark Zuckerberg, or Elon Musk.

As readers will discover, technology entrepreneurs have never made particularly good role models. Atari founder Nolan Bushnell essentially invented the role of the twenty-something Silicon Valley CEO almost a half century ago—and he may have been the baddest bad boy that the Valley has ever seen. His protégé, Steve Jobs, was not much better. At the same time, this new nerd culture is the best possible news for our collective future, given the awesome challenges ahead. Soon there will be nine billion people crowding this warming planet, and each one will come equipped with a supercomputer in their pocket. So I'm optimistic, bullish even. Who better to inherit the Earth, at a time of crisis, than a generation obsessed with science and engineering?

It's pretty clear where this new nerd culture came from—it came from the same place that the money did: Silicon Valley. And what is a culture? There's no mystery there, either. A culture is simply the stories that define a people, a place. It's the stories we tell each other to make sense of ourselves, where we came from, and where we are going.

And here those stories are, between two covers. Together they comprise an oral history of Silicon Valley.

Adam Fisher
Alameda, California
June 2018

Silicon Valley, Explained

*The story of the past, as told by the people of
the future*

Silicon Valley is a seemingly ordinary place: a suburban idyll surrounded by a few relatively small cities. So how is it that the Valley keeps conjuring up the future? Well, take it from this historian: The difference lies in how the people here tell their story. They see their own history in a way that's subtly different from the way history is generally taught. It's not the historical materialism of Karl Marx—the story of exploiters and the exploited. Nor is it the romanticized history of bards and poets—one tragedy after another. History, to Silicon Valley, is the story of the new versus the old: how one technology is vanquished or subsumed by the next. Traditionally, history itself starts with the written word, the ur-technology, the first medium. But in the Valley we fast-forward to the invention of the computer—the meta-medium that absorbs all media before it. Hard on the heels of the computer comes the invention of the internet, the network of computer networks. Next up is the always-on, always-connected smartphone—the internet in your pocket. And so on. In this telling, history is not something that happens to people. It's made by people. And, in our era, it's made in Silicon Valley.

Steve Jobs: I do think when people look back in a hundred years, they're going to see this as a remarkable time in history. And especially this area, believe it or not.

Steve Wozniak: Creativity is high here—it's okay to have dreams and think about them and think maybe you could make them. Here more than other places.

Ron Johnson: There's the Bay on the east and the foothills on the west, and they're about five miles apart, and the entire Valley kind of runs from

Stanford to the south toward Cupertino, and to the north toward San Francisco.

Jamis MacNiven: The first gold rush: That's what launched San Francisco. It was a sleepy little town of twelve hundred, and then three years later there's three hundred thousand people in the Bay Area.

Scott Hassan: Back in the 1870s, California, for some reason, decided to enact a law that prohibits employers from suing former employees for going on to a competing company.

Brad Handler: The reason goes back to the Spanish control of California—through Mexico. The law in Spain, through Mexico, to the California territory did not allow what we now call "covenants not to compete." Most other states will enforce a covenant not to compete. But in California covenants not to compete are nonenforceable, period.

Scott Hassan: That's a big deal. A lot less innovation happens on the East Coast because they believe in these noncompetes.

Jamis MacNiven: We had railroads, real estate, aviation. We had oil. Hollywood was a giant rush. Oranges. So we're used to that gold rush mentality, and we've had a lot of rushes.

Rabble: It's easy to see the web companies or this generation of companies and think that the stuff going on in Silicon Valley is new, but it's not. The reason there is stuff going on here with technology is because this is where radios were designed during World War One.

Dan Kottke: Lee de Forest! He invented the vacuum tube. It was called the Audion and it was the first amplification device. It was a huge step forward. His company was called the Federal Telegraph Company, and there is a bronze plaque in downtown Palo Alto where his lab was. Lee de Forest, he's right up there with Edison in my mind.

Jamis MacNiven: The vacuum tube allowed for amplification of sound, which led to the music industry and also to Adolf Hitler—Adolf Hitler was big on the radio. But they also discovered that the vacuum tube could be used as a switch: On. Off. On. Off.

Steve Jobs: Before World War Two, two Stanford graduates named Bill Hewlett and Dave Packard created a very innovative electronics company—Hewlett-Packard.

Jim Clark: You can name probably thirty, forty companies that started out at Stanford now—big ones.

Ron Johnson: Stanford really is the epicenter of the Valley, and most of the companies, the people of these companies, had a connection to Stanford.

Jamis MacNiven: Stanford was the first major university that reached out, in a big way, into the business community and said, "Hey, we have this open-door policy. Come on in, do business: Go out and do business!"

Jim Clark: Contrast that with the Ivy League. They had their nose in the cloud: "We are beyond business. Business is dirty. We are not talking about applications. We are talking about advancing knowledge and research."

Steve Jobs: Then the transistor was invented in 1948 by Bell Telephone Laboratories.

Jamis MacNiven: And that allowed us to turn switches on and off even faster, faster-faster-faster. In the binary world, on/off is very important.

Steve Wozniak: William Shockley invented the transistor and that was going to be the growing industry.

Andy Hertzfeld: In some ways Silicon Valley itself was an accident of William Shockley being born here.

Steve Jobs: Shockley decided to return to his hometown of Palo Alto to start a little company called Shockley Labs or something. He brought with him about a dozen of the brightest physicists and chemists of his day.

Po Bronson: At Shockley Semiconductor some people who felt mistreated, who felt mismanaged, left. They went over to a new funder, Fairchild, who said, "Yes, we'll take you guys over here."

Steve Jobs: Fairchild was the second seminal company in the Valley, after Hewlett-Packard, and really was the launching pad for every semiconductor company in the whole semiconductor industry which built the Valley.

Po Bronson: Then Bob Noyce and Gordon Moore left Fairchild to start Intel, and that just didn't happen anywhere else in the country. Labor laws were different in other states.

Brad Handler: It's just a public policy difference from hundreds of years ago.

Jamis MacNiven: Then Moore created this law he's famous for: computer power doubling every eighteen months.

Alvy Ray Smith: Moore's law went into play in 1965, and from then on the more powerful computer you have, the smaller they got.

Gordon Moore: By making things smaller, everything gets better at the same time. The transistors get faster, the reliability goes up, the cost goes down. It's a unique violation of Murphy's Law.

Jamis MacNiven: And Moore keeps saying, "Well, that's about to end," and then it doesn't end.

Steve Wozniak: Transistors were subject to Moore's law, and boomed into chips, and bigger chips, that could do more and more over time. Silicon Valley is called "Silicon Valley" because of the material silicon that makes hardware chips. The area was growing economically.

Alvy Ray Smith: And then the venture capital idea came along. And as far as I can tell that idea had never been tried before. There were bankers who could loan you money, but they wanted guarantees that could get it back. Venture capitalists expect failure. They discovered that they could gamble big with maybe ten firms…

Marc Porat: Just throw a lot of things into the funnel, attract a lot of smart people, and make them interact in unmanaged ways. Go chaotic on the whole system of innovation. Just let it happen. Cull the good stuff from the bad stuff. Find a relatively efficient—not perfect, but relatively efficient—solution. The thing blows up? Nobody cares. If it goes big, then they want to stuff as much money into it as possible.

Alvy Ray Smith: And if one of them hit, it would pay for all of the failures. That was a new idea.

Steve Jobs: People started breaking off and forming competitive companies, like those flowers or weeds that scatter seeds in hundreds of directions when you blow on them.

Po Bronson: That is pretty legitimately the true origin of the culture here. The I-need-to-leave-my-big-thing-to-start-something-small story.

Ezra Callahan: It's a place where people with an idea and some talent can make something huge out of something small. It's the success stories of young entrepreneurs with no particular business experience who, within a matter of months, become industry-creating technological celebrities.

Po Bronson: That became an imprint that got repeated and repeated and repeated.

Steve Jobs: And that's why the Valley is here today.

Carol Bartz: The future doesn't just happen.

Ev Williams: Sparks happen and then it just erupts. I think a lot of it has to do with networks. Networks—they get bigger.

Ray Sidney: There's this network effect: Sharp people come to be with other sharp people. People come to Silicon Valley just to be with these movers and shakers and brilliant engineers and product designers and marketers and sales folks and whoever.

Lee Felsenstein: The story of Silicon Valley is the story of networks. There was never any centralized place. They were all what they call local maxima, little mounds of people here and there, and people move between them, that's the important thing. So it's a decentralized set of networks with mobility among them.

Marissa Mayer: I've heard both the founding stories of Google and Yahoo, and for both those companies, the founders didn't even have to get into a car. They could literally go to the law office, the venture capitalists, the bank…on a bike. It's all that close together.

Orkut Büyükkökten: The network in Silicon Valley helps you to connect with people who can create that magic, right? So if you have a great idea, you can meet up with a good designer or engineer and then maybe with an angel who would support it and then you can make it happen.

Jerry Kaplan: And for every one of them there are a thousand people who came here with a good idea and burned through their savings without being successful.

Marc Porat: People fail in one, but they learn enough so that they succeed in another. And success breeds success.

Orkut Büyükkökten: I think all over the world people have great ideas, but they don't necessarily have the means to implement them. In Silicon Valley it's a lot easier to make an abstract idea and turn it into reality.

Carol Bartz: There are more people attracted to the concepts and there are more tools. Computing has gotten stronger, faster. Data collection? Stronger, faster. Video? Stronger, faster. And so we have the basics now to make increasingly fast changes.

Andy Hertzfeld: Once you have a pipeline going, the pipeline wants to be filled.

Biz Stone: The infrastructure is here: the real estate people, the legal people, the you-name-it people. They get start-ups, so it's easier: "Oh, okay, you're a start-up. So, here you go." It's just easier to do start-up stuff, because everyone in the whole ecosystem knows about start-ups.

Guy Bar-Nahum: Silicon Valley is not one thing. You have layers. You have the engineers and you have the bankers. But the people who are active ingredients, the troublemakers, the Tony Fadells, the Steve

Jobses—they're really lusting for fame. They want to be relevant, they want to be recognized, they want to be famous, and when they get that it all ties to money, of course. Big money.

Jeff Skoll: We're living in an age of ever-growing celebrity status for entrepreneurs.

Jerry Kaplan: That's what makes it tick. Without that nobody comes, nobody invests, and it doesn't work.

Rabble: One of the things that I find fascinating is that people from the outside perceive it as supercompetitive. They see it as a kind of über-capitalism. But when you look at people who work in it, the identity and the communication, the vibration, is more with communities, networks of people, than the companies. You really often see teams of people who will go from one company to the next company.

Aaron Sittig: The best way to think about Silicon Valley is as one large company, and what we think of as companies are actually just divisions. Sometimes divisions get shut down, but everyone who is capable gets put elsewhere in the company: Maybe at a new start-up, maybe at an existing division that's successful like Google, but everyone always just circulates. So you don't worry so much about failure. No one takes it personally, you just move on to something else. So that's the best way to think about the Valley. It's really engineered to absorb failure really naturally, make sure everyone is taken care of, and go on to something productive next. And there's no stigma around it.

Scott Hassan: It's like a huge, huge, huge company, where there's no one running it. And you can do whatever the fuck you want! And the only thing that matters is whether the market cares about what you produce or not. That's the only thing that matters. And you will go away if you're not producing something of value, period, and nobody decides it for you other than this nameless market. Fortunately as the internet grows, the market gets even bigger and bigger and bigger.

Aleks Totić: We have had a lot of help from Moore's law.

Chuck Thacker: Moore's law is like compound interest. It gave us exponential growth.

Aleks Totić: But that doesn't explain it. The rest of the world gets the same help from Moore's law.

John Battelle: At its core, Silicon Valley is a culture.

Dan Kottke: Silicon Valley was a nerd-driven culture, and the engineers were the priesthood.

John Battelle: It was a culture of hardware hackers and software writers, who were generally older, midthirties to fifty, generally bearded, generally grimy, super fucking smart—what we would call on the spectrum now, but back then there wasn't a term for it—generally that was the heart of the Valley culture.

Jim Warren: For the most part folks were doing it as a labor of love, and it was about the excitement of pushing the frontier.

Tiffany Shlain: There was something incredibly exciting about being at the frontier of this new medium and this new culture that we were creating.

Steve Jobs: In the seventies and the eighties the best people in computers would have normally been poets and writers and musicians. Almost all of them were musicians. A lot of them were poets on the side. And they went into computers, because it was so compelling, because it was fresh and new. It was a new medium of expression.

Po Bronson: You have a culture, riding on top of a technology that's accelerating superfast. And so that culture gets replicated, just as much as the technology does. And you put those two together and you get something distinctly unique.

Steve Jobs: This is the only place in America where rock and roll really happened, right? Most of the bands in this country outside Bob Dylan in the sixties came out of here: from Joan Baez to Jefferson Airplane to the Grateful Dead. Everything came out of here: Janis Joplin, Jimi Hendrix, everybody. Why is that? That's a little strange when you think about it. You also had Stanford and Berkeley, two awesome universities drawing smart people from all over the world and depositing them in this clean, sunny, nice place where there's a whole bunch of other smart people and pretty good food. And at times a lot of drugs and a lot of fun things to do, so they stayed.

Fred Davis: As Detroit was to the car, the Bay Area was to LSD. This is where it was all coming from. It was definitely part of the culture. Acid was a university lab product since the fifties.

R. U. Sirius: Then the Youth International Party—or Yippies—a group combining psychedelic culture with the revolutionary New Left, sprang up and became very powerful in 1968. And the phone phreaks were into the Yippies. And the Yippies were into the phone phreaks. And Captain Crunch somehow figured out that he could make a free long-distance phone call anywhere.

Captain Crunch: You blow a whistle and if it's just exactly 2600 hertz that's the tone that operators use. With that you have the same power that an operator has.

Steve Jobs: It was miraculous. Blue boxing, it was called.

R. U. Sirius: It was an idea that Steve Jobs and Steve Wozniak and Lee Felsenstein at the Homebrew Computer Club loved. That's the connection between the hacker culture and the counterculture going all the way back to the seventies. That's not really widely known.

Brad Templeton: As a personal computing nerd, I think of Nolan Bushnell as the first round. Yes, there was much stuff before: There was Fairchild.

Steve Wozniak: Engelbart! When you get to computers, Engelbart is great.

Brad Templeton: But I count Atari actually as the beginning.

Don Valentine: Steve Jobs was the son of Nolan Bushnell. Not literally, but he evolved the same way, and Apple was in many ways an evolution of Atari. A lot of Steve's original thinking came from Nolan.

Steve Wozniak: Atari, yes, they started an industry of arcade games, but what was the first arcade game ever that was software and what was the first time it was color? The Apple II. That was a huge, huge step.

Larry Brilliant: Steve used to write me when he started Apple. I would get these letters, and then one day he just called me. He said, "Do you remember when we would say 'power to the people'? That's what I'm doing, I'm giving power to the people. I'm building a computer that every person can put on their desktop, and I'm going to get rid of the high price of the mainframes."

Bruce Horn: There was so much discussion and so much talk and so many brilliant ideas and everyone just came together.

Kristina Woolsey: But we've also got our hippies, our beatniks, our gays... I'm an old-timer around here, and that's just in my lifetime.

Ray McClure: The sharing of ideas and merging of alternative cultures, alternative lifestyles, people's ability to express themselves, whoever they are: I think that that's the real heart of the creativity. That's what has made it a fertile place for new ideas.

John Battelle: It was the people who kind of first gathered around the Atari and the Apple II, and later around the Mac, and then later around the CD-ROM revolution and the multimedia revolution of the late eighties and early nineties. And they were the same people.

Jamis MacNiven: We had the *Whole Earth Catalog* as bible, then it went electronic, The Well, and it became one of the backbone models for the internet. These guys around here invented the new world.

John Battelle: They also gathered around science fiction, right?

Steve Wozniak: Science fiction leads to real products, but first you've got to deal with the laws of physics, and ask, "What's it going to cost?"

John Giannandrea: I'm fond of this book by Arthur C. Clarke, *Profiles of the Future*. He splits the book into two sections. One section is things that will happen, like self-driving cars. The other is things that will be surprising if it happens, like time travel. And the thing is the kind of people who worked at General Magic or Netscape would be like, "Show me the technical detail that makes this possible—and sign me up!"

John Battelle: They were these really smart engineering types who had nonstandard political views. They leaned more libertarian, sometimes anarchist. They were far more left than right, but at its core, it was a culture of people who believed that something really fucking big was going on: that it was bigger than them, that it was bigger than the products that they were working on, that there were the seeds of a true movement happening.

Kristina Woolsey: It wasn't about being able to talk on a less-expensive phone line or being able to do video-chats or having a cell phone with the power of a computer. The bigger question was always "How is the digital revolution going to change society? And people?"

Jamie Zawinski: We weren't building a toy. We were building a communications medium. We were letting people connect to each other in a way that they hadn't been able to before. It was the opposite of television; it was giving people a voice.

John Battelle: But it was not until the Web 1.0 boom that there was a sense that this was only a Silicon Valley thing. It was just that this is where most of the companies were.

Jordan Ritter: The defining moment was the rise of Google. If there was ever a paragon of the engineer thumbing their nose at conventional management, conventional software development, conventional everything—it was them. That's the point where it shifted: when the engineers were treated as valuable as they really are.

Ev Williams: One of the interesting shifts that has happened is the move to San Francisco. Silicon Valley until the dot-com boom wasn't in San

Francisco. It was in Palo Alto and Mountain View, because tech companies weren't built in the City.

Steve Wozniak: The hardware industry has sort of settled down and all the new things that catch our attention—the apps and Ubers—they've kind of chosen to move up to San Francisco. Silicon Valley includes San Francisco now.

Biz Stone: San Francisco is the cool place to be.

Ev Williams: My theory is that as these tech companies have become more culture makers, then they gravitate toward the City, because that is who makes culture—people who live in cities. And now it's maturing to a degree where you can be in tech and be a pure businessperson and there's lots of them. And lots of lawyers, and now there's lots of artists and those people are part of tech, too. And they live in cities and they live in San Francisco.

Louis Rossetto: The nineties marks this transition where all the original technical breakthroughs started to bite in society at large. There is a switch happening. The blog gave way to the tweet and to Facebook.

Jim Levy: To me, by the nineties it wasn't fun anymore. Though I'm sure a lot of people had a lot of fun. I'm sure Mark Zuckerberg had a lot of fun putting his thing together. I know the Google guys had a lot of fun. They are still having a lot of fun.

Steve Wozniak: The iPhone is the coup de grâce. When you use your iPhone you still have a personal computer—but your PC is in a data center. It's taking data off hard disks, analyzing it and determining what to show, and it sends it to your iPhone to show you, so you really have a PC but you don't see it, it's not yours, so it's not personal. You have a computer but it's not personal.

Sean Parker: And then it becomes the post–social media era. It's all the people who would have become investment bankers who want to go start internet companies, and it's a purely commercial, purely transactional world. It's just become this transactional thing, and it's attracted the wrong type of people. It's become a very toxic environment. A lot of people have shown up believing, maybe correctly, that they can cash in. But that's Silicon Valley the ATM machine, not Silicon Valley the font of creativity and realization of your dreams.

Biz Stone: Only in Silicon Valley can you be like, "Yeah, we would like $10 million and we'll sell you a percentage of our theoretical company that may one day have lots of profits and if we lose all the money we don't have to give it back to you. And maybe we'll start something else." In what crazy world does something like that exist? Like wait, you can just blow the money and then you don't owe it back, just wash your hands clean, done? "Sorry about that. Sorry, I spent all your money. Oh, well." I mean it's just crazy. And not only that, here's another scenario. Here's what some people do. They say, "We need $25 million, but you know, just so we can stay focused my cofounder and I each need $3 million of that money in our bank accounts. Then we don't have to worry about bills so we can really focus. Okay?" And then they blow the money and they say, "Oh, well, that didn't work out, but we're still keeping the $3 million each, so now we're rich." What the hell? That is crazy. So, it's a crazy world. This is like some kind of nutty place where you can do that kind of stuff.

Steve Perlman: We, today, in Silicon Valley, are at a similar level of maturity as the automobile industry in 1950. What happened in Detroit was that the winners ended up getting consolidated to the Big Three, and there was no other player that could enter that industry that could possibly compete. The people that succeed in these large companies are the ones that make the safe bets. The people that don't rock the boat, the people that get along with everyone. The people that don't do things which show up the other lifers there, and disrupt their opportunities to advance in the organization. Those are the people that gravitate to and stay at the larger companies.

Steve Wozniak: That turns me off a lot. Now your importance is measured by what company you are in and what you do. I don't like that.

Steve Perlman: Is there any way that Silicon Valley can recover what it's lost?

Andy Hertzfeld: The most important thing really is the motivation. Why are you doing what you're doing? That seeps into the product at every level, even though you think it might not. Your basic values are essentially the architecture of the project. Why does it exist? And in Silicon Valley there are two really common sets of values. There are what I call financial values, where the main thing is to make a bunch of money. That's not a really good spiritual reason to be working on a project, although it's completely valid. Then there are technical values that dominate lots of places where people care about using the best technique—doing things right. Sometimes that translates to ability or to performance, but it's really a technical way of looking at things. But then there is a third set of values that are much less common: and they are the values essentially of the art world or the artist. And artistic values are when you want to create something new under the sun. If you want to contribute to art, your technique isn't what matters. What matters is originality. It's an emotional value.

Chris Caen: This is a little clichéd and eye-rolling—and it's not true for all the entrepreneurs—but I think for a large portion of the entrepreneurs here they generally kind of make the world better. Look at all these entrepreneurs who are living five to an apartment, who are not making a lot of money, and may never make a lot of money. They can get a six- or seven-figure job at a bank or a large company but they choose to do this. At the end of the day it's not money, it's this emotional connection: "I am somehow connected with this world in a different way, and I can act upon that." I think that is unique to Silicon Valley. Even back to the original days of Atari, there was this idea that you can have an emotional-professional career, and that's okay. You can say, "I want to do this because I want to do this," and that's accepted.

Andy Hertzfeld: Woz might not say that he is driven by artistic values, but if you look at the work—that's what it is. All that crazy creativity in the Apple II was art. Steve was fundamentally motivated by artistic values. I had artistic values. The artist wants to spiritually elevate the planet.

Ev Williams: Larry Page and Sergey Brin are driven to create. I saw this at Google. They are like this.

Steve Jobs: The people who really create things that change this industry are both the thinker and the doer in one person. The doers are the major thinkers. Did Leonardo have a guy off to the side that was thinking five years out about the future? About what he would paint or the technology he would use to paint it? Of course not. Leonardo was the artist—but he also mixed all his own paints. He also was a fairly good chemist. He knew about pigments, knew about human anatomy. And combining all those skills together—the art and the science, the thinking and the doing—was what resulted in the exceptional result.

Ev Williams: Tons of other businesses are driven by people who like to transact, to "do business" or "make money"—I mean you can be on Wall Street and just transact. What's different about Silicon Valley are those people who are driven by the creation.

David Levitt: People make a huge difference and in a crazy way the technology is just incidental.

John Battelle: Technology is just the artifact, it's not the culture. Culture is the values we shared. And it's no longer an outside-the-mainstream cultural movement, it is our culture. What's different now is not the Valley, it's every major city in the world.

Aleks Totić: In the rest of the world, technology didn't matter. Twenty years later we won. It's worldwide. There are internet cafés in Zambia; they have cell phones; they know Google; they know Apple: The Valley has utterly conquered the mind share.

Marc Porat: If you say "New York," people around the world generally know what it is, and now if you say "Palo Alto" they kind of know what it is, too: That's kind of weird—it's a village!

Aleks Totić: Now what happens here really affects things everywhere else. It's very different. Now we attract normal people. Before we were like a tribe and now everyone's in it.

Brenda Laurel: It's ironic: Silicon Valley has gone from a countercultural kind of thing that grew out of hippie ethics and the *Whole Earth Catalog* to a mainstream belief system shared by young entrepreneurs around the world.

Steve Wozniak: And they are usually young people. Look at the people who started Apple, look at the people who started Google and Facebook—very young people, just out of college.

Jim Levy: I don't believe that the guys who founded Google at the time they founded it thought it was going to be what it is now. Certainly Zuckerberg didn't think Facebook was going to be what it is now. But over time so many of those things happened that now you find the Valley to a great extent populated by people who expect that that's what they are in it to do—another Google, to do another Facebook. That's not taking anything away from Silicon Valley as the engine that it is now. It is what it is, right?

Jeff Skoll: And it's interesting to think about where it will go. I don't think that the Valley has ever tried to use its political power in the way that it might. Certainly now it has the money. It certainly has a stranglehold on world dynamics in the way that it just didn't ten or fifteen years ago.

John Battelle: I don't know how I feel about that, to be honest.

Jeff Skoll: I get this image in my head of these very powerful overgrown kids who don't quite know what they have, and the rest of the world is just sort of at their disposal, and so what happens next?

BOOK ONE

AMONG THE COMPUTER BUMS

The best way to predict the future is to invent it.

—ALAN KAY

The Big Bang

Everything starts with Doug Engelbart

*D*oug Engelbart was the first to actually build a computer that might seem *familiar to us, today. He came to Silicon Valley after a stint in the Navy as a radar technician during World War Two. The computer then (and, in postwar America, there was only one computer) was used to calculate artillery tables. Engelbart was, in his own estimation, a "naïve drifter," but something about the Valley inspired him to think big. Engelbart's idea was that computers of the future should be optimized for human needs—communication and collaboration. Computers, he reasoned, should have keyboards and screens instead of punch cards and printouts. They should augment rather than replace the human intellect. And so he pulled a team together and built a working prototype: the oN-Line System. Unlike earlier efforts, the NLS wasn't a military supercalculator. It was a general-purpose tool designed to help knowledge workers perform better and faster, and that was a controversial idea. Letting nonengineers interact directly with a computer was seen as harebrained, utopian—subversive, even. And then people saw the demo.*

Doug Engelbart: In 1950 I got engaged. Getting married and living happily ever after just kind of shook me. I realized that I didn't have any more goals. I was twenty-five. It was December 10 or 11. I went home that night, and started thinking: *My God, this is ridiculous.* I had a steady job. I was an electrical engineer working at what is now NASA. But other than having a steady job, and an interesting one, I didn't have any goals! Which shows what a backward country kid I was. *Well, why don't I try maximizing how much good I can do for mankind?* I have no idea where that came from. Pretty big thoughts.

Stewart Brand: There is a sense about Doug that he is trying to express what humanity needs.

Doug Engelbart: Well, then I thought, *I'm an engineer and who needs more engineering? What are the things the world really needs?* So then I started poking around, looking at the different kinds of crusades you could get on. Someplace along there, I just had this flash that, *Hey, the complexity of a lot of the problems and the means for solving them are just getting to be too much.* The time available for solving a lot of the problems is getting shorter and shorter. So the urgency goes up. The product of these two factors, complexity and urgency, had transcended what humans can cope with. It suddenly flashed that if you could do something to improve human capability to deal with that, then you'd really contribute something basic. That just resonated.

Jane Metcalfe: Everything starts with Doug Engelbart.

Doug Engelbart: It unfolded rapidly. I had read about computers. I think it was just within an hour that I had the image of sitting at a big CRT screen with all kinds of symbols, new and different symbols, not restricted to our old ones. The computer could be manipulating, and you could be operating all kinds of things to drive the computer. The radar technician training let me realize that a computer could make anything happen on a display screen and that the engineering was easy to do. That was the spring of 1951. I got accepted to graduate school at Berkeley, because they were building a computer there.

Steve Jobs: And in those days—again, it's hard to remember how primitive it was—there was no such thing as a computer with a graphics video display. It was literally a printer. It was a teletype printer with a keyboard on it. And so you would keyboard these commands in and then you would wait for a while. And then the thing would go *ta-dah-dah-dah-dah*, and it would tell you something out.

Bob Taylor: Computing was, in those days, believed to be for arithmetic. That's it. Data processing, calculating payroll, calculating ballistic missile trajectories, numbers...I wasn't interested in numbers and neither was Doug.

Doug Engelbart: I finally got my PhD and was teaching and applied to the Stanford Research Institute, thinking if there was anyplace I could explore this augmenting idea it was there. Stanford was a small engineering school then. Hewlett-Packard was successful, but still small. By 1962 I'd written a description of what I wanted to do, and I began to get the money the next year.

Bob Taylor: There was this proposal called "Augmenting the Human Intellect" by someone at SRI, whom I had never heard of. I loved the ideas in this proposal. The thing that I was most attracted to was the fact that he was going to use computers in a way that people had not: to, as he put it, "augment human intellect." That's about as distinct a phrase as I can think of to describe it. I got in touch with this chap, and he came to see me in DC, and we got him started on a NASA contract, which was quite a bit larger funding than he had had previously. It got him and his group off the ground.

Doug Engelbart: I got the money to do a research project on trying to test the different kind of display selection devices and that's when I came up with the idea of mouse.

Bill English: That happened in '63. We had a contract with NASA to evaluate different display pointing devices, so I collected different devices—a joystick, a light pen—and Doug had a sketch of this "mouse" in one of his sketchbooks. I thought that looked pretty good, so I took that and had the SRI machine shop build one for me. We included that in our evaluation experiments, and it was clearly the best pointing device.

Bob Taylor: The mouse was created by NASA funding. Remember when NASA was advertising Tang as its big contribution to the civilized world? Well, there was a better example, but they didn't know about it.

Doug Engelbart: We had to make our own computer display. You couldn't buy them. I think it cost us $90,000 in 1963 money. We just had to build it from scratch. The display driver was a hunk of electronics three feet by four feet.

Steve Jobs: Doug had invented the mouse and the bitmap display.

Jeff Rulifson: An SDS 940 was used for the demo.

Butler Lampson: The SDS 940 was a computer system that we developed in a research project at Berkeley, and then we coaxed SDS into actually making it into a product. Engelbart built the NLS on the 940.

Bob Taylor: Doug and his group were able to take off-the-shelf computer hardware and transform what you could do with it through software. Software is much more difficult for people to understand than hardware. Hardware, you can pick it up and touch it and feel it and see what it looks like and so on. Software is more mysterious. It's true that Doug's group did some hardware innovation, but their software innovation was truly remarkable.

Bill Paxton: Remember—put yourself back. This was a group, an entire group, sharing a computer that is roughly as powerful…If you measure computing power in iPhones, it was a milli-iPhone. It was one one-thousandth of an iPhone that this group was using for ten people. It was nuts! These guys did absolute miracles.

Don Andrews: We knew from the onset, based on Doug's vision, what we were trying to do. We were looking for ways of rapidly prototyping new user interfaces, and so building a framework, an infrastructure, that we could go into and build something on top of, over and over, very quickly. We knew that things were going to change very quickly, and essentially we were bootstrapping ourselves.

Alan Kay: In programming there is a widespread theory that one shouldn't build one's own tools. This is true—an incredible amount of time and energy has gone down that rat hole. On the other hand, if you can build your own tools then you absolutely should, because the leverage that can be obtained can be incredible.

Bill Paxton: Being immersed in the group where everybody was using the same tools and using them on a day-to-day basis, and where the people who were developing the tools were sitting next to the people who were using the tools—that was a really tight loop that led to very rapid progress.

Doug Engelbart: By 1968 I was beginning to feel that we could show a lot of dramatic things. I had this adventurous sense of *Well, let's try it then*, which fairly often ended in disaster.

Bob Taylor: In those days there were always, at these computer conferences, panel discussions attacking the idea of interactive computing. The reasons were multitudinous. They'd say, "Well, it's too expensive. Computer time is worth more than human time. It will never work. It's a pipe dream." So, the great majority of the public, including the computing establishment, were not only ignorant and would be opposed to what Doug was trying to do, they were also opposed to the whole idea of interactive computing.

Doug Engelbart: Anyway, I just wanted to try it out. I found out that the American Federation of Information Processing conference was going to be in San Francisco, so it was something we could do. I made an appeal to the people who were organizing the program. It was fortunately quite a ways ahead. The conference would be in December and I started preparing for it sometime in March, or maybe earlier, which was a good thing because, boy, they were very hesitant about us.

Bob Taylor: Even within that community of people who were doing work on interactive computing, there was probably a pecking order of some sort. There always is. Doug's group, at that time, before this demo, was probably at the bottom of that pecking order. At SRI, in those days, Doug was also having problems. His manager, early in the NASA support, came to see me, which is unusual. He came to see me at my office in Washington and he said, "I want to talk to you about Doug. Why are you funding him?" I said, "Because he's trying to do something that's very important that nobody else is trying to do and I believe in it." I got the sense that this fellow felt that Doug was ethereal in some fashion or another and he wanted to see if his funding source was also ethereal. If the funding source disappeared, then the manager would be in trouble. That's just the way it was, folks. It was very frustrating.

Bill Paxton: At the time 90 percent of the people thought he was a crackpot, that this "interactive" idea was a waste of time, and that this wasn't really going anywhere, and that the really good stuff was artificial intelligence... There were a few people like Bob Taylor who picked up on the idea. And it eventually fed out into Xerox PARC and then Apple to take over the world. But, at the time, Doug was a voice crying into the wilderness.

Bob Taylor: Doug and I talked about doing this demo in early '68, and I was strongly encouraging Doug to do it. He said, "It's going to cost a fortune. We're going to bring in this huge display, we're going to have online support between San Francisco and Menlo Park, and it's just going to cost a ton of money."

Alan Kay: And basically, when they approached Taylor about doing this, Taylor said, "Look, spend what you need, but don't do it small—and be redundant enough so the thing really works."

Bob Taylor: I said, "Don't worry about it. ARPA will pay for it." ARPA was created by the Department of Defense at the instigation of Eisenhower. The idea was to launch an agency that would support high-risk research without red tape so that, hopefully, we would not get surprised again the way Sputnik surprised us.

Doug Engelbart: It was a time when we were just sort of on a good-friends basis and you interact. *How much should I tell them?* I told them enough so that they got the idea of what I was trying to do and they were essentially telling me, "Maybe it's better that you don't tell us." We had a lot of research money going into it, and I knew that if it really crashed or if somebody really complained, there could be enough trouble that it could blow the whole program. They would have had to cut me off and blackballed us, because we had misused government research money. I really wanted to protect the sponsors, so I could say that they didn't know. So that's the tacit agreement we had between us. As a matter of fact, Bill English never did let me see how much it really cost.

Alan Kay: I believe ARPA spent $175,000 of 1968 money for that one demo. That's probably like a million bucks today.

Doug Engelbart: A lot of money.

Bob Taylor: Bill English was the miracle worker for most of that demo.

Doug Engelbart: Actually, the demo really never would have flown if it weren't for Bill English. Somehow he's in his element just to go arrange things.

Alan Kay: Even good ideas are cheap and easy, but we also had Bill English and his team of doers who were able to take this set of ideas and reify it into something.

Bill English: It was a challenge to get everything from SRI to the Civic Center. I mean that was thirty miles away! What we did was lease two video circuits from the phone company. They set up a microwave link: two transmitters on the top of the building at SRI, receiver/transmitters up on Skyline Boulevard on a truck, and two receivers at the Civic Center. Cables of course going down into the room at both ends. That was our video link. Going back we had two dedicated 1,200-baud lines: high-speed lines at the time. Homemade modems.

Doug Engelbart: We needed this video projector, and I think that year we rented it from some outfit in New York. They had to fly it out and send a man to run it.

Bill English: We used an Eidophor, a Swedish projector, a complex machine. It was a large machine—almost six feet tall—an arc-light projector. And what it did was focus the arc light on a spherical mirror. The mirror was wiped with the windshield-wiper blade that smeared oil over it, between each frame, and the electron beam actually wrote the image in the oil! It was an incredible method.

Alan Kay: There wasn't just one Eidophor there, there were two. They were both borrowed from NASA. And then the question was, "Well, what if our redundancy there doesn't work?"

Stewart Brand: They were on the screaming edge of what this technology was capable of, in terms of bandwidth and reliability and all the rest of it.

Alan Kay: So then they went to Ampex, which had just started doing high-res video recording, and got a huge Ampex recorder and they did the entire presentation on it and it was running while the live thing was going on just in case something fucked up.

Doug Engelbart: There were boxes that could run two videos in and you turn some knobs and you can fade one in and out. With another one you can have the video coming in and you can have a horizontal line that divides them or a vertical line. It was pretty easy to see we could make a control station that could run it.

Alan Kay: Bill was the one who designed this whole thing, who made this whole display system and everything else. Bill was the coinventor of the mouse; he was not really a second banana.

Doug Engelbart: Bill just built a platform in the back with all this gear.

Alan Kay: Engelbart was the charismatic one. Bill was the engineer.

Doug Engelbart: The four different video signals came in and he would mix them and project them. There was no precedent for that, that we had ever heard of.

Alan Kay: The scale of that demo—it's just unbelievable.

Bill Paxton: I was the new guy and pretty much clueless, so what I was most impressed about was that Stewart Brand, of the *Whole Earth Catalog*, was our photographer. Talk about somebody to loosen up the atmosphere!

John Markoff: Stewart was really the person who, more than anybody, shepherded psychedelic drugs from the spiritual and the therapeutic to the recreational. He was the vector of the counterculture.

Stewart Brand: Some people through Engelbart's office had been paying attention to the stuff I was doing with the Trips Festival and so on. They thought I might bring some production values, or knowledge about how to put on a show, to the demo that they were planning. So, they invited me over.

Alan Kay: In those days the *Whole Earth Catalog*, which was actually a store as well, was located right across the street from SRI.

Stewart Brand: I remember walking over there thinking, *This could be interesting and maybe even important.* When I saw what they were doing, it seemed all very swell and obvious. *Of course you would want to do the kind of things that they were doing with computers!* They invited me to several of their meetings planning the show.

Alan Kay: Stewart was just involved. I met him through Bill English, at a party. A lot of the *Whole Earth* people were there.

Stewart Brand: When I went into Engelbart's lab for the first time, there was a big poster of Janis Joplin, which is kind of an indication that they were feeling like part of the counterculture.

Alan Kay: In that whole area—University Avenue in Palo Alto and then El Camino going all the way into Menlo Park—the counterculture was going on.

The NLS debuted at the national computer conference at Brooks Hall in San Francisco's Civic Center in December 1968. When the lights came up, Engelbart sat onstage with a giant video screen projected behind him, and a mouse at his fingertips. Then, in what has become known as "the Mother of All Demos," Engelbart showed off what his computer could do.

Stewart Brand: I participated in the demo part of the show itself.

Bill Paxton: Stew was behind the camera, and one of the first things he did was focus it on a monitor and zoom in so that the image filled up the entire screen. He was getting this great feedback loop going. Do that at home. It's really cool—very psychedelic. This is basically where things were behind stage.

Stewart Brand: I had been a professional photographer, and that was why they said, "Oh great, you handle the camera." It was pretty much just point and focus. But the demo was astounding.

Doug Engelbart: It was the very first time the world had ever seen a mouse, seen outline processing, seen hypertext, seen mixed text and graphics, seen real-time videoconferencing.

Alan Kay: We could actually see that ideas could be organized in a different way, that they could be filtered in a different way, that what we were looking at was not something that was trying to automate current modes of thought, but that there should be an amplification relationship between us and this new technology.

Engelbart's NLS terminal had a screen and keyboard, windows, and a mouse. He showed off a way to edit text, a version of e-mail, even a primitive Skype. To modern eyes, Engelbart's computer system looks pretty familiar, but to an audience used to punch cards and printouts it was a revelation. The computer could be more than a number cruncher; it could be a communications and information-retrieval tool. In one ninety-minute demo Engelbart shattered the military-industrial computing paradigm, and gave the hippies and free-thinkers and radicals who were already gathering in Silicon Valley a vision of the future that would drive the culture of technology for the next several decades.

Bob Taylor: There was about a thousand or more people in the audience and they were blown away.

Andy van Dam: I was blown away to see this professional system with this unbelievable richness and complexity. It was an otherworldly experience, and in fact, I couldn't quite bring myself to believe that it was all for real.

Bob Taylor: Nobody had ever seen anyone use a computer in that way. It was just remarkable. He got a huge standing ovation after it was over.

Alan Kay: The thing that we loved was the scope of it—Engelbart was really cosmic.

Butler Lampson: It was pretty spectacular.

Stewart Brand: I've seen a lot of demos since then, and been part of demos, at the MIT Media Lab and so on. But I've never seen anything that was so dangerous, and undertaken with such bravado. We never did a full rehearsal of the event; there were partial rehearsals, but that was a real-time improvisation that people saw. I don't think people realized that it was improvisation, but it was improvisation. It gave a certain extra high-wire-act quality to the thing that may have come through. And I have to say, Doug was pretty spectacular at managing all of that, and being such a master of the medium itself, and having become the master of the completely kluged-together communication system he was operating with, he was totally unflappable up there on the stage. When Bill would whisper in his ear, "Stall for a couple of minutes, we can't get the..." whatever it was that wasn't working, Doug would just pause and discourse on something else until he got word that "Okay, we're good to go." And, we lucked out. It all worked enough to take the day.

Andy van Dam: At the time, I had been working with Ted Nelson on our first hypertext system with a team of three part-time undergraduates. We were working in the hammer-and-chisel phase of this industrial revolution, coding in assembly language, and we were pretty good at it. But, here these guys had invented machine tools. They had built tools to build tools: This whole recursive "bootstrap" idea, starting with the system itself, and working all the way up through augmenting the human intellect, was just mind-boggling. It informs us today, still.

Steve Jobs: We humans are tool builders. We can fashion tools that amplify these inherent abilities that we have to spectacular magnitudes. And so for me, a computer has always been a bicycle of the mind.

Ken Kesey: It's the next thing after acid!

Steve Jobs: Something that takes us far beyond our inherent abilities.

Bill Paxton: A lot of people in the audience that day were profoundly impacted by it and went out and said, "How can I do this?"

Ready Player One

The first T-shirt tycoon

*T*he earliest computers were war machines, and in the beginning Silicon Valley ran on Defense Department contracts. Yet it was a "war
machine" of a totally different order that gave Silicon Valley's entrepreneurial culture the character it has today. Spacewar, *the first compelling computer
game, was written for an early computer called the DEC PDP-1. One of the
PDP-1's big selling points was the screen: It had one. The PDP-1 was also
"inexpensive"—which meant that universities could afford to buy them. Nolan
Bushnell, an engineering student at the University of Utah in the midsixties,
played* Spacewar *in his school's computer lab and it left a lasting impression.
After graduating he moved to Silicon Valley, and in just a few short years
figured out how to bring computer games to the masses with* Pong. *It was a
massive hit and Bushnell, still in his twenties, suddenly found himself in charge
of what was, arguably, the most important company ever to rocket out of the
Valley. Not only did Bushnell single-handedly create an industry around a new
American art form—video games—he also wrote what has become the quintessential Silicon Valley script. The story goes like this: Young kid with radical
idea hacks together something cool, builds a wild free-wheeling company around
it, and becomes rich and famous in the process.*

Michael Malone: Nolan Bushnell is hugely important. He's the first T-shirt
tycoon. He's the first modern Silicon Valley entrepreneur. He's not building heavy-duty hardware. He's not doing silicon. He's doing consumer
electronics.

Nolan Bushnell: I was at the University of Utah. Evans and Sutherland
headed the computer science department at the time, and yielded a whole
bunch that we called the Utah Brats: Jim Clark, John Warnock, Alan

Kay, Ed Catmull. I was in their thrall. I thought that the work they were doing was truly extraordinary. They were graduate students and I was an undergraduate.

Alan Kay: Nolan would sneak up and do things on the computers. We didn't go out and drink beer, but Nolan is a good guy.

Nolan Bushnell: My fraternity brother said, "There's something you got to see," and we stealthily went into the computer lab. He had jammed something in the lock mechanism so we could get in. We didn't turn the lights on. It was just us and the green display glowing back at us. It was a big round radar screen, a converted radar display left over from the Korean War.

Al Alcorn: Nolan saw *Spacewar*, a game that ran on the PDP-1.

Nolan Bushnell: It was a life-changing experience for me.

Michael Malone: The great breakthrough was the realization that semiconductor technology would be getting to the point that you could bring intelligence to consumer products. And Nolan's the guy.

Nolan Bushnell: To put it into context, I was putting myself through school working at the local amusement park, Lagoon. It was a dollar-an-hour job, sucker pay. "Come throw a ball! Win a stuffed animal! Let me guess your age, weight, and occupation!" The job was really selling at a very basic level, and I found that I was pretty good at it. I could regularly bust my quota and make a lot of commission. The following year they made me assistant manager, and then manager, and so at twenty-one years old I had 150 kids reporting to me. So I knew intimately the economics of the coin-operated game business.

Al Alcorn: I think that Nolan will proudly say that he graduated last in his class at the University of Utah, but he also worked in an amusement park, so he saw the connection: *If I could put a coin slot on that million-dollar PDP-1, I could make a few hundred bucks.* He knew it was economically unfeasible, but he also knew there was play value there.

Michael Malone: Nolan was the first guy to look at Moore's law and say to himself: *You know what? When logic and memory chips get to be under ten bucks I can take these big games and shove them into a pinball machine.*

Nolan Bushnell: When my wife and I were driving from Utah to California I told her, "I will have my own business in two years."

Ted Dabney: I knew Nolan back when he had a junky little car, he owed money on his school, he could barely afford his rent...early on he was a really neat guy. We had fun, we'd go around to pizza parlors, he'd tell me all about his brilliant ideas, so I knew him then. I had no reason to doubt he'd ever lie to me. Or to think that he would lie to me, but it turns out he did.

Nolan Bushnell: In 1968 Silicon Valley was really just starting in a lot of ways. Everywhere there were these large tracts of prune orchards and every month another orchard would be taken down and they'd pile up all the cut trees next to a sign that said "free firewood" and soon there would be a concrete tilt-up there. The center of gravity was Mountain View and Sunnyvale, where Fairchild and Intel and National Semiconductor were.

Bob Metcalfe: Intel was just getting started. There was no Apple. There was a Hewlett-Packard, but they barely made computers, just barely. They were old and respectable, but a tiny little company.

Al Alcorn: The Valley at the time in the sixties and seventies was dominated by military stuff: Lockheed and all these big places—and semiconductor companies. Also, the Del Monte fruit factory; they were making fruit cocktail right there behind Videofile in this giant factory that smelled like a fruit cocktail in the summer.

Nolan Bushnell: I worked for a division of Ampex called Videofile in Santa Clara.

Steve Mayer: Coincidently, Larry Ellison and Oracle came out of Ampex, also.

Al Alcorn: I was an undergraduate at Cal Berkeley in the work-study program, where you work for six months in the industry and then go to school for six months. I got a job at Videofile, working in the camera group. Nolan was in that same group under a guy named Kurt Wallace, the boss of all of us. Nolan was a brash, young, tall engineer, right off the bus from Salt Lake City, and he basically showed up and said, "I want to work here." Kurt was impressed and hired him.

Ted Dabney: Videofile was a way of recording documents and video on a very large rhodium disk, which was kind of like the disc you have in your computer, only now it was huge and you could have instant access to video and pictures.

Steve Mayer: It really brought computers and video technology together for the first time in a significant way.

Ted Dabney: Nolan and I shared an office. We were very close.

Al Alcorn: At Ampex in the late sixties, you had to wear a suit and tie. You had to dress properly, neatly. You had to be there at eight a.m. You took a one-hour lunch and left at five…I don't recall anybody staying after work. It was pretty straitlaced. You stayed there. You retired there. You got the golden wristwatch and your pension and all that. That was the plan.

Nolan Bushnell: As an engineer I wore a coat and tie to work every day; however, we all had our hippie costumes and we would go to San Francisco and pretend we were flower children. The ultimate posers, we were! The hippie culture was fascinating to us.

Al Alcorn: Dabney was the old man. At the time, he was probably in his late thirties or something, and served in the military at one point. He certainly was not a hippie by any stretch of the imagination.

Ted Dabney: Nolan was supposed to be an engineer, I mean, he was hired as an engineer but I don't think he was capable of doing any engineering work. He didn't have the background; he didn't have any training in it. He just wasn't very capable. And a lot of that I found out later. I didn't know that at the time.

Al Alcorn: Nolan was an okay engineer, but he was such an entrepreneur. Nolan wanted to play the market but didn't have enough money. He actually got together a stock investors' club—a group inside of Ampex— to invest. Back in those days, one could not buy stock unless one was of the right class—you couldn't afford to open an account. It was not like today, where everybody buys stock. But if you pulled about five or six guys together, then you could buy stock. Nolan was that kind of guy. He worked adjacent to me. We were friends.

Ted Dabney: Nolan knew games. He was really into that kind of stuff.

Nolan Bushnell: I became infatuated with the game of Go and I would drive to San Francisco every other weekend and spend basically most of Sunday morning playing Go at the old Buddhist church by Bush Street. There I met a guy who worked at the Stanford Artificial Intelligence Lab and one day we got to talking and he said, "Have you ever heard of *Spacewar*?" I was like, "*Spacewar*? The first time I played *Spacewar* was at the University of Utah, in 1965 or 1966." We left and drove up to the Artificial Intelligence Lab and played until all hours of the night. *Oh yeah! Oh boy! Bingo!* And that in some way was the recatalyzation of the idea for the video game. I can remember coming back to work next day and regaling Dabney with it.

Ted Dabney: He wanted me to take a look at this game on a computer over at Stanford. It was called *Spacewars* or something. It was a neat game but it was on this big computer—a million-megabyte kind of thing—and he said, "Hey, we should be able to do that with a smaller computer and, you know, TVs."

Nolan Bushnell: Coincidently there was an ad for a Data General Nova minicomputer. It had an eight-hundred-kilohertz clock cycle, and it looked like you get a stripped-down version of it for about four thousand bucks. I thought, *A-ha! The time is right!* It was time to form a company. I asked Dabney if he wanted to be part of it, because he was really, really good at analog circuitry, and I knew I need an analog interface for the television. He said, "Sure!" I was the digital guy—I could put integrated circuits together. This was probably October of 1969.

Ted Dabney: Nolan and I were sitting around my living room one day trying to think of what we wanted to do. We had decided to come up with a partnership, and we each were to put in $100 to kind of get this thing started. We knew that wouldn't be enough, but at least it was a place to start. So I started a bank account and put in my hundred dollars, Nolan put in his hundred dollars.

Nolan Bushnell: So I said that my addition will be to figure out a way to have a regular television set driven by this computer. We had to take the TV apart, cut a trace, solder a wire on it—you'd get a hell of a shock if you did it wrong. In a TV the signal is very fast, up to 3.5 megahertz. And the chips in those days didn't like to go more than a megahertz. I tried to solve this timing problem over the Thanksgiving holiday. I built a little circuit that would bring up the stars, I built a little circuit that would put the score up, various things to off-load tasks from the computer. But soon I found myself thinking that *This just isn't going to work, the computer just isn't fast enough.* I abandoned the project for about two days. Then I had an epiphany: *Let's not use the computer at all… I can do it all in hardware!* We never did buy that computer.

Al Alcorn: The story I heard was that Nolan was going to send $10,000 to Data General for a Nova minicomputer, for several of them—and his wife just refused—so she never sent the check. And by the time the computer should have shown up, they had figured out that the minicomputer, although wonderful for its time, was nowhere near fast enough to do anything. The hardware they were building was doing more and more, and pretty soon the stuff that was left to the computer was so trivial that they didn't need the computer.

Ted Dabney: Nolan really, really worked on it. I helped him with the circuitry and then he would say, "Well, how do I this? And how do I do that?" and I would show him, you know, the basics, and then he'd go and turn it into a circuit that actually worked.

Nolan Bushnell: Both Ted and I had daughters, and he was working in his daughter's bedroom, and I was working in my daughter's bedroom, and when we brought this stuff together we got this electric feeling. The first time we had a little rocket ship going, flying on the screen, it was just one of those wow-we-thought-we-could-do-this-and-now-we-did-it moments. Productizing it happened that summer.

Ted Dabney: At some point, and I don't know when it was, he decided to contact Nutting Associates, mainly because there's no way we could have done anything with it, no matter how good it was. So, he went and talked to Nutting.

Al Alcorn: Nutting was the only coin-op manufacturer west of the Rockies—they had something called *Computer Quiz*, which is a filmstrip game—fairly simple, nothing big.

Ted Dabney: Nutting had a game called *Computer Quiz* that they had been milking for years. So Bill Nutting was really kind of desperate for something.

Al Alcorn: Nolan did a deal with them.

Nolan Bushnell: I called him up and said, "Would you guys be willing to manufacture this new game?" And I went to lunch with him and showed him the thing and they were all jacked up and said, "Yeah, we can do this!" So we negotiated the licensing agreement and they said, "You are going to have to come on as our chief engineer as well," and I felt good about this, because I literally doubled my salary. Not only that, but they agreed to my salary so quickly that I said, "And a company car." So I ended up doubling my salary and got a company car and left Ampex. About June I talked Nutting into hiring Ted as my assistant.

Dick Shoup: The first time I met Nolan Bushnell, his desk was in the hall of Nutting Associates. I heard what he was doing and came by. He was making video games. Then occasionally I would run into him at the Dutch Goose having a beer. I thought what he was doing was pretty interesting. He had some great ideas.

Al Alcorn: Conceptually, the idea, from an engineer's standpoint, of making a device, that was digital, that generated a video signal, and worked on a display, was fairly novel. That was interesting—very impressive…

The game that Dabney and Bushnell created was Computer Space, *a* Spacewar *knockoff, but more importantly: the first arcade video game.*

Ted Dabney: I spent most of my time building a cabinet, because we needed to have a cabinet to display this thing in. Nolan actually did the design work and I worked on the cabinet.

Nolan Bushnell: It was totally rounded, it had a screen, it had a pedestal that organically grew out of the base and four buttons which were illuminated, and it looked like it was from outer space.

Ted Dabney: Totally Nolan's idea. The guy is brilliant! He's absolutely brilliant! He just happens to not be a particularly good engineer. But his imagination and his ideas and…I mean, all the ideas that came around were his ideas.

Nolan Bushnell: We put it in the Dutch Goose, which was a hangout for Stanford students. It was an immediate success—just cash dropping through! Now people say that *Computer Space* was a failure, but it did about $3.5 million, which was a lot of money in those days. The royalties deal allowed me to start Atari, so it was a winner for me.

Al Alcorn: *Computer Space* did modestly well, so Nolan said to Nutting, "Give me some stock in the company and I'll be the VP of engineering." They were contractors, not employees.

Ted Dabney: We owned the game. That was the whole deal. We owned the game. They were going to manufacture it and they were going to pay us a royalty. And they're going to pay us a salary while we're building this thing up. You know, developing it.

Nolan Bushnell: I said, "I'm clearly making this company, I want 15 or 20 percent." I think they came back with an offer of 5 percent of the company.

Ted Dabney: Nolan tried to negotiate with Nutting, you know, for an ownership of the business and all that kind of stuff, which didn't work for Nutting at all. Bill Nutting was not the sharpest pencil in the box.

Al Alcorn: Bill Nutting basically told Nolan, "You are not a businessman. I am a businessman. You're just a worker." The fact is, Bill Nutting's claim to fame was his wife, who was wealthy. He had kind of an attitude.

Nolan Bushnell: I got on a flight to Chicago and called on Bally and said, "Would you like to license my next game? A driving game? It's going to cost you this much of money in development." They said, "Yeah, we'd love to do it!" And so I had my cash flow in hand. I went back to Nutting and said, "I'm going to have to leave."

Ted Dabney: We wound up getting this contract, $4,000 a month for six months to develop a video game and a pinball machine for Bally.

Al Alcorn: Nolan then goes and hires me, employee number three, and we go off and start Atari.

Atari is a Japanese term borrowed from the game of Go. It's the equivalent of the word "check" in chess.

Al Alcorn: It was a flyer. It was a big risk. But so what? I had nothing to lose at that age. I reasoned that if it did fail, which it probably would, I could get my job back at Ampex or someplace.

Lee Felsenstein: In 1972 Ampex was beginning to implode.

Al Alcorn: I figured Atari would be a valuable lesson, even in failure. Much to my surprise, that didn't happen.

Nolan Bushnell: My idea was simply to be a design shop for the big companies that had the factories and cash flow. Because even with all those royalties, we still were nickel-and-dime compared to what was necessary to really get into manufacturing.

Al Alcorn: Looking back on it, it was really risky, because we didn't have some slug of money or something like that. Nolan and Dabney didn't have much at all. Banks wouldn't talk to us because we were in the "coin-operated entertainment business," which meant jukeboxes and vending machines, which meant mob controlled, and no way were they going to give us money for that. We had no track record, no money.

Nolan Bushnell: That spring I heard that somebody else had a video game which was, you know, scary, particularly somebody like Magnavox.

Ralph Baer: The Magnavox caravan was a vehicle that traveled around with an original Magnavox Odyssey before it became public.

The Magnavox Odyssey TV Game System was not a coin-operated arcade game, but rather a gaming console—a box that hooked up to a TV so one could play computer games at home.

Nolan Bushnell: So I find out where it is and I drive up and it's at the Burlingame Marriott or something and I go in, sign the guest book, and I see the thing and I look at it and I think, *Ahhh . . . no competition here!*

Ralph Baer: Nolan played the game. His opinion from the get-go is negative. That's fine that he had that opinion. It was mostly based on not learning how to play it properly. I bet you dollars to donuts he never found the English knob on it!

David Kushner: You could turn it to put a little spin on the ball.

Nolan Bushnell: The game was fuzzy, it wasn't that fun, it had no screen, no scoring, no sound. It was just really what I considered to be a marginal product. I drive back from the thing feeling relieved.

Ralph Baer: Maybe it wasn't as fancy, it didn't keep score... but people have played table tennis for 250 years. How do they score? They call out the scores. No big deal. I really resent that Nolan treats me like an engineer that didn't know what the hell he was doing, like it was a piece of junk and he's the great hero. Bullshit! All you have to do is look at the record: 350,000 Odysseys sold in the first year. I didn't go before the president of the United States to have the National Medal of Technology hung around my neck because I am just an engineer. I invented video games!

Nolan Bushnell: It was Al's first or second day and I needed a training project. *Computer Space* was a pretty complex project, and Bally's driving game was going to be a hard project, and I thought, *Okay, I'm going to pitch him this learning project: this ping-pong game.*

Ralph Baer: My ping-pong game. *Pong*, they called it.

Al Alcorn: Nolan described *Pong*: one moving spot, a score, a net, and a ball. It couldn't be any simpler.

Stewart Brand: *Pong* was clearly a brain-damaged version of an interactive game, but it was one that you could play on a simple computer.

Al Alcorn: *Pong* was the simplest thing Nolan could think of. In his mind he thought that the game he really wanted to do was something more complicated than *Computer Space*. He wanted to do a driving game, something like that. But, he didn't tell me that. He told me he had this deal with General Electric.

Nolan Bushnell: I told him that I had a contract with General Electric because I find that people don't like training projects or dead-end projects. It was a little fabrication.

Al Alcorn: It never occurred to me that this could be bullshit.

Ted Dabney: *Pong* was an exercise for Al Alcorn in figuring out how to use this motion circuit that Nolan and I had developed.

Al Alcorn: I had never designed a video game before. Nobody had but Nolan. So we just talked about how you get the ball, the spot, to move all over the place with no memory—there's no memory as such in the thing. The trick of how you do that is a brilliant insight. Once I understood the Bushnell Motion Circuit it was, "Okay, no problem." I went right ahead and did it.

Ted Dabney: Al Alcorn was a good engineer. He didn't need anybody's help at all.

Nolan Bushnell: Al basically had the thing going in a week, it might have been a week and a half—and it was fun! I thought, *Maybe Bally would like* Pong *instead of the driving game?* Al built one which had a modulator so I could hook it up to a regular television set, and I took it to Chicago and presented it to Bally. They were troubled by the fact that it was a two-player game and at that time you needed to have a one-player game…so they thought. That was the entrenched wisdom at the time.

Ted Dabney: Bally paid for it. They paid $24,000 to us for that game, but Bally kept not accepting the game, kept not accepting it. Nolan and Al and I were sitting around looking at each other: *What are we going to do?*

Al Alcorn: So Nolan said, "Put it out on location." We put it next to a *Computer Space* at Andy Capp's Tavern. I remember the day we put it in. Nolan and I popped it in one day after work and went and bought a beer and watched until somebody played it. I never thought anybody would play it! Think about it. There's no instructions. It just says *"Pong,"* which meant nothing. There's two knobs and a coin box. What's the motivation here? So, some guy plays it. Nolan went up to him afterward: "What did you think of that?" And the guy said, "Oh yeah, I know the guys who made this machine." And I'm thinking, *Save the bullshit for the ladies.* So we left.

Not long afterward, Alcorn receives a phone call from Andy Capp's Tavern.

Al Alcorn: The machine stopped working. It didn't surprise me. It was just thrown together quick and dirty. It was never meant to run on location. So, I went there after work and the attract mode was working, so that tells me most of the thing is working. *So what's wrong with it?* I open up the coin box—it was a Laundromat bolt-on coin box—to give myself a free game to see what was going on, and when I opened that coin box up it was just jammed full of quarters. *I can fix this!* I told Nolan, "Wow, this is interesting…"

Nolan Bushnell: *Pong* was earning $300 a week: a huge amount of money! And so my greedy little mind thought *Wow, this thing's a gold mine!* It was so much better than *Computer Space* in terms of earnings. So I figured out how to get into the *Pong* business.

Al Alcorn: We were sitting, I'm pretty sure, at Andy Capp's after work having our beers and Nolan said, "We want to build this. We want to be in the manufacturing business," and Ted and I were saying, "This is not what we signed up for. We don't have any manufacturing capacity, nothing." And Nolan, just basically through force of will, said, "We're going to do this."

Nolan Bushnell: I figured out that I could build 'em for about 350 bucks. I priced them at $910. And I figured out this financing model where the manufacturing would self-fund. I negotiated thirty to sixty days from our vendors, and if we could build the machine and ship them in less than a week, the company would operate in positive cash flow. Because we had no capital. Venture capital? I didn't even know what it was at the time. But we had the tiger by the tail: more orders than we could fill. I remember telling Alcorn and Dabney that we were going to move production up to a hundred a day and they looked at me like I was just stark raving mad.

Al Alcorn: I would come home and tell Katie, my wife, "Nolan is crazy. He wants to build a hundred machines a day." We didn't have the money, the capacity, the experience. This is insane, but I'm going to go along with the gag to see how far it gets.

Ted Dabney: We started out in seventeen hundred square feet, but when we started building *Pong* we needed more room, and it turned out that the guy next door to us had moved out in the middle of the night. He didn't pay his rent. So I cut a hole in the wall, literally cut a hole, and moved into his area and took over that seventeen hundred square feet, too. The manager came around and said, "You can't do that!" And Nolan said, "We did it! You just figure out how much it's going to cost us." But even that wasn't big enough. But then this roller rink down the block became available: ten thousand square feet! I mean we were just jam-packed and we had people on roller skates actually running around on the roller-skate rink building *Pong*s.

Nolan Bushnell: Understand that we had no purchasing department, no manufacturing skills. We had no procedures. We had no quality control. We had nothing, you know?

Al Alcorn: We made all kinds of mistakes. It was absolutely crazy. We went to the unemployment office to hire people—parolees and very colorful people, and we were being ripped off pretty heavily, but we were making so much money on this thing that it didn't kill us.

Nolan Bushnell: What we didn't know is that when you want employees you don't go down to the labor department, 'cause that's where all the druggies are. We hired a bunch of these druggies and pretty soon we noticed some theft, we lost six or eight TVs before we fired the guys that were kind of, you know—dodgy.

Al Alcorn: When we started Atari I was twenty-four. Nolan was twenty-six or something. Because we were so young and so inexperienced we didn't try to put in rules. Punching a time clock wasn't the point, it was getting the job done. If you could do it without showing up, so be it! Some people had to show up and be there to get their job done, but there were some people that could just make it happen and they would work anytime they wanted.

Nolan Bushnell: People talk about the party atmosphere, but what they don't realize is that it was all based on hitting quotas. We had an extremely young workforce.

Chris Caen: I started at fifteen as a summer intern. By the time I was eighteen I was a product manager. I had a hard-walled office, was making good money, and thinking to myself, *Why do I need to go to college?*

Nolan Bushnell: The girls that were stuffing the computer boards and doing the testing were in their early twenties or eighteen or nineteen. The guys who were muscling the big boxes around, packing and shipping them, were all in their early twenties.

Chris Caen: People had dropped out of high school and dropped out of college to work at Atari; they met their spouses there.

Nolan Bushnell: What everybody wanted was a party and some beer and some pizza and they ended up going home with each other.

Al Alcorn: Atari at the time had all these people and the smell of dope burning in the back and whatnot.

Nolan Bushnell: It was the hippie lifestyle.

Chris Caen: There was something magical about working at Atari at the time, which I have never experienced in any tech company—large or small—since. This sounds eye-rollingly naïve now, but it really was a family.

Nolan Bushnell: We were working hard and playing hard, and everybody was happy.

Al Alcorn: I remember when we started, nobody would talk to us. Then, all of a sudden, we became big.

David Kushner: In terms of the American landscape, Atari was almost like punk rock or something like that. *Pong* was just so minimalist and so compelling, and it managed to hook an entire generation.

Clive Thompson: *Pong* was so bare to the metal that you really felt like you were interacting with the substance of computation: This thing hits that thing and we calculate a new trajectory.

David Kushner: Just like the Ramones could take a few chords and just grab you by the throat, *Pong* had a joystick, a button, and a bunch of blocks on the screen—and you would spend the entire day there, playing it.

Clive Thompson: I was completely mesmerized. *Oh! Finally, we are directly looking at Newton's concept of physics: trajectories in a frictionless world.* There was something unbelievably beautiful and unsettling about it, in this sensual and tactile way.

David Kushner: If you were a geeky teenaged guy in that period of time, you were blowing your lawn-mowing money on Atari. But *Pong* also became a status symbol: It was such a phenomenon that Hugh Hefner had to have one in his bachelor pad in Chicago!

Al Alcorn: Then Atari was a big thing and everyone wanted to talk to us, and they believed what we said! The more staid the company, the more outrageous Nolan would behave.

Michael Malone: It's hard to capture just how crazy Nolan was in those early days. He was a wild man. He was young. He lived high. He had the Rolls-Royce. The code name for each new product was named after some hot girl on the assembly line. He really did have that master-of-the-universe thing. I mean there was coke with the assembly-line girls in the hot tub. This guy did the whole Cash McCall thing. He was just throwing out sparks in every direction.

David Kushner: Nolan really was the Merry Prankster of Silicon Valley, and so he attracted all the other Merry Pranksters who had nowhere else to go.

Al Alcorn: It was fun.

Ralph Baer: They made a couple hundred *Pong* games during 1972 toward the end of the year. And it was thirteen thousand *Pong* games the next year. The competition made more than that, everybody was making knockoffs.

Al Alcorn: Bally copied it, but it was with permission because we had a relationship with them. Nutting just stole *Pong* and made a game called *Computer Space Ball*, because everything they had was *Computer* this— *Computer Quiz*, *Computer Space*, *Computer Space Ball*…Clever! Why didn't they ask us? Everybody else stole it except Ramtek. When I say "stole it," I mean just copied the circuitry that I designed. Ramtek actually looked at it and made their own electronic version. They made a copy, but they didn't take the schematic.

Nolan Bushnell: Our copiers were essentially fast followers: They'd buy one of our games, they'd Xerox the printed circuit board, they'd ramp up manufacturing and go for it. They were jackals. It pissed us off.

Ted Dabney: They probably had somebody planted that, you know, stole for them. You know, industrial espionage is big business, it always has been.

Al Alcorn: The guy who was making our boards for us, I will not mention his name, was also making extra copies and selling them to the competitors. It was just a direct theft. You don't need to understand the circuitry, you just need to put the parts in. Everybody was stealing our games to the point where we were the advanced development group for the entire fucking industry. It was kind of frustrating.

Nolan Bushnell: We started countermeasures. We were getting to be a big semiconductor customer, so we asked them to privately mark some parts that we were using, and we designed some games so if the copiers just read the parts and ordered the parts and plugged them in they wouldn't work. Because they were the wrong parts! That put one of our competitors out of business. I can remember us having a little champagne party on their front lawn when they declared bankruptcy. By 1974 we had gotten rid of most of the jackals if not all of them. We had a huge market share at that point, we were dominant, but that summer was very dark.

Al Alcorn: Nolan's first attempts at selling games overseas were an utter disaster beyond recognition.

Nolan Bushnell: We didn't realize that Japan was a closed market and so we were in violation of all kinds of rules and regulations of the Japanese, and they were starting to give us a real bad time.

Al Alcorn: And Ron Gordon came in as a consultant and fixed all that for us for a huge commission.

Nolan Bushnell: I was twenty-eight years old and really trying to operate on a global stage with really no clue about what it was all about. There was never an Atari business plan; we were just making it up as we were going along.

Al Alcorn: Nolan had read this book about a company's growth and he learned from this book that the team that gets you to $1 million in sales can't get you any further. You've got to get pros to get any further.

Ted Dabney: He hired this president of the company that was a real yahoo, I mean a real yahoo, okay? He also hired a vice president of engineering that could not—would not—make a decision! And then he hired this salesman, as vice president of marketing, who didn't even know how to spell *marketing*.

Al Alcorn: They were like B players out of Hewlett-Packard: a marketing guy, a finance guy, and we hired an Ampex engineer to be VP of engineering, so we had a manufacturing guy. So Nolan had these people in and they really didn't understand start-ups, and they basically ruined the company.

Ted Dabney: So I says, "Nolan, you and I have got to talk." So we got on our motorcycles, headed over to a pizza parlor, and we sat down, and I say, "You've got to get rid of these guys, you've got to get rid of them."

Al Alcorn: Nolan couldn't fire people.

Nolan Bushnell: I drove into the parking lot of our Winchester Avenue facility, and it just kind of occurred to me that my company was paying the payments of everyone's car. It was just one of those things.

Ted Dabney: Nolan said, "All these people depend on us, don't they?" And I said, "Yeah." I said, "And their landlords, and the grocery stores, and everything. They all depend on us."

Al Alcorn: Then engineering made a key mistake—one of the mistakes that we made.

Nolan Bushnell: A part in our driving game had failed. And remember, we were operating on positive cash flow.

Al Alcorn: The games we did ship, we had to take back. It was so bad!

Nolan Bushnell: So all of a sudden we had a floor full of machines that couldn't be sold, which stopped our cash flow.

Al Alcorn: The peak of frustration was him in tears. Nolan was in tears. You could see Nolan was thinking, *This is over.* The company was going to die; we were going to shut it down.

Nolan Bushnell: We got sued for nonpayment of our bills. The sheriff had come to attach our assets—our bank accounts—so we had to switch bank accounts every week!

Al Alcorn: And then what happened was that Ron Gordon saw that his goose laying the golden eggs was dying.

Nolan Bushnell: We were just coughing blood.

Ted Dabney: Atari was going down.

Al Alcorn: And Ron came back and fired all those guys, talked to the banks to get our bank line going, and revived the company. And *boom!* He got us going again.

The Time Machine

Inventing the future at Xerox PARC

*I*n the wake of the young engineers who were drawn to Silicon Valley to work at the old-school electronics shops (like Ampex and Hewlett-Packard) and in the newfangled semiconductor foundries (Fairchild and Intel) came the blue-chip corporations looking for their own future—most notably, Xerox. In the early seventies the company had decided to build a research and development lab in the heart of the Valley: the Palo Alto Research Center—also known as Xerox PARC. The idea was to prototype the computerized office of the future. Inspired by Engelbart, the engineers at PARC designed and built a breakthrough computer, the Alto. Among the Alto's many innovations was the graphical user interface: The Alto had overlapping windows, menus, icons, and fonts. From those elements the first modern word-processing, e-mail, and paint programs were built. Other computers built by PARC could create psychedelic video and cartoon animation—in full color. Underlying all of it was the so-called bit-map display. Each pixel on PARC's many computer screens was mapped—connected—to a bit in memory.

Butler Lampson: The pitch was "Xerox is starting this new lab and it's a long way away from Rochester, and the charter is to invent 'the office of the future.' And nobody quite knows what that means, so we should be able to make it anything we want. And there's plenty of money!"

Chuck Thacker: What we did was very simple. We spent a lot of money to simulate the future.

Butler Lampson: Our game was to build time machines. It was extremely clear that the machines that we were building in the early and midseventies would be economical in the eighties. And it was extremely clear that they were wildly uneconomical at the time we built them.

Alan Kay: Through money—and assuming Moore's law—you can basically put a supercomputer on a desk, which is what we did.

Dan Ingalls: It was almost a free ride, because so many things were just ready to be invented: The entire world of bitmapped graphics was there to be invented.

Bob Metcalfe: And it was California, so we rode bicycles. I remember we used to ride our bikes along Arastradero Road over to Alpine Road and go to the Alpine Inn at lunchtime. We would drink beer at lunch, and that meant the afternoon was gone and that work might resume that evening after dinner. It was very laid-back. There was only one meeting per week. We'd sit in these beanbag chairs, and we would discuss what was going on at the lab. It was a very idyllic time.

Dick Shoup: We were all trying to figure out what to do.

Steve Jobs: They were doing some computer research, which was basically an extension of some stuff started by a guy named Doug Engelbart, when he was at SRI.

Bob Taylor: Engelbart's work influenced me lot. Some of our people spent summers on occasion working in Engelbart's group, before there was a PARC.

Alan Kay: I date PARC from 1971. Basically the five years through 1975: That's when most of the stuff got done.

Chuck Thacker: The first thing we did was try to figure out what we would use for computing resources. And we looked at SDS: because Xerox had just bought Scientific Data Systems.

Butler Lampson: The standard computer for computer science research at that time was the DEC PDP-10. We needed to be able to run the PDP-10 software, because that's what all the other researchers were running.

Charles Simonyi: But DEC was a competitor to Xerox, so it was completely politically incorrect for Xerox to have a competitor's machine for their research.

Chuck Thacker: And we realized that it would really be unseemly if we bought a PDP-10 from DEC—so we built one: the MAXC.

Alan Kay: It stands for Multiple Access Xerox Computer.

Chuck Thacker: I designed the memory, Butler designed the processor. And you could do that in those days.

Charles Simonyi: There was this wonderful tradition that new people had to do a shit job—that's what it was called in the vernacular. So my job at Xerox was building the MAXC.

Bruce Horn: I was the guy who would do the backups at night, and I'd go and, you know, put the tapes on the thing and run all the backups and everything. So that's kind of fun. Not bad for a fourteen-year-old.

Dick Shoup: We all were playing with *Spacewar*. There was really nothing else to do then, because we were building things.

PARC's first outside visitor of note was Stewart Brand, fresh from editing and publishing The Last Whole Earth Catalog, *and newly famous as a result of its countercultural success. He came to PARC for the tour in 1972.*

Stewart Brand: I went to Jann Wenner at *Rolling Stone* and said, "I want to do this story about what's going on with computers," and he said, "Fine, go ahead and do it." He was doing it based, totally, on his good feelings about the *Whole Earth Catalog*.

Alan Kay: Stewart and I knew each other a bit. He contacted me and said he was going to do a piece basically on *Spacewar*.

Stewart Brand: What I had seen with *Spacewar* was that it drove the technology. It drove the human interface faster than any other application.

Alan Kay: The game of *Spacewar* blossoms spontaneously wherever there is a graphics display connected to a computer.

Stewart Brand: Bear in mind that I had seen people playing *Spacewar* as early as 1962 in the computer labs at Stanford. They were young, randomly dressed people, in a back room of a nice Stanford office crying out with joy and excitement about something—and that something was the game.

Steve Russell: There were games on computers before *Spacewar*, but *Spacewar* was very influential, because it circulated through lots of colleges and universities, so everyone who was interested in computers in the nineteen sixties all knew about it.

Stewart Brand: That was my first inkling that there was something going on around computers that was not the standard New Left take on computers, which was that they were dire instruments of control.

Lee Felsenstein: The line was that they were instruments of the future in terms of automating production. The New Left didn't think that was such a great idea.

Stewart Brand: Amongst the general flow of hippie romanticism, there was an opposition to technology and, by implication, an opposition to science. And I thought that was dreadful.

Lee Felsenstein: Those who sneered at computers had no involvement with them. It was a convenient thing to do from a distance. But if you had any degree of involvement with them you came to love the magic of it, "the romance of it," as Alan Kay put it.

Bob Taylor: When Stewart Brand showed up, I had no idea he was coming. I didn't know why he was there. I didn't know who put him up to it. He said, "Well, we just want to talk to a few people." Stewart Brand didn't just want to just talk to me. He wanted to talk to a lot of people. He wanted to find out what was going on. I said, "Okay, fine." So I encouraged him and he did his thing. I saw nothing wrong with it.

Stewart Brand: Alan Kay was definitely my avenue into understanding the Xerox PARC and a great deal more. He just introduced me to his perspective on what was going to become personal computers, and probably he used that term offhandedly, which I'd then put in print. It became the standard usage. He was very articulate and operated on a level of abstraction, in terms of having a theory of what's going on, that was reminiscent of Doug Engelbart. Though Alan talked and acted like more of an engineer with a strategic vision and Doug acted more like the visionary with the visionary vision, I suppose.

Alan Kay: Stewart wrote a very good piece. Unfortunately *Rolling Stone* was a rag in those days—a West Coast rag, not an East Coast rag.

Stewart Brand (writing in *Rolling Stone*): Ready or not, computers are coming to the people. That's good news, maybe the best since psychedelics. It's way off the track of the "Computers—Threat or Menace?" school of liberal criticism but surprisingly in line with the romantic fantasies of the forefathers of the science…These are heads, most of them. Half or more of computer science is heads.

John Markoff: One of the terms of art then was *head*, meaning "acid head."

Alvy Ray Smith: Corporate Xerox was three-piece button-down suits—businessmen in Western New York State. Then they found out that there was this place out in California that was all beanbag chairs and hippies riding bikes and wearing sandals!

Dick Shoup: They felt it made Xerox look bad and we shouldn't have let it happen.

Stewart Brand (writing in *Rolling Stone*): Until computers come to the people we will have no real idea of their most natural functions. Up to the present their cost and size has kept them in the province of rich and powerful institutions, who, understandably, have developed them primarily as bookkeeping, sorting, and control devices. The computers have been a priceless aid in keeping the lid on top-down organization.

Bob Taylor: Xerox thought themselves to be a very responsible, upright, solid-citizen corporation. They thought *Rolling Stone* was a rag magazine of degenerated hippies and would have nothing to do with them and didn't want *Rolling Stone* to have anything to do with Xerox.

Alan Kay: And so Xerox went batshit when they saw it.

Stewart Brand (writing in *Rolling Stone*): Alan Kay is designing a hand-held stand-alone interactive-graphic computer (about the size, shape, and diversity of a *Whole Earth Catalog*, electric) called "Dynabook." It's mostly high-resolution display screen, with a keyboard on the lower third, and various cassette-loading slots, optional hookup plugs, etc. And that is the general bent of research at Xerox, soft, away from hugeness and centrality, toward the small and personal, toward putting maximum computer power in the hands of every individual who wants it.

Alvy Ray Smith: He had written up Xerox PARC as a hippie place.

Bob Taylor: Xerox corporate had apoplexy.

Alvy Ray Smith: They went nonlinear.

Stewart Brand: They put an armed guard at the door and they didn't let reporters in ever after.

Dick Shoup: But of course from our point of view we were proud to be on the leading edge and trying to move Xerox and anybody else who would listen into the future, because we could see what was coming.

Stewart Brand (writing in *Rolling Stone*): When computers become available to everybody, the hackers take over.

Bob Metcalfe: Right after we finished MAXC we started work on the Alto computer, arguably the world's first personal computer.

Chuck Thacker: And we finally started listening to Bob Taylor, who had been telling us what to do for a long time but we didn't understand it.

Bob Taylor: The design center of computers in those days was the arithmetic unit. Everything, all the design, was focused on making that as efficient as possible. So I said, "Look, the eyeball is the connection between the brain and the computer. The computer therefore has to be centered on the display. And furthermore, it has to be personal. We want one for every user."

Chuck Thacker: Taylor had this bee in his bonnet. He thought that computers were not really just for computing. They were communication devices. And he kept trying to explain it to us, why this was very important, and we didn't understand it. And finally we did, and one of the things we realized was that if you wanted to communicate with it, it kind of had to be personal.

Bob Metcalfe: Can you imagine that? A computer on every desk? Wow. Very controversial in 1973. Why would you want a computer on your desk? What possible use could there be for such a thing? I remember people had that discussion.

Charles Simonyi: Alan Kay had a clear vision of the Dynabook, and he was talking about it all the time.

Alvy Ray Smith: His idea was that the computer should be simple enough that a kid could use it. He saw it all. He was very clear about it.

Alan Kay: Because of my experience with Seymour Papert I got converted from thinking about computers as tools for adults to thinking about them as media, like reading and writing. And once I got that idea, you have to make it usable for children—which means that the user interface has to be very different.

Larry Tesler: We had a diagram with an idea of what the user interface might look like. It didn't look at all like the Mac or Windows; it looked more like a 3-D room: a desk, a file cabinet in the corner, things on the desk, and a trash can next to the desk.

Alan Kay: What we wanted was something like a room: an area for a project where we keep all the tools.

Larry Tesler: The idea was that you could point to stuff that looked semi-realistic, representational, but it was not equal-sized icons like we have today. Alan Kay loved it.

And so Alan Kay, PARC's software theorist, got together with PARC's resident hardware wizards, Butler Lampson and Chuck Thacker. Together they created the Alto.

Chuck Thacker: For a while we actually called the Alto the Interim Dynabook, because it allowed Alan to do a lot of the software things that he wanted to do on the Dynabook. And so he actually paid for the first twelve machines or something like that.

Terry Winograd: With a lot of this stuff at PARC, Alan had the vision, and the tech guys like Butler Lampson and Chuck Thacker actually did it.

Alan Kay: Chuck basically designed the whole Alto from start to finish in just a little over three months. That's why Chuck got that Turing award. It was magic.

Dan Ingalls: So the Alto was a beautifully designed minicomputer that had a removable disc pack and a bitmap display, and lots of computing power for creating images on a black-and-white display.

Chuck Thacker: And so the bitmap display was one of the key good ideas in the Alto. The idea was that you could represent the picture on the screen as a pile of bits in memory and that any program could manipulate those bits and produce a picture.

Alan Kay: The bitmap display acted as "silicon paper" that could show any image and this led directly to bitmap painting, animation, and typography.

Butler Lampson: We thought a really important aspect of interactive computing was to be able to simulate as many of the properties of paper as possible. It's a technology that's been around for a long, long time.

Bob Taylor: We saw that if we had a personal computer with a bitmap display that we could then use it for a lot of different things that people had not been using computers for—and that's what we did.

Butler Lampson: But initially the Alto wasn't that much of a hit because we had no interest in software development. It was only after we started to have things like Bravo that it really got hot.

Charles Simonyi: The Alto had a black-and-white bitmap display. It had a mouse. And so it was clear that quite a beautiful editor could be written for the Alto. The Alto needed an editor and I needed a PhD thesis. So that's why the name Bravo. It was the second experiment for my PhD thesis.

Butler Lampson: Bravo was a joint project between Charles Simonyi and myself.

Charles Simonyi: We finished it in about three months; the first version didn't have formatting yet, but it was attractive enough that Larry Tesler kind of picked up on it.

Bruce Horn: Larry Tesler is one of the inventors of the modeless text editing concept, where you clicked and typed at the insertion point.

Larry Tesler: Bravo was fairly low on modes at the beginning, but if you were in command mode and not typing mode and you typed the word EDIT—which would be a common thing we would type because we were building editing systems—E would select the entire document: E was shorthand for "entire." And then when you typed the D in edit mode, it would delete the whole document. Then when you typed I it would go into insert mode, so when you typed T it would insert the *T*.

Charles Simonyi: So all your work disappeared and T appeared on the screen.

Bruce Horn: Larry would wear a T-shirt that said NO MODES on it. We got a big laugh out of that.

Charles Simonyi: The first really useful version was in early '75. It really worked like a modern word processor.

Chuck Thacker: Bravo morphed into Microsoft Word. Laurel was an almost recognizable e-mail system, recognizable today. PostScript came from us. We had a lot of drawing applications. So it was a good run.

Alan Kay: If you had come to PARC four or five years after it started, you would have been in the midst of something. There were probably 150 Altos by then, Ethernets all over the place, laser printers were starting to appear…

Charles Simonyi: About that time, we had a lot of visitors: the first time probably in history that nonprofessionals, especially spouses of researchers and friends of researchers, came in at night to use computers. That never happened before.

The most persistent visitor was Alvy Ray Smith. He had dropped out of academia after realizing that what he really wanted was to be an artist, and he was searching for a new direction in life. One of the PARC researchers, Dick Shoup, was an old friend and invited Smith to the lab. Reluctantly, Smith came. And when Smith saw what Shoup was up to, he stayed.

Alvy Ray Smith: I was a weekend hippie. I got my PhD at Stanford while I was hanging out in the parks. Dick Shoup has long hair but he was never a hippie. He never dropped acid or anything like that. He's out on his own strange limb of the universe.

Dick Shoup: I've always been into parapsychology and psychic phenomena and strange phenomena. And UFOs are even more of a third rail than psychic phenomena. And as usual there's a tremendous amount of noise and hallucination and just plain misidentification of something in the sky, stuff like that. But there's also plenty of very, very good evidence. I think there's really something going on.

Alvy Ray Smith: And we kind of hit it off at a conference and then he takes off to Xerox PARC. And so we stayed in touch. Even at PARC, Dick was always kind of out on his own branch.

Dick Shoup: There were lots of ideas about what the group should do, and then people had individual things they wanted to do. And I had some ideas about what I wanted to do: graphics. I always wanted to do graphics.

Alvy Ray Smith: He decided that he wanted to build this artist machine. And he built the whole thing. The computational guts of it was a Nova minicomputer—not an Alto. Altos were being made all around, but Dick needed more horsepower.

Bob Taylor: Dick had opted to use a Nova instead of an Alto. Well, that put him at odds with everybody else in the lab.

Alvy Ray Smith: They were always trying to get him to come into the mainstream. And then he built this picture memory, as he called it—a frame buffer. It was a magical device that had enough memory to hold a video picture, so roughly five hundred by five hundred pixels, and then display it using the standard NTSC-compatible video. All this is new.

Dick Shoup: Not too many people were doing color. And in particular within PARC and within Xerox it was anathema. Because the Alto was just black-and-white pixels. And a lot of what other people were doing was just black and white. But I thought that if we were going to do video—to do anything remotely related to television as opposed to office documents— it was important to have color.

Butler Lampson: Which is much more expensive, because you need at least eight bits for each point instead of one.

Dick Shoup: I was diverging a lot from what most of PARC was doing, because they were orienting themselves around office systems and documents and the laser printing and document creation and editing, text editing, book editing, publication stuff, and so forth—for which black and white was much more appropriate. I thought that color was interesting personally, and I thought that Xerox ought to damn well be interested in that, not just black-and-white memos.

Bob Taylor: Then Dick brought in the so-called artists who were not computerists and therefore had little in common with the rest of us.

Alvy Ray Smith: I didn't really understand what I was walking into. I walked into the great heyday of PARC, and Dick took me into his lab and blew my mind.

Dick Shoup: I was making a train wreck of these two giant technologies—computers and television.

Alvy Ray Smith: I went, *Oh, that's what he's been talking about!*

Bruce Horn: I remember the room was always kind of dark. And so there was a big monitor and it was like, *Look at this thing!*

Alvy Ray Smith: Now everybody just grows up knowing what a paint program is and that if you move the mouse this thing happens up there, but nobody understood it at the time.

Bruce Horn: It was psychedelic. It was mind-blowing.

Alvy Ray Smith: SuperPaint was a whole video graphics system: hardware and software. Dick Shoup could do both—he built all the hardware and wrote all the software. Dick's machine was the first to have eight bits of color—256 different colors, enough to be interesting. And I went nuts.

Bruce Horn: It was very, very unlike all the stuff we were doing, which was on the Alto, and in black and white.

Alvy Ray Smith: Nobody had ever seen full-color and graphics ever before, and he had it. You could paint in full color! And I saw it and I thought, *That's my future, that's it!* It might've been the second visit where I stayed for like fifteen hours or something. Anyhow, basically that's how we bonded on a permanent level.

Jim Clark: Alvy and Shoup were always the artists. I do not know how you define *artist*, but I know one when I see one. I never was one.

Bob Flegal: Alvy didn't get hired. I don't think he was even paid; he just came in on his own. We got him permission to come in.

Alvy Ray Smith: One of the first things I did on Dick's machine was the walk cycle—a silly pirate walking across the screen. I started teaching myself from this Preston Blair book by animating on this machine. Blair was this great animator who did the dancing hippos for *Fantasia*. The book had the classic walk cycle and the run cycle and it had a striptease—all the stuff.

Adele Goldberg: You learn by playing. And Alvy just always had that twinkle of someone who played. There was a passion, and that passion was shared by a network of colleagues who he would bring in: creative people—and Alvy is certainly one of those people.

Bob Flegal: Alvy was such a character!

Dick Shoup: People were doing things in the middle of the night, painting and maybe smoking something.

Alvy Ray Smith: I would come home at like four in the morning typically, because I'd stay until I dropped. I'd come home and crash and be up as soon as I could and go back in and keep going and keep going and keep going. It was so much fun. It was just a thrill a minute. Every day I was just flipping out. It was hard to go to sleep because it was so much fun tearing the world apart. Every day everything you touched had never been seen before or never been thought of before or codified before. Just everything that happened was new. We used to sit around and talk about how this was what it felt like to be with Balboa in Panama or something. You know: the first guys ashore, the first Europeans ashore. And you get to name everything. It was the early days, right?

Dick Shoup: Some very strange images came out from Alvy and others.

Alvy Ray Smith: I had a friend, David DiFrancesco, that would come down at night. He was a video artist, that's what he did in the city. And we would jam. I would make pictures for a while and then just turn over to him and he would take what I'd left and he would work away on it and then he turned it back to me. I mean none of it was worth saving, but it was lots of fun—just crazy stuff. And meanwhile we are talking about all the fun stuff that's going on and the hippie thing and drugs and so forth. We really hit it off.

Dick Shoup: Was it a fancy-pants salon art rave? It was getting there.

Alvy Ray Smith: People would be dropping by all the time to see. They'd heard about this machine and they'd drive by. Boy, everybody came to look at it.

Dick Shoup: I'm sure if we had let it get bigger it would've turned into a real party. It was pushing the boundaries and it could've gone much farther, and some people think we had gone too far already. But I didn't think so. It was pretty modest, really, compared to what was going on in the world as a whole at that time. So it wasn't all that outrageous.

Adele Goldberg: Bob Flegal, who's a marvelous artist, and Dick Shoup told me a fabulous story that I bought into because I'm a little naïve. They said that they were making a monster movie. I think what they really told me was they were making some monster sex movie and I left that to them.

Bob Flegal: I was doing a fair amount of drinking at the time so who knows? We had a lot of fun there, all of us.

Alvy Ray Smith: I just sat around and made art all day, that's what I did, I just made art. And it was only after I was there for a while I realized, *Hey, I'm a programmer, I can make it do what I want it to do.* So I started creating programs and I would record the output onto video. And then eventually I edited all of it together into a piece I called Vidbits, and that was my entry into the New York avant-garde art scene.

Bob Taylor: Alvy loved to interview. He loved to be a spokesman for something. And he would talk about PARC as though he were an insider and gave people the impression that he was an expert on PARC. So a lot of people would go to him and say, "Tell us about PARC." And he would, but he didn't know much about PARC. He was a self-promoter.

Alvy Ray Smith: They were trimming back anything that was hippie-like or avant-garde, and that meant me. They didn't want my art associated with them.

Dick Shoup: Taylor got fed up with Alvy and the wacky graphics we were doing and the late nights and whatever else was going on, and he said we shouldn't do that anymore. And he wanted to just cancel the project as a whole.

Alvy Ray Smith: One day Bob Taylor came into the lab—Dick's lab—where I worked every day, and he was puffing on his famous pipe.

Bruce Horn: Bob was into smoking his pipe.

Alvy Ray Smith: And he said, "Alvy, do you agree with me that over here in the corner is more the direct line of computer graphics than what Dick Shoup is doing?" Off in the corner was an Alto, with a black-and-white display, just doing really crude one-bit graphics. I've got eight bits and full color! Who cares about one-bit graphics? All of a sudden inside my heart just sunk and I went, *Oh my God, this guy hasn't got a clue.* He's basically not supporting Dick Shoup, who's got the greatest idea of all. So although I've heard great things about Bob Taylor, my impression of the guy was he just didn't get it.

Stewart Brand: Bob was a close student, actually, of computer games. He took the same attitude I did, that this was where he could see a lot of the most important trends in interactivity that people would have with computers. But he got slapped and constrained by Xerox Corporate back East.

Alvy Ray Smith: I get the call from Jerry Elkind, my boss. He says, "We're going to let you go." And I went, "Well, why?" And he said, "Well, we've decided not to do color." I said, "But Mr. Elkind, the future is color. It's obvious. And Xerox owns it completely!" He said, "That may be so, but it's a corporate decision to go black-and-white." Okay, bye.

Dick Shoup: Alvy got the boot.

Bob Taylor: Alvy was an asshole.

Alvy Ray Smith: Several years later Dick Shoup and Xerox PARC got a technical Emmy for SuperPaint. And what was used to seal the deal was Vidbits—my calling card. And of course Xerox PARC tried to deny any association with this thing.

Stewart Brand: Xerox PARC went from freewheeling and freethinking and openly connected to shut off and under suspicious corporate oversight. It was the beginning of the end.

Breakout

Jobs and Woz change the game

*W*hen Xerox PARC and Atari were both just getting started, so were two Silicon Valley whiz kids: Steve Wozniak and his best buddy and sometime business partner, Steve Jobs. In the spring and summer of 1972, they built and sold "blue boxes" door-to-door in the UC Berkeley dorms. The highly illegal electronic gizmos emitted tones that could be played into a telephone to trick Ma Bell's switching equipment into letting the bearer of the blue box place free long-distance calls. Selling blue boxes was a lucrative sideline, but ultimately too risky. By January 1973 both Steves had dropped out of college and went looking for real jobs. Woz landed at the staid and respectable Hewlett-Packard, while Jobs wound up at the hippest new start-up in Silicon Valley: Atari. After a few years of feverish after-hours hacking by Woz and, for Jobs, a mind-expanding trip to India, the two friends founded their own computer company.

Steve Jobs: I don't think there would have ever been an Apple computer had there not been blue boxing.

Dan Kottke: The blue-box article came out in 1971.

Captain Crunch: Ron Rosenbaum wrote "The Secrets of the Little Blue Box" in *Esquire* magazine.

Ron Rosenbaum: It was only my second magazine story.

Steve Jobs: It was about this guy named Captain Crunch who could supposedly make free telephone calls.

Captain Crunch: You can do what a telephone operator can do, even more—including calling overseas and deciding whether you want to go through cable or through a satellite.

Ron Rosenbaum: Woz's mother sent Woz the *Esquire* piece and Woz couldn't believe it could be real.

Captain Crunch: You can route the calls through different cities, hiding your location. They can trace the calls all they want, they'll never find you! That's how, for instance, I was able to call the CIA crisis hotline into the White House. I told Nixon that we needed toilet paper—that there was a toilet paper crisis!

Steve Jobs: We were captivated. *How could anybody do this?* And we thought it must be a hoax. And we started looking through the libraries, looking for the secret tones that would allow you to do this. And we were at Stanford Linear Accelerator Center one night, and way in the bowels of their technical library, way down at the last bookshelf in the corner bottom rack, we found an AT&T technical journal that laid out the whole thing. And that's another moment I'll never forget. When we saw this journal, we thought, *My God, it's all real!*

Steve Wozniak: So I designed this little box and Steve said, "Oh, let's sell it."

R. U. Sirius: And then Jobs and Woz manufactured those blue boxes.

Captain Crunch: And I met Steve Wozniak, Steve Jobs, and one or two more individuals who were UC Berkeley students. Steve Wozniak was going there for his engineering degree.

Steve Wozniak: We both sold it to people in the dorms for a year.

Ron Rosenbaum: It was the beginning of the Apple partnership, even though as far as I can tell they weren't very good at it then.

Steve Jobs: We built the best blue box in the world! It was all digital.

Captain Crunch: Their blue boxes were not pure. You use those blue boxes, you go to jail. I drilled it into Woz. After I met him in the dorm, I said, "Steve, you really don't want to be selling these. I'm going to give you some advice. I'm talking to you from an analog engineer's point of view, not from a digital engineer's—like you are. I deal in analog signals, not digital. Analog signals are much more complicated and they are not as cut and dry as digital signals." I told him, I said, "When you put those tones together by whatever means of injection into the phone line, you are going to drop a trouble card. They will notice it immediately." I made that very clear to Steve. He didn't seem to care.

Steve Wozniak: I liked exploring the network and being able to convince a Tokyo operator that I was a New York operator and get her to put it over to London and around the world. And I'd call one phone and speak into my phone and it would come out the other one a second later.

Steve Jobs: And you might ask, "Well, what's so interesting about that?" what's so interesting is that we were young. And what we learned was that we could build something ourselves that could control billions of dollars' worth of infrastructure in the world. And that was an incredible lesson.

But after being robbed of a blue box at gunpoint in the parking lot of a Sunnyvale pizza parlor while trying to close a sale, Woz and Jobs decided to shut down their illegal business. The venture wasn't making much money, and besides, Silicon Valley had just spawned something even more exciting than phone phreaking—Atari had just invented the video game industry.

Steve Wozniak: I had a friend that worked at Stanford Artificial Intelligence Lab, so I'd ride my bike over there and it was just open. You know smart people are always open thinkers, and they don't lock doors. So I walk right in. *Spacewar* was running on a PDP-11 there. And wow! You had the spaceship going around, being pulled by gravity into the center. So that was the idea of what could come in arcade games, but it was so expensive. No person could ever afford it.

Nolan Bushnell: The big X-Y displays were about $20,000 at the time.

Steve Wozniak: But then I saw a real arcade game, and that was *Pong*, and I was stunned: *My God—the television set solves the problem of the cost!* The movement toward the games we have today was largely one of how do you do it at a reasonable cost that people can afford. That was a big challenge, you know. I thought, *Oh my gosh. I know how TVs work. I know all their signals for drawing lines in drawing frames and putting dots on the screen.* So I built a little device—twenty-eight little one-dollar chips—and I built my own *Pong*. But it was hardware. See, nowadays you would just write a game of *Pong* in software and if you kind of know what you are doing it might even take you a day or two days. Back then it was hardware. That was a different world.

Steve Jobs: I decided I wanted to travel, but I was lacking the necessary funds. I was looking in the paper and there was this ad that said, "Have fun and make money." I called. It was Atari.

Al Alcorn: I enjoyed getting these people that were really bright that wanted to come to work and do fun things. The semiconductor business wasn't much fun. It was like that Ampex model. Atari was probably the most fun place to work at the time in the Valley.

David Kushner: Atari was the company that established Silicon Valley's casual culture as we know it today. Just the idea of showing up to work in jeans and a T-shirt? Prior to Atari the Valley was the era of Intel and essentially men in suits. With Atari it became smelly hippies in jeans smoking weed. Atari was the counterculture come to Silicon Valley. And so it was no coincidence that one of the smelly hippies that walked into Atari was Steve Jobs.

Steve Wozniak: I did a copy of *Pong* in twenty-eight chips, and two of the chips put a four-letter word on the screen when you missed the ball—I'm into humor—and I showed it to my friends at Hewlett-Packard and then Steve came into town, and he saw it, and he took my board to Atari.

Al Alcorn: One day the personnel lady came in and said, "I know you like to see these walk-ins. I've got one for you: an eighteen-year-old hippie." I said, "Bring him in."

Steve Wozniak: I don't know if he was telling them that he had designed it or he and another guy had designed it or what, but I'm sure they were impressed by it.

Al Alcorn: Jobs had a résumé which was pretty much nothing. He dropped out of Reed College. I go, "Well, was Reed an electrical engineering school?" "No. It's a literary school." He's not an engineer, but he had spark and enthusiasm and he could solder. I needed a tech, so I hired him.

Nolan Bushnell: Jobs was a technician—basically someone who did a lot of the actual soldering.

Steve Jobs: I was, like, employee number forty. It was a small company. They had made *Pong* and two other games.

Nolan Bushnell: He came to me one time and said, "No one in this company knows how to solder!" I looked at some of his work and it's just pristine. He was extraordinary. I said, "Well, teach everybody." So he did. He was not kind in doing it, but I actually think he upped our game somewhat.

Lee Felsenstein: In 1974 I found myself standing in front of Al Alcorn's desk at Atari looking for a job, and it was Jobs who conducted me into that office. Jobs was wearing a nice little white shirt. He might've had a tie. He had a scuzzy beard. It wasn't the mythical Jobs, let's put it that way. He didn't exist yet. I think this was before his India trip.

Al Alcorn: He had this weird diet. He would pass out occasionally. And he said, "Don't call 911 if I pass out. Just push me under the table." *Oh, okay.*

Dan Kottke: Steve had found the Mucusless Diet Healing System, which is a fruitarian diet for healing. It's not like it cured anything. In fact it's not a great diet; it's just sugar.

Nolan Bushnell: Steve was brash. He was difficult. He didn't bathe. And subsequent to Al hiring him, Steve and I got to be friends. Steve was very, very up to speed on Eastern philosophy. I was very up to speed on Western philosophy, and we would have these interesting conversations which I really liked.

Dan Kottke: I don't think that Steve had a spiritual phase—not really. The whole thing for Steve was that he was always looking for a mentor; he was actively looking for a mentor.

Nolan Bushnell: Steve had one speed—full-on—and I was always impressed by that.

Al Alcorn: Jobs was impressive, but not that impressive.

Dan Kottke: Kobun Chino Otogawa was the roshi—or teacher—at the Los Altos Zendo. I'm not sure when Steve first met him, but the very first time I came down from Reed to visit, Steve brought me to the zendo and we did the meditation. Steve was big on doing that a few times a week. Kobun was his mentor.

Kobun Chino Otogawa: Steve always say, "Make me monk. Please make me monk." I say, "No."

Nolan Bushnell: I think Steve perceived himself to be a deep thinker on philosophical issues.

Dan Kottke: Eastern literature, Hare Krishna food, and then psychedelics, that was part of the mix.

Steve Jobs: This was California. You could get LSD fresh made from Stanford. You could sleep on the beach at night with your girlfriend. California has a sense of experimentation and openness—openness to new possibilities.

Dan Kottke: And I just assumed that Steve had taken LSD in high school. But all these years later I pretty much know who all his friends were then—and I can't think of any of them who would have taken LSD; certainly not Woz.

Steve Wozniak: I never used LSD—or even pot!

Dan Kottke: So I might have turned him on.

Steve Jobs: I'd been turned on to the idea of enlightenment and trying to figure out who I was and how I fit into things.

Dan Kottke: Steve was looking for a guru who would tap him on the head and make him enlightened.

Al Alcorn: I remember when he said, "I'm going to go to India to meet my guru." I said, "Well, great!"

Dan Kottke: I dropped out of Reed to go to India. Steve generously offered to buy my ticket, $800. I was carrying a stack of books. Steve was looking for his enlightenment experience. There was one pilgrimage we took, where we went to Kainchi, which was where Neem Karoli's ashram was.

Larry Brilliant: Neem Karoli Baba is the guru of Baba Ram Dass and Danny Goleman and a lot of other people you've heard of a little bit. Ram Dass wrote a book called *Be Here Now*, which was about Neem Karoli Baba. He had a hundred different names: Neem Karoli Baba, Blanket Baba, Maharaj-ji. He was a guru, probably in his eighties, and he had an ashram or a monastery. I say "monastery" sometimes just to make it seem more familiar, but it didn't train monks, it trained people who were interested in being enlightened. By the time they got there Neem Karoli Baba had died.

Dan Kottke: It was deserted, completely deserted. It had been a huge scene only a year before.

Larry Brilliant: Steve wanted to find a guru—there were other gurus in the hills—and he became interested in one called Hariakhan Baba.

Dan Kottke: And then we took this pilgrimage to see Hariakhan Baba.

Larry Brilliant: And Steve followed Hariakhan Baba through the forests, barefoot, and Hariakhan Baba told him to cut his hair, and so he shaved his head.

Al Alcorn: And he came back a few months later. I remember Ron Wayne came in and said, "Hey, Stevie is back." And I said, "Steve who?" "Steve Jobs." *Oh that kid, yeah.* "Oh, bring him in." And I wish I had a camera. He was wearing a saffron robe, shaved head, barefoot, had a Baba Ram Dass book, *Be Here Now*. Steve gives me the Baba Ram Dass book and says, "Can I have my job back?" "Sure."

Nolan Bushnell: Steve was creating too much havoc in the engineering department, so I decided to put him on the night shift knowing that I would get two Steves at the price of one.

Steve Wozniak: I got to play *Gran Trak 10*—the first car game—before it was out.

Steve Jobs: I would just let him in at night and let him onto the production floor and he would play *Gran Trak* all night long.

Al Alcorn: I had the responsibility to get these games out, coin-op games. The number one responsibility is to keep the factory going, right? But Nolan had a very short attention span, and he'd come into engineering and change the product halfway through. *Well, they're never going to ship if you keep doing that!* So I had a pager installed, and when Nolan went into engineering they'd buzz me. That way Nolan could talk all he wanted, but later I'd come in and I would get everything set back on course. I had to do that. I had to get things to ship.

Nolan Bushnell: I had designed this game that was called *Breakout* and I couldn't get any of our engineers to do it. We had this thing where engineers could bid on projects, and the perception was that ball-and-paddle games were over. We'd done *Pong*, *Pong Doubles*, *Quadrapong*, what have you—about as much ball-and-paddle as you can do. And so everybody thought ball-and-paddle was over, and *Breakout* was a ball-and-paddle.

Al Alcorn: And so Nolan went around me and he cornered Steve Jobs to go do *Breakout*. Nolan didn't know that Steve wasn't an engineer. That's a fact. He thought Steve was an engineer, and Jobs never dissuaded him of that notion.

Nolan Bushnell: I thought, *Okay, I'll put Steve on it*—knowing that Wozniak was actually going to do the stuff, and they knocked it out in amazingly short time.

Steve Wozniak: Steve Jobs came to me and he said that Atari wanted me to design *Breakout* and I had only four days to do it. This was a half a man-year project! It was all in hardware. Four days! I didn't think I could do it. But I'd already done *Pong*, so it was really just an extension of that game. It was just putting in the reflection counting when you hit bricks. We went four days with no sleep. Steve and I both got mononucleosis, the sleeping sickness. But we delivered a working *Breakout* game.

Al Alcorn: By this time Jobs was working for Harold Lee, and Jobs shows Harold this *Breakout* game that wasn't even on the list of things to do. Steve alleges to Harold that "I, Steve Jobs, designed it." And Harold looks at the design and was like, "What the hell! I mean I've never seen a design like this." And he said, "If Jobs did this, I'm really blown away."

Steve Wozniak: Supposedly the Atari engineers couldn't understand my design. It was just so beautiful and advanced. I never got to talk to them. I don't know if they knew that I did it.

Al Alcorn: Here was *Breakout* in, like, forty integrated circuits. I couldn't do *Pong* in less than seventy!

Steve Wozniak: Atari was getting tired of their engineers designing games with 150 chips, 160 chips, 190 chips in them.

Al Alcorn: And Nolan had said, "If you make the game with anything less than fifty chips, you'll get a thousand per chip bonus"—pretty well knowing there is no way you could do less than fifty.

Steve Wozniak: Steve had asked me how many chips my design was. I said, "It looks like forty-five chips." He said, "If we get it under fifty chips, we get seven hundred dollars, but if we get it under forty chips, we get a thousand." He was motivating me to get it down.

Nolan Bushnell: They came in at, I think, forty-five chips. The deal was that they got a bonus.

Steve Wozniak: They paid Steve Jobs and then he paid me half the money, supposedly. He told me that we would get paid seven hundred bucks.

Nolan Bushnell: It was about five grand.

Steve Wozniak: Then he wrote me a check for 350. So, whatever. Steve should have been more open and honest with me. He should have told me differently because we were such close friends. But the fun of doing it overrides anything like that. Who cares about money? Right after we finished the game he went up to Oregon, and bought into that orchard or whatever it was.

Dan Kottke: It was apple harvest time and Steve may have stayed there longer than me, but I was there for like a week doing the apple harvest. We were fasting on apples. It was our fruitarian experiment. And that's why the name Apple was in the air.

Also in the air, thanks to Stewart Brand's 1972 Rolling Stone *article, was the idea of a "personal computer." Brand coined the phrase after meeting Alan Kay at Xerox PARC. There were lots of personal computers at PARC—the Altos—but it wasn't until the release of the Altair, a build-it-yourself computer kit, that an ordinary hobbyist could get his or her hands on one. It was all the excuse the hackers in Silicon Valley needed to get together. The first meeting of the Homebrew Computer Club was held in 1975, and no one was more excited about it than Woz.*

Steve Wozniak: I realized that finally the day had come when you could buy low-enough-cost memory chips, low-enough-cost microprocessors that did enough to build an affordable computer. And I thought, *Wow! This is what I've wanted for ten years. I've got it! I'm there.*

Steve Jobs: The clubs were based around a computer kit called the Altair.

Jim Warren: The Altair was what really started the hustle-and-go around microcomputing. But the Altair was a pretty shoddy design.

Steve Jobs: You didn't even type; you threw switches that signaled characters.

Steve Wozniak: A bunch of switches and lights and you can push a button and some ones and zeros go to memory? That's geeky computer stuff; that's not usable.

Jim Warren: And Woz, he looked at the Altair and thought, *Okay but it doesn't do exactly what I want it to do. So hey, why don't I design one?*

Michael Malone: People started realizing that this stuff could be done by people who weren't businesspeople. And they weren't PhDs in electrical engineering. It could just be done by smart kids.

Lee Felsenstein: Woz was there from the beginning with his high school cadre, Randy Wigginton and Chris Espinosa prominent among them.

Randy Wigginton: We knew that everything was changing: Mainframes were not the future, and even minicomputers were just continuing to shrink very quickly. And that's why everybody was so excited.

Steve Wozniak: I'd already built a terminal that talked to faraway computers. And I just did it for fun; it was a hobby.

Lee Felsenstein: The terminal was mentioned in the first meeting. Woz was there at the first meeting and just kept coming.

Randy Wigginton: Then MOS Technology started selling the 6502 for twenty-five bucks each.

Dan Kottke: The 6502 was way, way cheaper than the 8080 in the Altair.

Chuck Thacker: At PARC we used a 6502 in our keyboards for digitizing the stream that went down to the computer. The 6502 was the keyboard controller.

Steve Jobs: It was the computer hobbyist community that first thought about making a computer out of these things.

Randy Wigginton: Woz bought some and then went home and figured out how to hook it into the terminal.

Steve Jobs: And so what an "Apple I" was was really an extension of this terminal—putting a microprocessor on the back end.

Randy Wigginton: That's how the whole Apple I came to be. And it was really fast.

Steve Wozniak: I was a hero at the club. I had built a computer. I had given the plans away for free. I helped others build it.

Lee Felsenstein: And then of course what's missing was any software to make it go, and that's why Woz was writing BASIC in less than 256 bytes.

Steve Wozniak: I called my BASIC "Game BASIC." And my whole idea was if you write a language that can play games, it can do all the things that computers do that I don't know about. Financial stuff? I don't know what companies use computers for. I only know what I like to use them for. And it's games.

Lee Felsenstein: Then this guy turns up at Homebrew in the spring of 1976 who doesn't say anything. I thought of him as "that rat-faced kid who hung around Wozniak and never said anything." He was the business guy.

Steve Wozniak: Steve came into town and I said, "You've got to come down and see this!"

Randy Wigginton: Steve was never very technical, even though he liked to say he was, but he really never was. I don't know that he could actually program or design hardware. I never saw any evidence of it. He was more interested in the Homebrew Computer Club as a way to make a business. I mean he just wanted to work for himself, he always wanted to be in control of his own destiny. He really wanted to be rich when he was young.

Steve Wozniak: Steve wasn't into social good. He was into "Do we have something that can make money?" He was always looking at that. He'd been selling my stuff for five years. He'd come into town about once every couple years and see what I'd created lately and he'd turn it into money. He was really money oriented at the start.

Dan Kottke: So Steve Jobs had nothing to do with the Apple I—Woz just completely did it on his own, showed it off to Jobs, and Jobs was thinking *ka-ching!* Cash registers.

Steve Wozniak: He suggested that we make a company.

Steve Jobs: I sold my Volkswagen bus and Steve sold his calculator, and we got enough money to pay a friend of ours to make the artwork to make a printed circuit board. And we made some printed circuit boards and we sold some to our friends. And I was trying to sell the rest of them so that we could get our microbus and calculator back.

Steve Wozniak: And for a while we were getting the parts on thirty days' credit with no money and we built the computers in ten days and sold them for cash at the Byte Shop.

Trip Hawkins: The Byte Shop was the first of the chain computer stores. In the Byte Shop basically you've got a whole bunch of card tables set up that had a bunch of circuit boards on it, and a bunch of really smelly geeky people talking jargon.

Steve Wozniak: So that was how we ran for, you know, a good year with the Apple I computer.

Trip Hawkins: It was not really a commercial product. It was a kit. They only made maybe 150 of those.

Al Alcorn: Mostly made with parts from Atari, by the way.

Nolan Bushnell: It was considered okay for people to do their own projects with Atari parts. We wanted engineers to work on their own projects, and we were happy to subsidize that.

Randy Wigginton: While Woz was doing the Apple I, he also started thinking about how he could do color graphics. Because what he always wanted to do from the very beginning was to write *Breakout* on his own computer at home.

Steve Wozniak: Back when I designed *Breakout* for Atari I was so tired—in and out of sleep. But you know what? That makes your mind creative. There was a TV set on the factory floor. They only use black-and-white TVs for their games. And this TV set wasn't playing a game but it had a dot going from left to right and right to left. And as it moved it was changing colors: red…green…blue…yellow. There must have been a Mylar overlay or something, I couldn't see it. And I just went back to my lab bench. And so I'm just sitting there thinking, *Color?* It was hypnotizing—like a psychedelic light show or something at a concert.

Dan Kottke: Something about the fuzzy colors in his blurred vision made him think about how if you shift the phase of the dot clock it would cause different colors, which is true.

Steve Wozniak: And here an idea popped into my head—a little way to put out a repeating digital signal of ones and zeros: one, one, zero, zero. It goes up and it goes down. If you think about it, ones are up, and zeros are down. Up and down, up and down. It's not a sine wave, but I know how televisions work: They are going to interpret the signal as red. And if I put the ones and zeros in at a slightly different point in time they are going to call it blue! Oh my God, I have sixteen different colors! Would it work? There had never been a book that ever talked about creating color digitally. It wasn't allowed. It wasn't done.

Dan Kottke: Color television was new and expensive then. And it wasn't like you could read *Popular Electronics* explaining how it worked.

Steve Wozniak: I knew the analog world of color televisions well, but I had crossed over to the digital world.

Andy Hertzfeld: Woz just kind of tuned the Apple II to the frequencies that the television worked on, such that it was synchronous with the color burst signal, the signal that tells the TV what color to display.

Lee Felsenstein: You are supposed to look at color as a two-dimensional vector of the same frequency with a phase relationship. That's an analog way to look at it. When you look at it on the oscilloscope it looks like a series of pulses, which ultimately is an approximation of a sine wave. Shifting it a little bit relative to another series of pulses gives it the color. Woz had looked at color as sort of a bit stream. So what he did was do that sine wave with digital data, and shift it with hardware. And you know everybody's jaw dropped and we thought, *My God! I guess you can do that . . .* It was really spare, sparse, and very imaginative.

Steve Wozniak: Two little parts—maybe twenty-five cents' worth of parts—and it almost converts a square wave into a sine wave.

Andy Hertzfeld: It was like a magic trick, getting color for free just by a sort of alignment with color TV technology. And once I saw that I thought, *Boy, there is genius at work here.*

Randy Wigginton: When he first started showing it at the Homebrew meetings people were amazed, because it was like this little tiny board running things.

Steve Wozniak: Color in those days was very complicated analog stuff—hardware circuits with feedback and resistors and capacitors and inductors.

Andy Hertzfeld: It was an incredible advance costwise, because to get a computer monitor to do that would have cost as much as the rest of the computer pretty much. But Woz designed it to use a standard color television set, which could be gotten very cheaply.

Steve Wozniak: It made it possible for a little one-dollar chip to generate color instead of a thousand-dollar color-generation board.

Lee Felsenstein: Nobody had to pay for that additional hardware to do all that vector generation and phasing and so forth.

Andy Hertzfeld: It was the single cleverest thing in the Apple II. That was one of the first revolutions.

Steve Wozniak: And then I thought, *I wonder if I can write a game that's playable with my slow BASIC?*

Dan Kottke: A video game has to move fast. Maybe the BASIC was fast enough. But most video games would not be written in BASIC.

Steve Wozniak: I had done *Breakout* for Atari. I knew *Breakout*. Would it be fast enough or would it be too slow? Because BASIC is a slow language. And I wrote a little program and put in a bunch of bricks in color. And I changed the color of them and I changed the color and I changed the color twenty times till I had what I liked. And then I programmed in a little paddle that would move up and down as you turn the knob. I built paddle hardware into the Apple II deliberately for the game of *Breakout*. I wanted everything in there. I put in the speaker with sound so I could have beeps, like games need.

Andy Hertzfeld: For the Apple II you could write programs that played music—but it had no real sound hardware. All you could do with the software was hit a memory address, which I still remember: C030. So if you hit C030 it would produce a "click" on the speaker. That was all the hardware could do. But if you did that a thousand times a second with software you'd get a one-kilohertz tone, if you did it three thousand times a second you'd get three kilohertz. And the Apple II was literally full of stuff like that, where by just using tiny little bits of resources it could do amazing stuff. In fact, just looking at the Apple II design gave you the feeling that anything was possible, if you were just clever enough. That's what the main lesson of the Apple II is: that it had infinite horizons.

Steve Wozniak: So I programmed *Breakout*, and in half an hour I had tried a hundred variations that would have taken me ten years in hardware if I could've even done it. So I called Steve Jobs over to my apartment, and we sat down on the floor next to the cable snaking into my TV that had the back off of it so I could get the wires inside. And I showed him how I could change the colors of things and change the shape of the paddle and change the speed of the balls. *So easy!* And he and I looked at each other and we were both kind of shaking, because we knew that the world of games was never going to be the same now that they were software. I mean up until then there were no software games in the arcades. But we now knew that animated games were going to be software. *Oh my God. A fifth-grader could program in BASIC and make games like* Breakout. *This was going to be a new world!* We saw it right then.

Al Alcorn: Jobs was telling me he's going to get this thing funded and all that. And he said, "Well, I want to pitch it to Atari, to you guys." I said, "Well, I'm not the one to do this." I pointed him to Joe Keenan. Joe was the serious, levelheaded businessman. Nolan would probably just say yes, just because he couldn't say no, and anyhow he couldn't focus on this. And basically the meeting did not go well. Jobs came out and, "How'd it go?" "Not too good." Joe had basically kicked him out of the office for mis-behaving. You just don't put your bare feet on a guy's desk! Joe was, you know, Joe was a great guy, but Steve was barefoot and this was just too much. This was like, *Geez.*

Steve Wozniak: The story of Apple is a little misunderstood. It's not like Steve and I did it ourselves.

Al Alcorn: We helped them get the account to get the 6502 microprocessor. Remember these kids were like eighteen or nineteen years old, they can't get a credit card let alone a trade account and a credit account. They'd show up and say, "Al, they won't talk to me!" *Of course not, you're a high school kid. You know, they've got a business to do.* And I said, "Well, I tell you what, I'll put in the word to my friend." And I said, "Talk to these kids, they've got something cute and they'll go somewhere." He said, "Okay." I get a call back a couple weeks later saying that was the funniest meeting he's ever had! "What? What happened?" He said, "Well, these two kids, Steve Jobs and this Wozniak come in, and they were trying to get parts from us and they got no accounts—no trade accounts, no references." And Jobs not only is trying to get an account, he's trying to negotiate a great price. And Woz is trying to tell him, "Just go get an account. Let's just take the parts and worry about that later." Jobs is trying to get Woz to shut up. So Jobs tries to hit Woz under the table and Jobs slides off the chair because he's so thin and he falls under the conference room table and they have to pull him out! "Okay, tell you what: We'll give you a ninety-day account and if you don't pay you're cut off. Okay?" "Okay." That was the biggest account they ever had.

Steve Wozniak: A guy funded us, an angel. And he joined us. He had made his money working in marketing at Intel and as an engineer before. And he was a mentor. He was kind of young but he was wealthy and he owned as much of Apple as Steve and I did. Same amount of stock. Mike Markkula was his name.

Mike Markkula: The two of them did not make a good impression on people. They were bearded, they didn't smell good. They dressed funny. Young, naïve.

Arthur Rock: Mike asked me if I'd be interested in investing in Apple Computer. And I met with Jobs and Wozniak and I really didn't think I wanted to be involved with them. Steve Jobs had just returned from six months in India with a guru or whatever you call them. And they didn't appear very well and they were bragging about the blue box they had invented to steal money from the telephone companies. And I didn't like that too much.

Mike Markkula: But Woz had designed a really wonderful computer.

Al Alcorn: That was the trouble with starting a company when you're under twenty-one. You're underage. You can't drink—but you can start a company.

The fledgling company made its first big splash in 1977 at the inaugural West Coast Computer Faire which—like the Mother of All Demos a decade before—was held at Brooks Hall in San Francisco's Civic Center. The confab was a coming-out party for the talented hackers of the Homebrew Computer Club, and Apple was just one company out of perhaps a half-dozen Homebrew hopefuls.

Jim Warren: I didn't know anything about producing a computer convention. I got on the phone and started calling people. Because I had been editor of *Dr. Dobb's Journal of Tiny BASIC Calisthenics & Orthodontia*, to say nothing of being very active in Homebrew, I knew all of the owners and founders of the start-up companies. So I called them up: "Oh yeah, we really want to exhibit in that. Oh yeah, that would be wonderful." I called up Steve Jobs: "Hey, we're doing this faire." "Hey, I want to be in on that. We want the front of the hall right at the main entrance." "Okay, send me some money." "Sure. How much you need?" "Okay, those are going to be high-priced. Why don't you send me this much?" "Sure. It will be in the mail."

Andy Hertzfeld: I remember walking into the West Coast Computer Faire, I was impressed with just how many people were there.

Randy Wigginton: There were thousands of hippies walking around, people who were more, I should say, freethinkers. Because it wasn't mainstream. It was a lot of just hobbyists and people who had heard something about it. It was more of a counterculture event than I think you would expect.

Steve Jobs: In the first several years of Apple, we were selling to people just like us.

Randy Wigginton: The industry had no direction, it was just sort of growing like mold, just sort of popping up. Nobody knew what was going to win, what was going to lose, what the direction was going to be. Nobody knew about Apple, for sure.

Al Alcorn: I remember thinking, *All those poor bastards making computers nobody is going to buy.* And nobody did, except for the Apple II.

Andy Hertzfeld: Somehow like a magnet I was just drawn to the Apple booth.

Randy Wigginton: Steve was so proud of that booth. I mean he just thought it was the greatest thing ever. I mean he couldn't stop talking about it. You know he was so proud of the whole sign, how beautiful it was, how it looked professional, and how he was going to make us look better than everyone else: That was just totally his baby. Because I mean honestly at that point in the company we had no idea what to do with Steve. Steve was just a pain. He would try to negotiate with vendors: "Ah, I think you need to go sharpen your pencil." It was code for "Give me a better price." Woz and I were making fun of him all the time for that, but he really found his calling developing the booth.

Lee Felsenstein: The Apple booth had a projection color display. No one else had that. And that projector said it all: "You guys can play around with your little alphanumeric terminals and so forth. We got color graphics on the base product!" And a lot of people got interested in that. Their booth was crowded.

Steve Wozniak: That, bringing color to the world: You're not in Kansas anymore! That's why we chose a six-color logo. We were the ones that rocked color, because nobody would have ever expected color on an affordable computer, much less the graphics that we had. We even had pixels so you could almost have photographs on the screen. No, it was so far ahead of its time that everybody else was going to have to sit back and figure out ways to do it.

Trip Hawkins: It had a bunch of innovative features. It had the bitmapped graphics. It had the color. It had a really crappy little speaker and it could make a few beeping noises. That was about it. But a lot of machines didn't even do that.

Andy Hertzfeld: I could tell right away there was something magical about it. Part of it, I guess, was the look of the case.

Al Miller: I remember Jobs standing up at a Homebrew meeting holding the very first plastic case. Somehow that symbolized an advancement—a move into the mainstream.

Trip Hawkins: It looked fabulous compared to all the other junk. I mean if you'd asked a bunch of women at that time, said, "Okay, which of these would you tolerate having at your house?" they would say the Apple II. And they would look at all the other and go, "That stuff is really hideously ugly." It was all this sort of geeky hard-core-looking stuff, and then the Apple II was this beautiful thing.

Andy Hertzfeld: At that point all the other ones were rectangular metal boxes that looked like they were industrial equipment. The message was "Oh, that means it's not for people." They were for manufacturing or business or something, but they didn't give off an inviting feeling, whereas the Apple II looked beautiful. It kind of looked like a futuristic typewriter.

Trip Hawkins: Oddly enough, Steve made a mistake there. Because he had a choice between a character-generator ROM chip that did a traditional upper- and lowercase characters set, and he went instead for one that was all uppercase, either regular or inverse—or flashing inverse, you could do that too. So our first-generation word processors looked like hell.

Alan Kay: But Apple was starting to get interesting. Not because there was anything interesting about the Apple II. The thing that was really interesting was the spreadsheet.

Steve Wozniak: VisiCalc was the killer app.

Alan Kay: We had almost invented the spreadsheet at Xerox PARC, but none of us were businesslike enough. All of us when we saw the spreadsheet, we thought it was just the best damn thing, just fabulous.

Charles Simonyi: My jaw dropped. Wow! Here is this primitive machine and it's doing something we were just dreaming of, just kind of hinting at.

This is big! If you can do this with this machine, imagine what could you do on a more serious machine—like the Alto or the machines that were coming.

Chuck Thacker: VisiCalc was a thing that was useful to businesspeople, and a lot of them. And did increase the productivity of people who did who use that kind of tool—enormously.

Butler Lampson: It was a success of the Apple II and VisiCalc that created the whole personal computer industry, really.

Steve Wozniak: The Apple II was the only one of those computers, the three of them that existed, that had enough memory to run VisiCalc. So they had to write it for this computer only. Everybody else had to go back to the drawing board and make computers that could add floppy disks and more memory. So that was a big leap for us. It was an accident, too; we hadn't really thought about how we were going to make sure that we were well ahead of the competition. We just lucked out.

Andy Hertzfeld: By the fall of '79, VisiCalc was making Apple sales triple every month.

Trip Hawkins: VisiCalc was the number one driver. Word processing was the number two driver. And that was a pretty potent one-two punch.

John Markoff: The Apple II was about getting out of the hobbyist ghetto and making personal computing accessible to the broader world. At its peak the Apple II was sold through Macy's—and a lot of the big marketing push early was educational, in the sense that this was an intellectual activity that you could give to your kids.

Clive Thompson: The Apple II was the first time that computers went really mass market. It was this moment when those great big huge room-scale computers suddenly shrunk down and got into the hands of teenagers. And the moment anything gets put in the hands of teenagers, it becomes part of mass culture. That's a trip-wire moment. That's what happened to the car in the fifties and sixties, it was the computer in the late seventies and eighties—and the Apple II was the vanguard machine.

Steve Jobs: The Apple II went on to sell maybe ten million units over its lifetime. It was the first really successful personal computer—by a mile.

Towel Designers

Atari's high-strung prima donnas

*A*t about the same time that Woz realized that the new low-cost microprocessors would make his dream of a personal computer a reality, Alcorn
and the others up in Atari's product development group realized that they could
use the same chips to power a new kind of machine—a home video-game console
with a slot on top in which one could plug in game cartridges. The Atari VCS
(later to be renamed the Atari 2600) brought the excitement of the arcade to
America's living rooms and established a new economic model for Silicon Valley. The VCS was a platform and the cartridges—software, in essence—were
a predictable revenue stream. Atari, after more than a few brushes with bankruptcy, had invented a golden goose—but there wasn't enough money to build
it. Nolan Bushnell knew he had to sell.

Nolan Bushnell: We'd been on this ride that was almost vertical, and I just
knew that if you just keep pushing all your chips out in the middle, at some
point you're going to lose your stack. And you're not just gambling with
your own future, you're gambling with the future of all the people that
are working for you. The number of times that I didn't have the cash on
Wednesday for payroll on Friday! I can remember one time I had nine
uncashed paychecks personally. We had been running the company with
tricks and mirrors from a cash standpoint—and it was very clear that the
launch of the VCS was going to consume a massive amount of capital. I
think I was tired.

Al Alcorn: We actually had a prospectus for a public offering. But the market went bad, so we couldn't do that and then we had to look for a buyer.

Nolan Bushnell: And then Manny Girard from Warner said that the chairman, Steve Ross, had been with his kids to Disneyland and played Atari

games for a full afternoon. He came and kind of kicked the tires and looked around a little bit and said, "We'll send the jet." I'd never been on a private jet before. I climbed up into this G2 and they said, "We hope you don't mind but we are going to have to stop in Sun Valley and pick up another guest." Clint Eastwood! So all of a sudden we're feeling like really big shits. There were four of us on the team and after we land in New York we get picked up by a limousine and swept off to the Waldorf Towers: not just the Waldorf, the Waldorf *Towers*! It was a nine-room suite that had a pool table and a grand piano. Our normal fare was Holiday Inn. They were clearly warming us up. The next day we go up to meet Steve Ross to sort of negotiate and we came to a handshake agreement in form and structure of the deal by probably four or five o'clock that afternoon.

Al Alcorn: Warner concluded the deal in November 1976. The VCS was already under development, underway.

Chris Caen: This was the very beginning of the whole synergy between Hollywood and Silicon Valley.

Al Alcorn: So we sell the company for $30 million to Warner and we thought we were the richest guys in the world.

Nolan Bushnell: My personal payday was about $16 million, which was considered to be a lot at that time.

Al Alcorn: When I sobered up about a month later I bought a Shelby Cobra. And I bought a house after that—and an airplane. So that was my splurging. Nolan goes off and buys himself a Learjet, starts a charter operation.

Nolan Bushnell: Corporate air transport: We leased jets. That was Nolan living larger than life and getting a little ahead of himself. It turns out that you cannot run a private jet economically. And getting two jets doesn't help at all.

Al Alcorn: Because I was a pilot and I love to fly, I would fly Nolan and the guys back to Grass Valley quite often.

Ron Milner: Atari's secret think tank in the mountains.

Al Alcorn: I go there every week.

Alan Kay: Atari had a group of inventors up in Grass Valley that had been involved in the original machines. They were kind of tinkerers.

Ron Milner: Their operation down in the Bay Area was so busy with production and making the stuff work that they didn't have time to develop the next new thing.

Al Alcorn: My title was VP of research and development. That's what was actually on my business card. Well, I wasn't that: There was no R&D per se—it was advanced product design.

The tinkerers in Grass Valley were developing the VCS; the initials stood for "Video Computer System." From an engineering perspective the VCS was a personal computer much like the Apple II—it was even powered by an almost identical microprocessor. But Atari marketed it as a home video game console for the masses.

Al Miller: The VCS was introduced in the fall of '77, and it was very well received.

David Crane: The VCS was actually designed so that they could sell one console and then sell a *Tank* cartridge and a *Pong* cartridge. Twice as many sales, right? So the hardware was designed to play those two games. And if it could do any other games? "Great! We can sell more cartridges." So that was the push: "Let's see if we do more arcade games on this home hardware."

Larry Kaplan: I saw their ad in the *Mercury News* and applied for the job. I was among a hundred applicants. They later told me they hired me because I had purchased an Altair. I started in August of 1976. I joined Atari so I could play *Breakout* for free.

Bob Whitehead: I was the second programmer hired at Atari. And then Alan Miller and Dave Crane came on my heels, just behind me. We were all brought on to program games for this new programmable cartridge-based system.

David Crane: Myself, Al Miller, Bob Whitehead, and Larry Kaplan, we generally worked together on our projects.

Bob Whitehead: We weren't the only guys being hired to program these things. What seemed to happen was just a group of us began to separate ourselves, professionally. It seems like a few of us just had a better knack for doing games. We became fast friends.

David Crane: Larry Kaplan made *Air-Sea Battle*, which showed that it could do more than just *Tank* and *Pong*. *Air-Sea Battle* had two little howitzer-like guns at the bottom, one player, two player. They had fighter jets and things flying over and you would shoot missiles at them, and there was a lot of action going on in the game—far more action than the designers of the VCS had originally expected. Al Miller did a game called *Surround*, which was originally an arcade game from Atari. Bob Whitehead went off the other direction and did an original title: a baseball game where you actually batted and pitched and all that kind of stuff. I did *Canyon Bomber* and *Depth Charge*, which were ported over from the Atari arcade games. The arcade versions cost like $4,000 each, and I put them both into a VCS cartridge that you could buy for $19 and play on your little $200 machine.

Al Alcorn: They were a huge hit. Warner got there at the right time.

Larry Kaplan: Within a year the company had enough money that they put in a sauna in the engineering section. We were spoiled, and we worked the hours we wanted to work, on any schedule, and a game was finished when we said it was finished.

Al Alcorn: We moved to a brand-new facility out in Sunnyvale next to Lockheed. Clearly the money was helping to pay for it. At first it was the same Atari.

Bob Whitehead: Atari began to change when Warner brought on Ray Kassar as CEO. Ray Kassar came from Ralph Lauren.

Al Alcorn: He has never been on the West Coast, didn't understand consumer products, electronic products, didn't understand games or anything like that—but he was a businessman.

Ray Kassar: When I arrived there on the first day I was dressed in a business suit and tie and I met Nolan Bushnell. He had a T-shirt on. The T-shirt said: I LOVE TO FUCK. That was my introduction to Atari.

Al Alcorn: Ray had an executive parking spot for his chauffeur-driven Rolls-Royce. He had a helicopter landing pad. He had an executive dining room so the executives did not have to rub shoulders with the unwashed. It was a term they liked to use: "the great unwashed." He was really tone-deaf to what was going on around him.

Al Miller: I was at one of the very first meetings that Ray had with the entire group. There were probably eighty to a hundred technologists in the room and somebody asked him, "Well, what kind of experience do you have dealing with creative people?" Because, you know, we were creating entertainment. And he said, "Oh, I've had a lot of experience dealing with creative people. I've worked with towel designers my entire career." And I don't know about the other people, but I was just flabbergasted when he said that, because it showed he was entirely clueless about the industry and what we were doing.

Alan Kay: Ray was a great guy and a very good businessman, but his former experience in computing was that he was an expert in Egyptian cotton.

Ray Kassar: I had not played video games before.

David Crane: Marketing circulated a memo: a single sheet of the sales of game cartridges for the previous year by percentage. It said number one was baseball or whatever and it did 22 percent of sales. And then number two was this and the percentage, and the percentage all the way down. And basically what the memo said was, "Here's what's selling, this is what's hot. Do more like that." They had no idea what it took to make a video game, what the effort was, and how you actually enticed people to play it and make it interesting, all those things. They just wanted more: more like the top of the sales chart than the bottom.

Al Alcorn: Miller, Crane, Whitehead, Kaplan—those guys could in three months design a video game cartridge that would generate ten to twenty million dollars in sales.

Bob Whitehead: The four of us began to look up the numbers and we found out just the four of us were responsible for in excess of $200 million in sales, just our games. That was after a year and a half or so of sales.

David Crane: Since we went out to lunch together and tended to look at these things together, we totaled up the sales of all the games that we did. And that was interesting, because about 60 percent of the total sales from the previous year were games that the four of us did. Twenty percent of the revenues from the previous year were people who had left Atari, just to go off and do other things. And of the thirty people left, they accounted for about 20 percent of the revenue.

Al Miller: And so I researched the music and the book publishing industries to develop a contract that I thought would be fair for my situation at Atari. And I submitted it to my management, saying this is the kind of relationship I want. I want recognition for my work and I proposed something like a 2 or 3 percent commission royalty, which was pretty low relative to the music and the book publishing industry. And I started having discussions with my management and then it bopped up to senior management, Ray.

David Crane: We went into Ray Kassar's office and we told them all about it.

Al Alcorn: They said, "Well, come on, give us a piece." They didn't want a lot, just five cents a cartridge.

David Crane: We said, "Look, obviously there's something that we have that is marketable, that makes value."

Clive Thompson: One of the things that are so interesting about those early games is how they wrestled with the great geopolitical anxieties of the times: not just global thermonuclear war, but also the idea that society was robotizing. *Space Invaders* was about these gibbering creatures from the id. You had games that literally made the subtext, text—like *Missile Command*. Games were responding to these floating senses of technological threat that were in the air. There is really something dreamlike about them. They were the poetry of the age.

Al Miller: We were doing creative efforts much like book authors. In that era it was single-person work. We did all the music. We did all the art. We did all the programming. We did virtually all of the conceptual design as well.

Chris Caen: You got to remember that back then you had one software programmer per product. It's not like the world now, where you have hundreds of programmers working on a decade-long plan to deliver a single software experience.

Howard Warshaw: Programming created this new phenomenon: the intellectual blue-collar workers that are as smart or smarter than the managers who are working with them. Ray and his staff weren't prepared to deal with this.

Al Alcorn: Ray's attitude was that the engineers were a bunch of high-strung prima donnas.

David Crane: And he said, "Well, you know this is a corporate product. It's an engineering product. There are hundreds of employees at Atari. And if that guy over there didn't do his job we wouldn't have sold $60 million. If that guy over there didn't do his job we wouldn't have sold $60 million. In fact if the guy on the assembly line hadn't put them together we wouldn't have made $60 million on the games. So you are actually no more important than the guy on the assembly line who puts them together." That was the end of that meeting. And in fact we were walked out by a senior vice president who just kind of chuckled at the way that meeting went and said, "Well, guys, it's been nice knowing you." Because he knew that we'd be gone soon.

Chris Caen: The idea that people would leave Atari to go start their own software company was baffling to the powers that be. It would have been like leaving Ford to start Edsel—why would anyone do that?

Bob Whitehead: We were starting to say, "You know what, enough is enough. We're not respected here. We can definitely make some big money elsewhere in the new emerging game business." And so we decided the easiest thing to do was write games in BASIC, the programming language BASIC. So one of us says, "I know a guy over at Wilson Sonsini." They were a pretty well-respected Silicon Valley law group.

Jim Levy: The mother church of venture law.

Bob Whitehead: We went over and had a meeting and one of the lawyers says, "You know what, I'm talking to a guy named Jim Levy who's kind of doing the same thing. He's from the music business, and he's been working with a group that has been distributing BASIC games and talking about splitting that off and starting his own company."

David Crane: He said, "You need to talk to this guy, because you guys have all the technical expertise but you don't have the business experience. So he's going to be your CEO."

Jim Levy: So, these guys pull up out in front of my house about two in the afternoon. They pile out of this car and they're all really young. I think the oldest was thirty, Larry Kaplan. But Al Miller, Bob Whitehead, David Crane were all like twenty-three, twenty-four, or twenty-five, somewhere in there. I wasn't that much older. I was thirty-four at the time. So they sit down and start explaining to me what they do. I had seen arcade games and I think I had brushed up against things like *Pong* and so forth, but I hadn't really seen or taken notice of the cartridge-programmable system, the VCS.

Clive Thompson: The Atari VCS games were definitely the rock and roll for that generation. It was a deep and intense culture for the kids.

Bob Whitehead: Well, it wasn't very long with our meeting with Jim where we started asking each other cool questions. It could have been prompted even by Jim himself: "Why can't we just do Atari games? Cartridges?"

Jim Levy: So they said, "Do you realize Atari has a million-plus systems out there and they are the only ones doing software for it?" I said, "Oh, okay." They said, "So would you consider doing your start-up, but starting with video games, Atari-compatible games instead of personal computer software?" And I said, "Well, that's an interesting idea. Tell me more."

Bob Whitehead: Within a pretty short period of time Jim put together a business plan.

David Crane: It all went pretty quickly after that.

Al Alcorn: Those kids went off with a lot of energy and a lot of venom and started a company.

David Crane: We ended up getting venture-capital funding and forming Activision in 1979.

Clive Thompson: There were video games before the VCS. There were arcade games. But once games became things that you could play over and over again while sitting in your basement there was a new economic logic. An arcade game wanted you to lose after about two minutes so that you would put another quarter in. But once a game became something that you could buy and take home with you, it created a different relationship with the game—you could explore a game that was longer and richer. So you started getting these games where you had to go roaming around and find things and complete this quest. It became storytelling.

David Crane: Programming for the VCS was the greatest technical challenge that has ever existed in video games. My goal was to do a game with an actual human figure in it. Back at the time your main character was a tank or a jet airplane. Creating a realistic-looking person was very difficult. And I would walk around the lab and freeze my motion and sketch where my legs were. Animators do that sort of thing; they often use themselves as models for their animation. And I would do that for hours and hours on end until I had a pretty nice character.

Jim Levy: David was really a computer animator and still is.

David Crane: I said, "Okay. There's my little running man. What's he doing? Well, he's running. He's probably running on a path." So I drew two lines that represented a path. "Where is this path? Let's put it in a jungle." I drew some trees. "And why is he running?" I put threats, treasures to collect, that sort of thing, and figured out these quicksand and tar pits and alligator heads to jump over and just tied a bunch of this stuff together. And, really, I had a sketch in about ten minutes which defined the whole game.

Jim Levy: *Pitfall* made David a rock star. *Pitfall* was our *Space Invaders*. It was a landmark game that had spin-offs: Saturday morning cartoon shows and an arcade game. I think it was the first time that a home game had spun out to the arcades instead of the reverse.

David Crane: *Pitfall* took me about a thousand hours sitting at a computer to do. Now that's a long time except it's only a few months. Six or seven months.

And it turns out that *Pitfall* earned Activision $50 million wholesale. So I made for them $50,000 an hour given the thousand hours I worked on it.

Jim Levy: The party for *Pitfall* was spectacular.

David Crane: Live animals hanging in cages from the ceiling in the jungle theme. Three live bands, a marimba band just in the entryway before you got into one of the three ballrooms. Those parties, everyone in the industry and then some would come. We'd have four thousand people coming to a party at Activision. Those were good times.

Larry Kaplan: Activision was so flush with cash, the programmers were traveling first class, had limousine service, company cars, a private chef, and a DO NOT DISTURB sign on the door and the phone.

Activision, one of Silicon Valley's earliest software companies, made millionaires and celebrities out of its programmer-founders. Meanwhile, things were starting to go downhill at Atari.

Al Miller: Atari just dropped the ball because senior management did not understand the technology. They had pissed off the really capable people.

Nolan Bushnell: What we were planning was a network for playing games online. It was going to be very simple; we were going to have a closet full of computers in every area code, because in those days a local call was free but long distance was very expensive. And then we were going to link the closets with a T1 line. Now when you think about that, that's fundamentally the architecture of the internet. Our IP stack was very, very similar to the internet's IP stack. And that was killed! Now think about if Atari owned the internet. That would have been massively cool!

Al Alcorn: There was a fear of introducing new products that might fail. Even though we were making the billions of dollars and if it failed it wouldn't even be a blemish, it would be just nothing. But to a big corporation it was, "What if we introduced this thing and everyone laughed at us? Thus if we don't introduce this thing they won't laugh at us." So that was going on. It was clear that Atari was not going to release any new products.

Nolan Bushnell: They started killing off the projects that we were working on because every prototype looked like a tiny business to them—not realizing that all businesses at the outset look like tiny businesses. They just thought they were geniuses. You know it was just a corporate culture that became very toxic.

Al Alcorn: So Nolan and I go off, well, actually Nolan goes off first. Nolan left in '80, I left in '81. And we were just going to start a bunch of companies.

Nolan Bushnell: I had a lot of ideas for businesses that were somewhat disparate, and so I got a facility, centralized Xerox machines, centralized health care. The idea was that if I funded an engineer to do something, they could go in, get a key, a stack of papers on their desk, sign their name sixty-four times, and within forty-five minutes they'd be incorporated, they'd have payroll set up and be working on their project without all this other noise. Catalyst was the first incubator. It was absolutely unique at the time.

Stan Honey: Upstairs at Catalyst, Nolan had a great big office suite and this enormous desk that had computer screens buried in it. The feeling of it was kind of tablet-like. It was a prediction of things to come, pretty interesting. Nolan has always been involved in a ton of stuff.

Nolan Bushnell: I did automobile navigation. I did this kiosk project called ByVideo, which was the first online shopping kiosk. I had a toy company. I had a little robotics company, and life was good.

The most notable spin-out was Etak, the world's first commercially available computerized in-car navigation system. Bushnell sold Etak to Rupert Murdoch for thirty million dollars—the same price that Warner had paid for Atari.

Al Alcorn: So now Ray was beginning to learn a lesson but he overlearned it, and now he's trying to coddle the engineers. The office was too noisy or whatever, and so they had to have a facility over at Santa Cruz or something.

Steve Perlman: I was an Atari intern. I remember going to the office where the software engineers were—I had to ask them something, I don't remember what. So I knocked on the door, and I heard from outside a voice say, "Who is it?" I say, "It's Steve, the summer intern." The voice says, "He's cool, man, let him in." So they open the door and let me in and blue smoke just comes flowing out.

Howard Warshaw: There was a lot of dope that was smoked at Atari when we were there.

Steve Perlman: They're passing around a joint, and the guy says, "Steve, I call this my number seven. Do you know why?" I say, "Why?" He says, "Well, it's got three bits in it: It's Maui Wowie—with hash—dunked in hash oil!" If you know binary, seven is one-one-one: three bits in binary. I'm like, "Great! Cool. I can understand how that can help with your creativity, but I've got work to do…"

Al Alcorn: And toward the end Ray got Larry Kaplan to jump ship from Activision and come back to Atari. He was going to run engineering, at least get the cartridges done. Larry was the first manager that could read the code, and he could see these guys were doing nothing. They were just goofing off, getting money, and nobody could assess them. So Larry came back and told them, he said, "You've got to start firing people." "Oh no, we can't fire them." "Look, they are out of control. They are not doing anything. We've got to show them who's boss because they don't take orders." And Ray wouldn't let him do it. So I think Larry gave up at that point.

Larry Kaplan: I left to start a game hardware company to build a replacement for the VCS, since no one had done anything comparable.

Al Miller: At Activision we kept waiting. "When is Atari going to introduce their next great game machine?" And they never did.

Alan Kay: Atari was greedy and they were making a shitload of money off really obsolete games toward the end there.

Al Alcorn: You know in Silicon Valley if you don't obsolete yourself somebody else will, right? The Warner guys didn't really understand that. They were from an East Coast company and thought that they had an evergreen kind of product. They thought that they would just sit back and mint money for the rest of their lives selling the Atari VCS for the next twenty years.

PARC Opens the Kimono

Good artists copy, great artists steal

*T*he *Apple II—which Woz designed in order to play* Breakout *at home— found its greatest success with businessmen who bought it to play with spreadsheets at work. Seeing this, Jobs concluded that Apple needed to turn away from games and reorient itself around the needs of business. The next computer should be an office-of-the-future machine. But what did that mean, exactly? The answer came when Jobs went to Xerox PARC to see the Alto. In December 1979 Jobs was ushered into the ivory tower to get the full demo and more. He saw the laser printer, networking, and something called object-oriented programming. But what impressed Jobs most about the Alto was the mouse. He was twenty-four at the time—too young to have attended Engelbart's Mother of All Demos—but he fixated on the mouse, which was Engelbart's most fundamental breakthrough. With the mouse one could point and click, cut and paste, doodle and paint. It was the key to the virtual desktop on the Alto's screen: the office of the future. After seeing what PARC had invented, Jobs knew exactly what he had to do.*

Bruce Horn: Bob Taylor's group at PARC was CSL, the Computer Science Lab. That group was very computer science heavy: real heavy hitters in the field, in the industry.

Bob Taylor: The people in CSL despised Steve Jobs, to a person. Steve Jobs was a college dropout who didn't know shit about computing. The CSL people were PhDs in computer science.

Alan Kay: The reason that the Homebrew-type stuff works is because some guys with PhDs put a lot of complicated electrical engineering into the chips.

Bob Taylor: We laughed at his Apple II! Compared to an Alto it was laughable.

Alvy Ray Smith: I knew this hobbyist stuff was happening. There were people at Xerox PARC saying, "Alvy, you've got to come over to the garage." But I just didn't give a damn. I was like, *Those are just toys.*

Larry Tesler: I went to Homebrew Computer Club meetings. And I was telling everybody that Xerox is going to miss the boat unless we get into this. This is really taking off! Xerox headquarters got concerned. And a guy named Roy Lahr, in the business development group in East Coast Xerox headquarters, came out to PARC. He said, "We have a task force—we're going to study these personal computers. We're going to come to some decision about what the company should do. And we want your input. But the decision is going to be made by the management of the company." So we all gave input. And as he went around meeting people, he found that there were three or four of us that thought personal computers were going to happen and that Xerox couldn't wait until PARC came out with something. And they wrote up their findings and recommended that they, in fact, partner with somebody like Apple to get into this market. And so they came up with a deal where Xerox would be allowed to invest in Apple.

John Couch: Steve said, "Well, you know, if you open the kimono of Xerox PARC, we'll let you invest." It was really an exchange of favors.

Larry Tesler: Apple would be allowed to see some PARC technology, which they've heard of from the graduate students that they'd hired from various places. And PARC would help Apple to do the kind of hardware they were doing with mice, and bitmap displays, and so on. They would give a limited amount of technical help, and more sourcing help, help them find manufacturers who would make things like the mouse for them. And Xerox's view was this would be a benefit to Xerox.

Alan Kay: By '79 we had shown this system to about three thousand people. Why did Steve Jobs decide he wanted to go there? His people had seen the demo, and they just wanted Steve to see it.

Adele Goldberg: Mind you, they could have come at the once-a-month demo day if they wanted to. But they wanted a special visit. So Xerox closed off the foyer and they set up a machine up there.

Alan Kay: There was a table with their display console and stuff and the demo was given.

Adele Goldberg: It was given by Dan Ingalls and Larry Tesler, with me in attendance.

Alan Kay: And our typical way of doing things is pilot/copilot things where people trade off. And Steve was sitting to the left and a couple of his people were there. Adele and I were standing in the back of the room watching this.

Adele Goldberg: In fact it wasn't one demo, it was two demos. They were maybe a week apart. But you've got to separate these because they were very different. The first demo was a management demo. So this was "look at multimedia as built on the Smalltalk system."

Smalltalk was Alan Kay's revolutionary object-oriented programming environment that was developed at PARC for the Alto. Thanks to the Alto's bitmapped graphics capability, Smalltalk was the first system to have a full-blown graphical user interface, or "GUI"—a feature that proved to be very influential.

Dan Ingalls: I gave a demo in which I was sort of showing off the programming environment.

Alan Kay: It was kind of like what we have today, but better. It was a completely integrated system that was not made up of applications but "objects" that you could mix and match anywhere. You've got a work area and you can bring every object in the system and you just start: You've got every tool, you've got every object, and you can make new objects. It was programmable by end users, and it had the famous GUI.

Larry Tesler: It was the first one that was graphically based—overlapping windows, a mouse, stuff like that.

Adele Goldberg: It was very much a GUI demo.

Dan Ingalls: I was scrolling up some text and Steve said, "I really like this display."

Alan Kay: At that time, when text scrolled it did so discretely: So jump, jump, jump, jump—like that.

Dan Ingalls: And Jobs said, "Would it be possible to scroll that up smoothly?"

Alan Kay: "Can you do it continuously?" Steve liked to pull people's chains.

Dan Ingalls: Because it did look a little bit jerky, you know?

Alan Kay: So Dan or maybe Larry just opened up a Smalltalk window and—

Bruce Horn: —changed a few lines of code in Smalltalk, and in the blink of an eye it could do a smooth scroll.

Alan Kay: And so it was like, "Bingo!" And so Steve was impressed but he didn't know enough to be really impressed. The other Apple people just shit in their pants when they saw this. It was just the best, best thing I've seen.

Dan Ingalls: I think that sort of blew Steve's mind. Certainly, just about anybody who worked on development systems responded really well to that particular demonstration. It really showed how the Smalltalk system could be changed on the fly and be extremely malleable in terms of stuff you could try out in the user interface.

Alan Kay: On Smalltalk you could change any part of the system in a quarter of a second. These days it's called live coding, but most of the stuff today—and virtually everything back then—used compiled code, so your programming and editing and stuff like that was a completely separate thing. You had to stop the system and rebuild.

Dan Ingalls: So you had a regular sort of programming environment sitting there in front of you with windows and menus and the ability to look at code and stuff. But the neat thing about it was that if you edited the code, it would actually change the code that you were running at that very moment. The object-oriented architecture made it really, really quick to make changes.

Bruce Horn: Even today if you wanted to change something at the level of the operating system in MacOS or Windows, it's simply impossible unless they have decided to give you that option to smooth-scroll or not. It's simply impossible to get to the source code to do that unless you're an employee—and even then it would probably take six months to get that feature in. Whereas within Smalltalk, you pop up the menus, accept the change, and it's working right now. You can't even do that today.

Larry Tesler: We were very proud of that, being in the Smalltalk group.

Bruce Horn: So long story short, that was a blinding insight for Steve. *Oh my God! This object-oriented programming stuff that PARC has is superpowerful!* He saw that power.

Larry Tesler: Steve was just really focused on the UI and the way it looked—the simplicity and the beauty of it.

Steve Jobs: I remember within ten minutes of seeing the graphical user interface stuff just knowing that every computer would work this way someday. It was so obvious once you saw it. It didn't require tremendous intellect. It was so clear. It was one of those sort of apocalyptic moments.

Adele Goldberg: Steve just sat there, didn't say a word. He sat there staring. And then they left.

Trip Hawkins: Steve had his private visit. So then he comes back to Apple and he goes, "Okay, we've all got to go back and see this again."

Steve Wozniak: There were about five of us.

Dan Kottke: It was the members of the Lisa team. The Mac at that point in time was a tiny little thing: just a processor, some RAM, and a video generator. It didn't even have a mouse.

Trip Hawkins: So we were the brain trust that had to go back and see what Steve was all excited about and decide what we were going to do about it.

Adele Goldberg: They came back with the entire development team: Steve Jobs and the entire Lisa programming team. This is what people don't understand.

Larry Tesler: We arranged a bigger demo where we'd show them more.

Adele Goldberg: I said, "You're blooming crazy!" You can't give a demo to the programming team. That's a "how" not a "what" presentation. That's a complete giveaway.

Bob Flegal: Adele did not want to do that demo. In fact, she turned on her Adele meter pretty high, like only she can do. She really caused quite a stir with management.

Adele Goldberg: I simply said, "You order me to do it, and I'll go do it. But it's a mistake." They didn't want to know why it was a mistake.

Larry Tesler: Nobody wanted to show them everything. We didn't even want to show them everything that was being done in Smalltalk, or everything that was being done on the Alto, let alone show them other hardware and other systems.

Adele Goldberg: Xerox inadvertently did a public disclosure at a level that freed the entire team, my entire research team, from its nondisclosure. That's the story that no one wants to talk about. A legal public disclosure is what happened.

Larry Tesler: We wanted to show them enough that they would build bitmap displays and mice and laser printers and other things that we could get at a consumer-ish kind of price, because they were buying it for a bigger market than we were. And that required showing them a little bit more. So we did.

Bruce Horn: Atkinson was just looking at it so closely—trying to figure it out.

Andy Hertzfeld: Bill Atkinson was the main graphics engineer at Apple, and he was doing the graphics on the Lisa.

Bruce Horn: He was nose to the screen, just trying to figure it out.

Trip Hawkins: We were not complete strangers to bitmapped graphics, because Apple II had them. It's just what you could do with them on an Apple II was kind of limited.

Trip Hawkins: What PARC had was completely innovative thinking about the entire user experience.

Steve Wozniak: Multiple windows on the same computer screen? When I saw that I said, "God, it's like you've got three computers in one! Once you have that you'll never go back." The Smalltalk language allowed them to write software in a different way than ever before.

Adele Goldberg: So we proceeded to have a Smalltalk language and implementation discussion, practically giving the answers to the team.

Larry Tesler: I was getting better questions from the Apple management than I ever got from the Xerox management. It was clear that they actually understood computers.

Adele Goldberg: We walked through all the details, all the details.

Trip Hawkins: So that was the seeds of it. Of course then we come back and we're thinking. *Yeah! So what do we do about this?* And by that time we already had a project.

John Couch: The Lisa.

Trip Hawkins: A dinky little R&D project, and Steve and a handful of engineers had got started on that, and we had a lot of debates about which processor to use. And there had been this debate where a lot of the engineers were thinking we don't want a mouse.

Steve Jobs: The problem was that we had hired a bunch of people from Hewlett-Packard, and they didn't get it.

Trip Hawkins: And this became a real bone of contention.

Steve Jobs: I remember people screaming at me that it would take us five years to engineer a mouse and it would cost $300 to build. And I finally got fed up. I just went outside and found Hovey-Kelley Design and asked him to design me a mouse.

Dean Hovey: I had no idea what a mouse was.

Jim Sachs: Steve Jobs said, "It's going to be the primary interface of the computer of the future." That's how he described it, which we thought was laughable. You could balance a checkbook without a mouse, and you could write BASIC programs without a mouse, and that was about all people were buying Apple IIs for. That and playing games, and there were already game paddles and joysticks, or keyboard input, and nobody could think of what a mouse would be used for in a game. We of course went back to the office and snickered, and thought, *Maybe he hasn't had enough meat in his diet?* But if he was willing to pay us $25 an hour to do this, we would design a solar-powered toaster for him. So we said, "Sure, Steve."

Jim Yurchenco: So the first thing we did was to go out and see what ideas we could steal, which is how engineers work—why reinvent the wheel?—and one of the things we came up with was a trackball module: a very large trackball used in Atari game machines.

Nolan Bushnell: We did a trackball for the coin-op business. We used it in several games: *Missile Command, Centipede, Atari Football*—that was a good one. But we really had problems with having it jam and all that, and so we solved a whole bunch of technical problems.

Jim Yurchenco: We looked at it, and what they were doing was interrupting beams of light with slotted wheels—a different orientation, and everything was at a much larger scale, and of course the ball was supported by the structure and was depending just on gravity, but it sure seemed like a real promising approach.

Jim Sachs: The mouse Xerox had had a mean time between failure of something like one week, at which time it would jam up irreparably, or the little wire fingers would break. It had a very flimsy cord whose wires would break.

Dean Hovey: It just shows, particularly in Silicon Valley, how you take a good idea and run with it and improve it. It's very rare that a lightning bolt strikes and you come up with something that's never been thought of before. It's a lot more taking from this, taking from that, and trying to make something work, and going for it.

Steve Jobs: In ninety days we had a mouse that could be built for fifteen bucks that was phenomenally reliable.

Jim Sachs: People often ask me, when they see the mouse patent on my wall, whether I invented the mouse. I point out that I did not invent the mouse. Doug Engelbart invented the mouse. His mouse was crude but effective, and remarkable because he demonstrated it, along with a graphical user interface and all kinds of other things, at Brooks Hall in 1968. Engelbart's mouse consisted of two rotating disks attached to potentiometers and a big, clunky wooden box.

Dean Hovey: Xerox had obviously implemented various versions of it, but the execution around it was still flawed. It wasn't something that would scale such that it could change the world as it had the potential to.

Jim Sachs: I credit Steve Jobs with having the vision that that is the way the masses would use computers. One would never have concluded that that would be "the primary interface to millions of computers in the future." But he was correct. I think he's not been recognized enough for that. I think people tend to spend more time thinking about the young, brash, rude, obnoxious Steve Jobs of the 1980s, not that he truly had a vision for seeing this gem in a lab in Xerox PARC, and saying, "They may not be able to commercialize it, but we can."

Dean Hovey: We had working prototypes in the late 1980, early 1981 time frame.

Jim Sachs: It's also interesting to note that in 1980, at Hovey-Kelley we didn't have access to a computer to plug a mouse into! So we were a little puzzled as to what this thing really was going to do.

Trip Hawkins: We gave it to Bill Atkinson, and he wrote a driver that did stuff graphically. And until that happened, there had been this debate where the mouse was a real bone of contention, but when Bill wrote that driver and it did something and at least half the engineers go, "Oh yeah. Yeah. I get it!"

Larry Tesler: And after that, it just kind of got away from Xerox.

Adele Goldberg: We started losing personnel to Apple over the next couple years.

John Couch: A lot of the people that were at Xerox PARC ended up joining my team, like Larry Tesler and others.

Larry Tesler: And Apple ended up getting all this technology, improving on it.

Steve Wozniak: Steve Jobs felt that Xerox had this great technology, but Apple was the one who could make it cheap and affordable—like Woz had done with the Apple II.

Dan Kottke: The graphical interface with the mouse and windows? That was pretty much taken from Xerox PARC.

Bill Atkinson: We tried a lot of things, we didn't just take what Xerox had done. In fact, many of the things that we did, they didn't have at all.

Trip Hawkins: For example, the Alto did not have icons.

Bruce Horn: And so the double-click to open, drag-and-drop, double-click to launch an app that had the associated file? So Smalltalk didn't have such things. There were no entities that were files in Smalltalk.

Andy Hertzfeld: The question for the sake of history is "How much did Apple get from Xerox PARC?" And so the biggest issue is, "Did we have multiple windows before the PARC visit?" Bill thinks we did; I think maybe we didn't. It's unclear. But definitely the mouse came out of the PARC visit. We did not have a mouse before then.

3P1C F41L

It's game over for Atari

*T*he video game craze peaked in the early eighties: MTV VJs talked about their love for Atari on air; "Pac-Man Fever," a novelty song, topped the charts; Hollywood was churning out thrillers like Tron and War Games. Tron's plot fictionalized the obsessions of the Valley's programmer class and one of its main characters, Alan Bradley, was modeled after Alan Kay. In the movie the Alan Kay character enters cyberspace to stop a computer program gone rogue. In actual fact, Kay was working on making Tron's fictional cyberspace real—at Atari. After PARC fell apart in the wake of Steve Jobs's visit, Kay was recruited by the high-flying computer gaming company to be its chief scientist. He was in charge of Atari's lavish new lab, and virtual reality was one of the main avenues of research. Unlike Xerox PARC, however, Atari Research never amounted to much. It was devoted to research, not development—R, not D—and just that one letter, that D, made all the difference.

Alan Kay: Things were not good at PARC in 1980. And somebody I knew said, "Well, as long as you are looking around, why don't you go down and talk to Ray Kassar at Atari?"

Al Alcorn: Ray Kassar was running Atari, and he would have the staff meeting just for the old-timers—me and a few others. We called it "the limp dick society" because we knew it was just to keep us amused. I left in 1981. So to replace me they hired the best R&D man in the world, Alan Kay.

Alan Kay: That year Atari's gross just by itself was $3.2 billion, and that was several hundred thousand dollars more than the entire movie industry in Hollywood. So at that time Atari was bigger than all of Hollywood: They had money coming out of their ears!

Chris Caen: It's funny. Now everyone talks about Apple, but people don't remember how big and pervasive Atari was. At one point Atari was twenty-seven buildings in six cities. You could almost trace the outline of Silicon Valley by connecting the dots. We used to call Highway 101 "Via Atari" because you'd be driving to meetings up and down 101, and all around you there are cars with Atari parking stickers. It was that first magical wave of Silicon Valley.

Michael Naimark: The company was just rocking and rolling.

Alan Kay: They used to send the corporate jet up the coast to get shrimp for the executive dining room, and the joke was you could tell how Atari was doing by the size of the shrimp in the executive dining room.

Michael Naimark: Jokes about garbage bags full of jumbo shrimp the size of lobsters.

Alan Kay: Ray offered me a job and the bribe was a huge budget.

Al Miller: We had heard rumors that they were spending $100 million a year on R&D, which I don't doubt.

David Levitt: Alan Kay was one of my big mentors, and suddenly Alan was chief scientist of Atari, this billion-dollar company that came out of nowhere.

Howard Rheingold: At Xerox PARC, Alan Kay had sort of been the kid to Bob Taylor, and now Atari was sort of the next generation. Alan Kay was the adult supervision there.

Alan Kay: My first act as chief scientist was to hire pretty much all of Nicholas Negroponte's graduate students from MIT. Because what Atari needed was something more like what Nicholas had been doing. And God, he had a great bunch.

Scott Fisher: I was working with Nicholas Negroponte, and Alan showed up at the Architecture Machine Group with a bunch of the Warner execs in tow. Basically they came to buy the lab. And Nicholas, of course, in his inimitable way was interested in that amount of money, but I think in the end he said, "This is MIT. You cannot buy the lab." So we were disappointed. But then Alan came back a little bit later and made offers to six or seven of us to come out to Sunnyvale and work for Atari. For me that was a hard choice, because so many fun things were happening with Nicholas.

Jaron Lanier: This was all wrapped up in the founding of the MIT Media Lab, which was essentially all the same people and which was getting going at the same time, although the Media Lab hadn't quite started yet.

Michael Naimark: All of a sudden by the fall of 1982 Alan had a critical mass, this amazing group of people. And I think we had a half-dozen key projects at the time.

Alan Kay: We did a bunch of things: virtual reality, communication...I mean Atari was basically a consumer electronics company, so my thought was that what we wanted to do was develop media.

Tom Zimmerman: Atari Research was close to paradise for an inventor. I was doing research on music, electronic music. The idea was to do a voice-controlled synthesizer, so you could hum into it and it would play any instrument. You could play a violin, trumpet, flute, just by humming. Very ambitious!

David Levitt: Media, music, artificial intelligence, all the things that we knew were going to be part of Atari's future.

Scott Fisher: Atari Research was meant to be a kind of resource for the whole company, all the operating divisions. And all these crazies that Alan hired that would then go get farmed out to work with different groups.

Michael Naimark: I hung out with the coin-op guys partly because they were doing optical video disc stuff, because we had done a lot of that at MIT, of course—just trying to get those into the arcade games they were building. So that was fun.

Brenda Laurel: It was me and a bunch of MIT kids and the most fun thing we did was we started writing memos in the voice of a guy named Dr. Arthur Fischell, also known as Artie Fischell.

Artie Fischell: Dr. Arthur Fischell is *artificial*. Get it?

Brenda Laurel: We actually made this guy up. We were a collective of people personifying this guy, who we claimed was the new director of the lab because Alan wasn't around a lot.

Michael Naimark: Alan, as chief scientist, was the head not only of Atari Sunnyvale Research Lab but Atari Cambridge Lab, which was next door to MIT in Tech Square. And the folks at the Cambridge Lab were mostly young artificial intelligence PhDs—and Jaron Lanier, the VR guy with the dreadlocks.

Jaron Lanier: Kay was spending most of his time at the lab in Cambridge.

Scott Fisher: I think this idea of having this additional member of Atari Research that was completely made up was mostly Brenda's idea, but it caught fire quite quickly.

Michael Naimark: Artie Fischell was something that we tried our best to have as our little secret from the rest of Atari. We wanted to see how far we could push things.

Brenda Laurel: We got him an employee number, an office.

Scott Fisher: He could get snail mail there. And we got to a point where Arthur had an e-mail account.

Michael Naimark: And a subset of Atari Research, not everybody, had his e-mail password to log in as him.

Kristina Woolsey: Realize that e-mail was somewhat new at the time. And we used it in the lab very fancifully.

Brenda Laurel: I mean memos were going around on e-mail and people were sending memos back to him, and I think he had a part-time secretary.

Scott Fisher: To get him to participate remotely in lab meetings, we had a pitch-shifting speech device so that we could take turns and always have kind of the same voice.

Brenda Laurel: I did his voice on the phone with an Eventide Harmonizer to lower it in real time. He had a British accent.

Michael Naimark: So he developed a personality as being this older, worldly guy.

Brenda Laurel: He had a whole family, a whole backstory. He had worked for the British Postal Service, he invented squid jerky, things like that.

Scott Fisher: Just seeing how far we could go in getting him into the physical world was a great challenge.

Michael Naimark: Eric Hulteen had submitted a purchase request under Artie Fischell's name for a Saturn V booster rocket.

Brenda Laurel: Alan added "a pack of gum" and signed it. So this is going through the system and I swear to God, these facilities guys show up and say, "Well, how are you going to launch this thing?" "How much space do you need?"

Scott Fisher: I was a little bit surprised that that actually went through.

Brenda Laurel: It culminated in a live videoconference with Arthur. We had laid cable all the way across the building down to the conference room so that it really could be live. And I played Arthur in drag with crêpe hair, and I wore Michael Naimark's jacket and stuff. And I'm at the desk and at every cutaway they put a different city in the window behind me. At one point some guy in Arab dress comes in and tries to give me an exploding pizza.

Scott Fisher: People had no clue that this was totally an act.

Brenda Laurel: And then we opened it up to Q&A and I'm answering the questions. And Douglas Adams was visiting that day, so he was in the audience.

Kristina Woolsey: Alan did a good job with bringing people through the lab. He would meet somebody and just bring them through.

Brenda Laurel: About four hours later I was at lunch with Ray and Alan and Douglas Adams, and Douglas was sitting next to me. They were debating whether or not it was live. Because they couldn't quite grok that it could be. And Douglas looks at me and he said, "It was you, wasn't it?" And I said, "I don't know what you are talking about." And he pulls this piece of crêpe hair off of me, a sideburn that I had missed. And he was the only person who got it. That was stellar. That was just the most fucking fun.

Scott Fisher: Artie Fischell kind of makes it sound like we didn't have enough to do. But in fact I think it was probably one of the most interesting research projects we did there. We were thinking about how do we as a group, author and give voice to a character?

Brenda Laurel: In those days people were struggling with building artificial intelligence so they could build good nonplayer characters. But we thought you didn't need to build a bunch of artificial characters to have a good interactive system—you could have people playing with people as opposed to people playing with made-up programmed characters. So by making this guy up and having multiple interactions with people at all levels, that was sort of a proof of concept. It presages so much, when you think about the value of the things we learned by doing that.

Alan Kay: Atari wasn't that interesting. Atari was us learning about corporations.

Scott Fisher: Arthur was us projecting the lab director we really wanted.

Michael Naimark: On the one hand, if you ask, "What actually came out of Atari Research?" The answer, I agree with Alan, is "Very little."

Brenda Laurel: We talked a lot about VR.

Kristina Woolsey: Virtual reality was not a well-known concept, but we had a big program and played with it for a while. We would spend a lot of time talking about it, writing, but we did not produce anything. We talked a lot, we thought a lot, we wrote a lot, we gave speeches a lot.

Michael Naimark: On the other hand, I truly seriously believe that the seeds of both virtual reality and multimedia came out of the work of Atari Research.

But Atari itself was going bankrupt: gradually at first—and then suddenly.

Bob Whitehead: Atari was generating a ton of cash. And unfortunately when you do that you tend to just start throwing money at things. There was a sort of disconnect: a misunderstanding of what really drives the creative process.

Brenda Laurel: Warner thought that a good license would be a good game.

Al Alcorn: Ray Kassar gets this phone call from Stephen Ross after getting off the jet with Steven Spielberg and Ross basically says, "I've bought the name *E.T.* for you." Ross sent Spielberg a check for I don't know how many millions, tens of millions.

Howard Warshaw: The licensing deal was for $22 million; if you pencil it out that's like the most profit you could ever expect to make on a game.

Nolan Bushnell: The only way that the *E.T.* deal made sense is if they sold a certain amount of cartridges. It was all based on the royalty that they would have to pay Mr. Spielberg. And so rather than being driven by the economics of the market, they were being driven by the economics of the deal. It went from the engineers being superstars to the marketing department. If you have a good product, any idiot can sell it. They didn't realize that.

Howard Warshaw: They bought *E.T.* as a loss leader to keep it away from other people. Back then Atari was the vast majority of the industry, but there was also Mattel, there was Coleco.

David Crane: They had to get it out by a certain time frame for Christmas.

Nolan Bushnell: Therefore the deal constrained the engineering time to six weeks.

Howard Warshaw: Five weeks and one day. But I didn't get to start until dinnertime the first day.

Al Alcorn: Ray was like, "What?" Ray had learned enough by this time to know that this was kind of crazy, but the deal was done and he had to do it.

Howard Warshaw: Nobody had ever done a game in less than six months on the VCS, and I had to do a game in five weeks. I was used to working under pressure, but this was just crazy. The CEO of Atari was betting a lot of his career on making this thing happen.

Al Alcorn: Throw in the fact that Ray didn't actually ever play games, so he had no way to know the game was a stinker.

Chris Caen: It was ugly. It really was unplayable. People talk about it as being the world's worst software game of all time.

Nolan Bushnell: People used to smuggle cartridges out to me: prototypes, that they would just burn onto ROM. And I can remember getting the *E.T.* cartridge and thinking that it was broken—literally!

David Crane: And what happened was that most people turned the game on, picked up the joystick, moved it to the right, and *E.T.* fell in a hole and they couldn't get out.

Jim Heller: The 1982 Christmas season was nowhere near as good as what had been projected because *E.T.* was pretty much a failure as a game. I guess it still sold probably two million cartridges. But that left close to three million cartridges that were unsold.

Ray Kassar: We made almost five million, most of them were returned.

David Crane: And so they've got millions sitting in the warehouses. Well, where do they go? Into the landfill.

Jim Heller: We only had three days' worth of dumping at the landfill before the weekend. When I came back I was told that the news media is going crazy as soon as I walked in the building. They're calling wanting to know what we're doing. Kids had come and scavenged the cartridges that we had dumped at the landfill, and they'd been arrested trying to sell the cartridges. No, this is not good. So I ordered six dump-truck-loads of concrete to be delivered to the landfill. And everybody thought it was crazy, but I figure if I covered them up with concrete, it would keep the kids out of the landfill.

David Crane: Same thing happened with *Pac-Man*. They tried to design *Pac-Man*, which was a well-known game and it had to work in a certain way. The VCS couldn't do that. So they made a subpar version of *Pac-Man*. And because they paid so much for the license they built millions of them. They sold none of them.

Al Alcorn: When they were hurting after the *E.T.* disaster they went to Alan Kay: "Where is the new thing to put in production?" He said, "You don't understand: research, not development." And Ray is like, "Well, what's the difference?" They didn't get it.

Alan Kay: Ray came to me and said, "Alan, I'm drowning. Throw me a rope." And I said, "I can't, Ray. You didn't pay for the rope factory."

Al Miller: Ultimately the only thing that came out was a very lame next-generation game machine based on the Atari 400 home computer.

Alan Kay: Atari had had their own version of a 6502 PC, which was in many ways better than the Apple II.

Al Miller: They had taken the same chips, added some terrible controllers which were just nonfunctional, and tried to sell that.

Chris Caen: The international sales team were notorious for having wine-and-cocaine parties. We used to joke that we had the only international sales team in the industry that imported more than it exported.

Michael Naimark: In '82 and '83, coke was everywhere.

Chris Caen: The home system was just bombing. But that's okay because we have this huge library of arcade games—this evergreen set of titles—and we'll just keep figuring out new ways to do them and bring them over to the home computer system. We will be fine.

Scott Fisher: By '83 it was kind of not so much partying as it was using the drugs to get these nineteen-year-olds to like hurry up and get this damn stuff ported over from the arcade game to the home machine by, like, tomorrow. "Here's more. You need more coke. Here. Anything, I'll give you anything if you get this done."

Brenda Laurel: A guy from biz-dev invited me into his hard-walled office. And he had the glass from the *Battleship* arcade game on his desk, and he cut a big line of cocaine and he said, "Here, snort this, it will help you work better." I'd never seen cocaine. I didn't know what it was. I'm a naïve hippie and I'm thinking, *Cocaine? You know they lied to us about grass. So this is probably fine.* And coke became a problem for me for about a year until I realized I was going to kill myself. So there was a lot of coke going around.

Scott Fisher: We were all watching this thing go down the tubes because of that. I think that's ultimately why things got so bad.

Chris Caen: In '83 the wheels are starting to come off of Atari in a dramatic fashion.

Alan Kay: I used to show them slides about what they were spending on, and ask, "How fast will it go when the thing flips?"

Kristina Woolsey: I was amazed at the exorbitant spending. And it did not taper off. It was exorbitant and then it was, "Oh, we are in trouble!"

Brenda Laurel: And this was right about the time they were burying the *E.T.* cartridges in the desert.

Kristina Woolsey: A group of us managers got pulled into a conference room, it was just plush, Hollywood plush. And a man in a suit—I think it was the CEO—said, "Guys, it's over. We are looking for a buyer." Basically he laid it all out. He had this wonderful line that caught my attention: "You cannot fool thirteen-year-olds."

Howard Warshaw: At the beginning of '83 or so, we're still at about ten thousand employees. By late '83, we were down to two thousand employees.

Chris Caen: You'd call someone in the morning, then you'd call them in the afternoon and the extension would be disconnected. The company was just being decimated on a weekly basis.

Scott Fisher: I remember sitting with Alan Kay in a park in Sunnyvale as a kind of retreat, just getting out of the office and going out to talk about what our plan was for the next twenty years. And then literally going back to the office and finding out that we—well, at least Brenda and me—had been fired and had fifteen minutes to get out of our offices. It just seemed so crazy to be talking about twenty-year-long projects and then have the security guards escort you out.

Brenda Laurel: It was insane. Like five thousand people in one go! I got it that day.

Chris Caen: Layoffs of fifteen hundred people? Now we're used to it, but when it happened in '83 there was no context for it. It was a psychic shock.

Alan Kay: Once a month or so Steve Jobs would come over and have lunch with me—and it used to drive Ray crazy. He'd come along and here I am sitting with Steve Jobs of Apple. In the course of our lunches Steve said, "You know, you should come over to Apple."

Michael Naimark: It was either in late '83 or early '84 that Alan Kay leaves Atari. And we are all thinking, *Okay, you know it's a little crazy out there, but the rest of Atari is still making lots of money, so what's the problem?*

Brenda Laurel: So the lab limped along for another half a year or so and then they finally shut everything down.

Michael Naimark: But after Alan left things started tanking very quickly. Atari went from like earning billions to losing billions. And I think the first quarter of '84 was the turning point.

David Crane: It was a whole disaster. There was a time when Atari lost a hundred million dollars in a quarter.

Alan Kay: It went awfully fast. In their first bad year I think they lost a billion.

Howard Warshaw: By mid-1984 we were down to two hundred people. So, in about a year and a half the company goes from ten thousand employees to two hundred employees, and I was still one of those people. It was a dark time.

Kristina Woolsey: We thought the video game business was over.

Brenda Laurel: At the very end, I'm told, it was like the fall of Saigon. People were dropping equipment into the trunks of their cars from the second story. You know it was just like chaos. Everybody moving out and people carting computers down the stairs.

Chris Caen: That's when that first magical wave of Silicon Valley ended. There was a shift: the idea that the company was a family disappeared with Atari.

Kristina Woolsey: If you are young and you don't know how the system really works and you believe in something, then it truly was horrible. It was heart-wrenching. It was socially destructive.

Chris Caen: There was a certain feeling of collective mission that broke, and all the corporate folks and off-sites in the world weren't going to put it back together. That sounds overwrought, but if you were there, it wasn't.

Kristina Woolsey: But this is how the Valley works—people come and go from job to job because start-ups are starting, and big companies can disappear, too.

Jamis MacNiven: Titans rise, titans fall—that's the nature of the world. It just happens faster in Silicon Valley.

Hello, I'm Macintosh

It sure is great to get out of that bag

While Atari was having its troubles, so was Apple. By the early '80s the Apple II was starting to show its age. Other competitors, including Atari but most importantly IBM, were coming out with newer, sleeker machines. Apple was trying to counter, first with the Apple III and then with the Lisa, but both machines were flops. The next machine in line to be released was the Macintosh, and it would make or break the company. The Mac was Jobs's exclusive turf. Although Jobs was the public face of Apple, he did not have actual control of the company. Jobs contented himself with being the (nonexecutive) chairman of Apple's board and directing what was, in 1980, a minor research effort: an experiment in making a computer for the common man. He and his hand-picked team of twenty-somethings had been hacking away at the idea ever since returning from PARC. It took four years, but by 1984 Jobs's vision became reality. The Mac put the mouse—and the PARC-inspired GUI that accompanied it—into the hands of the masses.

Dan Kottke: The IPO was November of '80, and Steve is raring to go. Because he had money to spend, right? The Apple III was launching, but Steve Jobs had already moved on. The Apple III didn't interest him, because there were too many other people involved and he wanted to be in charge.

John Couch: Steve says, "I want to run Lisa," because it was the newest, newest product.

Randy Wigginton: But they weren't listening to him.

Steve Jobs: I thought Lisa was in serious trouble. I thought Lisa was going off in this very bad direction.

David Kelley: Lisa—the Apple IV—was really mixed up.

Dan Kottke: Steve was trying to bully the Lisa group.

Randy Wigginton: Jobs was angry at the Lisa group because they named it "Lisa" after his daughter to make fun of him. Steve lived with a girl for several years, Chrisanne. And she got pregnant and insisted that the baby was his. He always claimed that she was sleeping with other people and it was probably someone else's, and no one inside of Apple believed him. And when the baby came she was named Lisa. Steve absolutely insisted that it was not his baby. And so that is why the Lisa was called "Lisa." It was a big fuck-you from the engineers and the people over there. So he hung around with all the Lisa folks basically until they drove him off.

John Couch: And so when they wouldn't let him run the Lisa project he went around looking for something else to do.

Michael Dhuey: And across the street was where the Mac group was, and the group had nobody in it except Jef Raskin.

Dan Kottke: Jef's outlook on the world was tiny, friendly machines—like a home appliance. Jef was completely obsessed with that.

Steve Wozniak: Jef is the one who brought that idea to us.

Andy Hertzfeld: The Mac was initially a skunkworks. At the time it was not an important project at Apple. It was a very minor thing.

Randy Wigginton: And Steve went over to Macintosh where Jef Raskin was, and he and Jef did not mix well.

Steve Jobs: Jef's a shithead who sucks.

Jef Raskin: Steve would have made an excellent king of France.

Michael Dhuey: They lasted about a week together, because Jef realized that Steve was going to immediately take over the group, and so Jef came back to the executives and said, "He's taking over my group, he's wrecking everything!" And then they said, "Tough. Steve wins."

John Couch: I don't think he asked Jef's permission; Steve just took it over.

Randy Wigginton: Steve hijacked it.

Andy Hertzfeld: When the Mac started, the big projects in Apple's future were the Lisa on the one hand and the Apple III on the other hand. Even the Apple II was diminished in their minds, not thinking it would last as long as it did.

Randy Wigginton: Remember Steve always had this insecurity about the Apple II, because he didn't actually make the Apple II, right? And Steve's idea for the Mac was basically the next-generation Apple II for the rest of the world.

Andy Hertzfeld: The Mac was going for well more than a year before it found what it really was: the volks-Lisa. We were taking the Lisa technology but making it for everybody.

Mike Murray: But there wasn't a lot out there that suggested the Macintosh would succeed. Could we do it with a smaller box?

Dan Kottke: The first prototype of a Macintosh was just a 6809 processor, some RAM and like a video generator, that's all it was. It didn't even have a mouse. And it wasn't good for anything other than just the most simple kind of a demo.

Steve Jobs: So we adapted the 68000 processor that Lisa had.

Trip Hawkins: It was just a rocket ship of a processor: elegant, sleek, fast—a simplified instruction set.

Andy Hertzfeld: But there was a technical problem. We could not use the 68000 because the 68000 wasn't able to work with eight RAM chips. It needed sixteen. And Burrell Smith, being the genius that he was, he came up with an incredible way to hook up the 68000 to eight RAM chips, what he called the bus transformer circuit. It was this sort of miracle. No one thought it was possible to hook up the 68000 to an eight-bit bus without losing any performance, to go just as fast. And suddenly when he had that, you had the Macintosh which cost about one-fifth of what the Lisa was projected to cost, and ran twice as fast. So that is lightning in a bottle. That's an amazing thing. Burrell, in an act of unprecedented genius, was able to do that. And when Steve saw that he thought, *Boy, this is the future*.

Dan Kottke: It was kind of like putting a big V8 in a tiny little sports car. It was way overpowered for this tiny little machine.

Joanna Hoffman: I knew the hardware was amazing. But when Andy joined the soul of the machine started to take shape.

Andy Hertzfeld: I noticed Steve Jobs peering over the wall of my cubicle saying, "I've got great news for you: You're on the Mac team now!" And I said, "Great! Tremendous!" This was late on a Thursday afternoon. I said, "How about I start Monday morning? Let me just get my DOS 4.0 notes together." I had worked on a new operating system for the Apple II for three weeks, so why not get it so someone else could pick it up?

Dan Kottke: Andy and I joined Mac in January of '81.

Andy Hertzfeld: Steve got mad. He goes, "The Apple II was going to be dead. Your OS is going to be obsolete before it's finished!" And then he said something like, "The Macintosh is the future of Apple and you are going to start now." I just wanted one day, really. It was Thursday afternoon. I said "Monday" because Monday seems to be a good time to make a new start. But he goes, "No! You're going to start on it now!" And he went and he pulled the plug on my Apple II, which had my code not saved. I was right in the middle of working on it. He just yanks the plug out, and then without pausing he picks up the Apple II and starts walking away.

Steve Jobs: We were on a mission from God to save Apple.

Andy Hertzfeld: I had no choice: I had to follow my computer. I couldn't work without it. That's why he took it, because he knew I needed it to do any work. So he just took it and he walked away! He goes outside—his car is always parked in the handicapped space so that it's close to the door—and he plops it into his trunk and goes, "We're going to the Mac building."

Bill Atkinson: We had a place that was behind the Good Earth Restaurant. And so they called it Salt of the Earth.

Dan Kottke: Burrell Smith had put a sign on the door that said "Danger! Contagious Algorithm Research Area!" That was Burrell's sense of humor.

Andy Hertzfeld: The Mac office looked like it should be a dentist office or something: wood paneled, but cheap. And he drops my computer on a desk and he says, "This is your new desk. Welcome to the Mac team." And then he walks out. I open the drawers at my new desk—the desk he put me at—because I don't even have my stuff, I just have my Apple II. And the drawers were all full. It was someone else's occupied desk. And I'm thinking, "What's going on?"

Randy Wigginton: Jef got shipped off to Siberia—you know Hooli? The roof? Jef was on the roof with the BBQ.

Andy Hertzfeld: I find out that it was Jef Raskin's desk, who was just exiled the previous day, he hadn't had time to get his stuff. That was my first day. At that point Lisa was still two or three years away from shipping.

The Macintosh group was a classic skunkworks operation: a breakaway engineering team tasked with the research and development of an alternate future. It's the engineering equivalent of a special forces unit being sent on a long-range reconnaissance mission.

Randy Wigginton: We had just gone through the horrible, abysmal Apple III. And Lisa just appeared to be going nowhere. So that's why we sort of went off the grid. We were our own ragtag group of people, and you know we just wanted to go off and do our own thing.

Dan Kottke: There was this whole thing that Apple was becoming bureaucratic, and so we were the renegade pirates.

Bruce Horn: Steve said, "We'd rather be pirates than join the Navy." Right? We'd rather steal great things than be bureaucratic and lockstep and follow the rules.

Randy Wigginton: We were not in the Navy. We hated the Navy, we hated the official establishment.

Dan Kottke: They made a pirate flag and they put it on top of the building.

Andy Hertzfeld: I was a lookout for installing the pirate flag. On a Friday evening we cased it. We climbed up onto the roof and looked around and on Sunday evening we put it there. But the most fun part was just seeing how people would react the next day. Steve loved it. But some other parts of Apple thought it was extremely arrogant.

Randy Wigginton: We were trying to be very counterculture even within Apple.

Bruce Horn: We were the rebel group, a little bit more willing to try crazy things.

Steve Jobs: Why can't we inject typography into computers? Why can't we have computers talking to us in the English language? Looking back, this seems trivial. But at the time it was cataclysmic in its consequences. The battles that were fought to push this point of view out the door were very large.

Bill Atkinson: I remember an interesting incident on the icons for the Lisa. We needed a way to show whether there was something in the trash: if you say "Empty the trash," is there something to empty? The very first version of the trash can that I wrote had little flies buzzing around it, but they got sanitized out. The Lisa user interface was a little bit hampered by who we thought it was for. We thought we were building it for an office worker, and we wanted to be cautious not to offend.

Randy Wigginton: On the Mac, we had the Cookie Monster in the trash can. When you dropped a file onto the trash can, all of a sudden the Cookie Monster would come up and go "munch munch munch" and then back down. Andy Hertzfeld and Susan Kare just did that as a demo. Everybody who saw it thought it was great.

Dan Kottke: The Mac was a very radical thing, because this was the age of CRT screens and big boxes, right? The Mac was going to be a computer you would take to bed with you, like you would sit on your bed to do your work.

Bill Atkinson: When the Mac was designed, we had a pretty clear picture of a fourteen-year-old boy using this thing, and we knew what they were like, and so we were going to have fun with that.

Bruce Horn: We were all young. Andy was in his late twenties. I was in my early twenties. Steve was maybe twenty-six or something. And we were motivated by this passion to make the life of everybody using computers much more fun, much more productive, and to give them a certain amount of creative license.

Andy Hertzfeld: That was why we were there. It's what we wanted. And the pressure wasn't external pressure, it was our purpose to make something great. The people working on it believed in it.

Bruce Horn: And that's how Steve was motivated. He thought of it as an artistic tool, among other things.

Dan Kottke: At the point of time when we were finalizing the custom plastic case, the Mac team was maybe only fifty employees.

Randy Wigginton: And everybody signed different places on the case diagram.

Andy Hertzfeld: Steve explicitly said, "We are signing the Macintosh because artists sign their work. We are artists."

Dan Kottke: Then Jerry Manock, the lead designer, shrunk it down and actually put it into the die for the case of the Mac, so that we all had our signatures in the plastic.

Randy Wigginton: Steve wanted us to be personally invested, he wanted us to be attached to it, and he felt that by having our signatures inside the case, we would be. It was pretty cool.

Andy Hertzfeld: But it wasn't until the fall of 1982 that the rest of Apple cared about the Mac at all. So the Mac, Apple began to take it seriously in 1982, once we had prototypes built. But then by the spring of '83, it became clear to everyone that the Mac was the future.

Mike Murray: Because the Lisa, which came out in January of 1983, was pretty much a failure.

Butler Lampson: It was a bust. It was too high-end, tried to do too much stuff, and it cost too much.

Trip Hawkins: Then Lisa basically gave birth to the Mac. Again this is a somewhat untold story, because Steve didn't want this to be known. Steve basically took all the bright ideas from Lisa and all the best people from Lisa. There were a whole bunch of huge pieces of Lisa that pretty much just shifted right over to the Mac.

Steve Jobs: Apple II was running out of gas, and we needed to do something with this technology fast or else Apple might cease to exist as the company it was.

Dan Kottke: The Apple II was paying the bills, but by '83 that was kind of the old generation and now the Mac was coming.

Andy Hertzfeld: If the Mac failed, Apple would fail.

Bruce Horn: There was a ton of pressure.

Andy Hertzfeld: We'd have retreats every six months and Steve would begin the retreat, with quotes. And for the January 1983 retreat one of the quotes was "Real artists ship," which meant "Hey, we've spent enough time developing." It was time to ship.

Bruce Horn: Steve Capps and I were working on the Finder. When you boot the system, you boot into the Finder. So it's the first thing you see. Capps and I worked all the time. I was twenty-three and I had just gotten braces and I'd have my headgear on at night, and Capps and I would sit and work at night. He was the only person who ever saw me with headgear. And we would listen to Violent Femmes and just hack all night. We had sweatshirts that said NINETY HOURS A WEEK AND LOVING IT.

Burrell Smith: Back then it was the joy of being absorbed, being intoxicated by being able to solve this problem. You would be able to take the entire world with its horrible problems and boil it down to a bunch of microchips.

Andy Hertzfeld: During the fall of 1983 the IBM PC came into its own and started selling. The IBM PC was introduced in the summer of '81. So it took two and a half years, but by the fall of '83 it was clear that it was an existential threat to Apple. That's when IBM came into focus as major competition to Apple.

Randy Wigginton: It was ninety hours and then it turned into a hundred hours, and we were getting really, really tired.

Bruce Horn: It was very intense because every time we needed to do something it would take a little bit more memory space. And then we would have to steal a little bit more from maybe the stack or something. And every once in a while it would break something.

Charles Simonyi: The first Mac, and it's not very much appreciated, was a little bit overpowered, so the performance was very high, but the memory was really low, it had 128K memory in the first version, but a lot of that was taken up by the bitmap and by the operating system, so the actual space for the application was pretty much minimal.

Bruce Horn: MacPaint would crash because MacPaint was like that, too. We were just trying to squeeze it in as tightly as possible and working up to that deadline date and just hoping we'd get there. Everybody was exhausted. I gave it one last push.

Randy Wigginton: It was truly insane pressure. I basically lived at the office for about three weeks before it shipped. I had a sleeping bag there and would go home and shower every so often, but it was terrible. It was not healthy. We were stupid—drunk tired—and at two o'clock in the morning when we integrated everything to prepare the release, literally nothing worked. I just lost it, I went totally hysterical, I just started laughing and couldn't stop. I had to walk out of the building and literally walk around the block just to get a grip, because it was not looking good. I mean the two o'clock release was just abysmal and we only had until six a.m. At that point we were like, "Okay, well, let's not try to do all that, let's roll back to stuff that works." And so we did, and it was a shockingly good release. God is good to idiots.

Andy Hertzfeld: There were actually two run-ups with huge pressure. The software that was in ROM—the system software—had to be finished before the end of September '83. So there was a crescendo there and then a party at Woz's house the day after the ROM was frozen. And then around the beginning of October came the buildup for finishing the disc-based software. And then the publicity had a similar thing, because the publicity was designed to reach a crescendo on the day of the announcement.

As the Macintosh was being finished—the circuit boards printed, the ROMs burned, the software debugged—Jobs's attention turned toward the marketing push that would accompany the computer's launch.

Lee Clow: Steve Jobs's simple challenge was, "I think Macintosh is the greatest product in the history of the world and I want an ad that's that good."

Mike Murray: In 1983 Chiat/Day had come up with this Big Brother/ George Orwell idea and they were shopping it to their clients. All good advertising agencies have a car company, a soda company, and, you know, a high-tech company. So they went to each one and then they came to us with some storyboards that showed this rough idea and I remember Steve just liking it, and I liked it immediately. We said, "That is great."

Lee Clow: Getting the spot approved wasn't that hard. I guess the biggest hurdle was that it required a pretty big budget. They had never spent one million dollars on a commercial before. But Steve loved it and John Sculley had just joined from Pepsi.

Dan Kottke: John Sculley was the celebrity CEO that Jobs had hired. He was like this glitzy marketing guy.

Lee Clow: John Sculley wanted to show off and prove that he was really smart, so he said, "Well, I love it too, because we used to do spots like that at Pepsi." He really never did but that's what he always said. Steve in his classic way basically said, "Go make it great." He didn't want to come to preproduction meetings or script consultations or casting or the shoot. He just said, "Go make it great."

Mike Murray: And they hired Ridley Scott.

Ridley Scott: When I first saw the boards, I thought, *My God, they are mad!* It was such a dramatic idea that it would either be totally successful or we'd all get put in the state pen.

Lee Clow: Ridley Scott had just finished doing *Blade Runner*, and so we were using a famous director and we were going to do something really special.

Ridley Scott: From a filmic point of view it was terrific, and I knew exactly how to do a kind of pastiche on what 1984 maybe was like in dramatic terms rather than factual terms.

Mike Murray: It was the first time that a filmmaker—a moviemaker—had been hired to make a commercial, because in those days that was considered dipping way down and becoming very commercially crass.

Ridley Scott: So I've always looked at each commercial as a film, as a little filmlet.

Lee Clow: Ridley is an amazing professional. We had some great discussions about everything from wardrobe to where the spot should be. At one point we were talking: "Maybe they should be marching through the mud from Quonset huts?" And that gave way to what you ultimately saw in the film.

Ridley Scott: Those two big black walls at the back of the main auditorium? We hauled in two huge 747 engines and just hung them on the wall. And they looked like...what did they look like? I don't know, they looked like ducts. So that is to me what I call "good dramatic bullshit."

Mike Murray: I remember Jay Chiat from Chiat/Day went over to London, and during the filming he sequestered himself behind a whole bunch of cardboard boxes because he was frightened. Because those guys in the ad who look like they're skinheads? They were.

Ridley Scott: We organized one of these rather frightening casting sessions where there were about three or four hundred youths, and we chose 150 skinheads out of that group.

Mike Murray: And whenever they'd have breaks in the filming these guys would all go out and get these bags with glue and they'd be sniffing glue! They were truly scary, scary people. So Jay told me afterward that he hid during the filming of this very famous commercial because he was afraid they were going to come and get him.

Lee Clow: It ended up being what it's become—a legendary commercial. The commercial's been talked about so many times over the years with different people ascribing as to what the genius of it was. I always suggested that it was a lot of things all happening just right with some serendipity thrown in: an amazing product, a client that wants an amazing commercial, a great director, and then the Apple board...

Mike Murray: So we get it done and we are told that we need to show it to the board of directors, and Steve is chairman of the board. It was Phil Schlein and Mike Markkula and Art Rock and Henry Singleton, and Steve and maybe one or two others. I remember very clearly going into that board meeting. We had a TV on one of these rolling carts and we send it out and we show it.

They roll tape: Big Brother is on a television of enormous size addressing a crowd of zombified skinheads. Suddenly a beautiful young blonde wearing a Macintosh-branded tank top and red hot pants runs to the front of the crowd while swinging a hammer. She throws it at the looming image of Big Brother. It's a bull's-eye, and the giant television explodes. As the action fades out, the words fade in: "On January 24th, Apple Computer will introduce Macintosh. And you'll see why 1984 won't be like 1984."

Mike Murray: Phil Schlein has his head down on the desk, on this big boardroom desk, head down and he's pounding on the desk: Boom! Boom! Boom! And I think, *He got it! He's with us! That's one vote!* And then Markkula turns to Jobs and says, "You really want us to run that?"

John Sculley: Then they all turned and looked at me because I was the adult supervision. "You are not going to really run that thing?"

Lee Clow: John Sculley—who originally approved it—then decided that he hated it because the board hated it.

Mike Murray: And it was such a horrible feeling, though, because all of us knew it was the epitome of the greatest ad of all time, and it was such a huge disconnect between us and our board, who were then on from our own point of view just a big bunch of schmucks.

Bill Atkinson: Apple management didn't want it to air. They ordered Steve to sell the time back.

Lee Clow: So we sold off all the time except for one sixty-second window during the Super Bowl which, not by accident, we didn't sell.

Mike Murray: So Steve talks to the board and the board says, "Well, then we will run an Apple II spot." So they went to the vault and looked at them but none of them were relevant. So reluctantly they said, "All right, run it." And it was shown on the Super Bowl.

Steve Wozniak: And there's Big Brother on the television saying, "Everybody has to think the same."

Big Brother (lecturing the skinheads): Today we celebrate the first, glorious anniversary of the Information Purification Directive! We have created, for the first time in all history, a garden of pure ideology, where each worker may bloom secure from the pests of contradictory and confusing truths. Our Unification of Thought is a more powerful weapon than any fleet or army on Earth! We are one people. With one will. One resolve. One cause. Our enemies shall talk themselves to death. And we will bury them with their own confusion! We shall prevail!

Andy Hertzfeld: It was clearly an allegory. Most commercials aren't allegorical.

Steve Wozniak: You know, kind of like IBM World: "Everybody has to have the identical thought. Contrary thoughts will not be tolerated." And then the young track lady with the red outfit throws a hammer at Big Brother on the big screen and it explodes and everybody's just gasping: A new world is coming...Oh my God, that was the most incredible thing I'd ever seen! An incredible piece of science fiction!

Ridley Scott: We nailed it.

Bill Atkinson: It won a Cleo.

Mike Murray: The following night on ABC, CBS, and NBC—those were the three networks then—it was shown as news. Not as advertising, not as PR. It's Dan Rather and Peter Jennings saying, "Apple Computer has introduced a new computer called the Macintosh." And they showed this outrageous ad and they would show it in its entirety.

The Macintosh itself debuts a couple days later at the Flint Center on De Anza Junior College campus in Cupertino. Jobs is the emcee. But before the big reveal, he delivers a short Valley-centric history lesson.

John Markoff: There had been leaks ahead of the Macintosh for years—literally—and so there had been tremendous buildup. There was the failure of the Lisa. Already the connection between the work at Xerox PARC and Apple technology was well known, so there was this expectation that this was the future of computing. All of that was palpable.

Steve Jobs (from the stage): It's 1958. IBM passes up the chance to buy a young, fledgling company that has just invented a new technology—called xerography. Two years later, Xerox was born and IBM has been kicking themselves ever since. It is ten years later: the late sixties. Digital Equipment Corporation and others invent the minicomputer. IBM dismisses the minicomputer as "too small to do serious computing," and therefore unimportant to their business. DEC grows to become a multi-hundred-million-dollar corporation before IBM finally enters the minicomputer market. It is now ten years later: the late seventies. In 1977, Apple, a young, fledgling company on the West Coast, introduces the Apple II, the first personal computer as we know it today. IBM dismisses the personal computer as "too small to do serious computing," and therefore unimportant to their business.

Randy Wigginton: Steve loved to point out how other people screwed up, and how he's amazing and stuff.

Steve Jobs (from the stage): The early nineteen eighties—1981—Apple II has become the world's most popular computer, and Apple has grown to a three-hundred-million-dollar corporation, becoming the fastest-growing company in American business history. With over fifty companies vying for a share, IBM enters the personal computer market in November of 1981 with the IBM PC. It is now 1984. IBM wants it all and is aiming its guns on its last obstacle to industry control: Apple. Will Big Blue dominate the entire computer industry? The entire information age? Was George Orwell right?

Randy Wigginton: The Mac group had the first three rows of the auditorium to sit in, and we were eating it up.

Bruce Horn: We really did think that this was going to liberate people from the tyranny of Big Computing.

Randy Wigginton: Like, "Here we are. We are freedom fighters, the Empire is about to win, the Death Star has moved into position, and we are the only ones there that can save the planet."

Steve Wozniak: The Macintosh was going to lead to big problems later on, but he convinced us all. I believed it was the future. John Sculley believed it. Steve convinced us all, and we believed it.

Randy Wigginton: We actually believed it!

Yukari Kane: Looking back, there's definitely an irony there. Apple is now an empire—the most valuable company in the world—and once you become an empire, you are the establishment.

Steve Jobs (from the stage): There have only been two milestone products in our industry: the Apple II in 1977, and the IBM PC in 1981. Today—one year after Lisa—we are introducing the third industry milestone product: Macintosh. Many of us have been working on Macintosh for over two years now—and it has turned out insanely great!

For his finale, Jobs pulls the Macintosh out of a bag, plugs it in, turns it on—and walks away so that Macintosh can "introduce itself" while sitting unattended on the podium at center stage.

John Markoff: Macintosh spoke when it came out of that bag.

The Macintosh computer (from the stage): Hello, I'm Macintosh. It sure is great to get out of that bag. Unaccustomed as I am to public speaking, I'd like to share with you a maxim I thought of the first time I met an IBM mainframe: *Never trust a computer you can't lift!*

Steve Jobs: When it was introduced, after we went through it all and had the computer speak to people itself and things like that, the whole auditorium of twenty-five hundred people gave it a standing ovation and the whole first few rows of Mac folks were all just crying.

Randy Wigginton: It was Steve at his finest. It was amazingly choreographed. It was a show to beat all shows.

Andy Cunningham: It was a hugely successful launch: I think it got more publicity than any launch that has occurred before or since, so it obviously worked.

John Markoff: Macintosh was Steve's marketing sensibilities coming into full flower. It was orchestrated in a way that he would later routinize—the onstage presentation format, the dramatic unveil, the "secret," and all of that. Everybody is still playing the same script that Steve invented, basically.

Mike Murray: So then the next board meeting, three weeks later, they summon the Mac senior team and when we all go into the boardroom they stand up and clap.

Bill Atkinson: And so, one of the biggest differences when the Mac finally shipped is it was the first computer that people really fell in love with. People actually would say, "Gosh, I want one of those. How can I justify it to my spouse?"

Jaron Lanier: I remember bringing one to MIT: "Look, there's a Xerox PARC machine that you're actually going to be able to buy in the store!" And they were like, "Wow!"

Fumbling the Future

Who blew it: Xerox PARC—or Steve Jobs?

*T*asked *with building "the office of the future," the scientists and engineers at Xerox PARC churned out a slew of ideas—most of which ended up in another company's computers. Nineteen eighty-four's Macintosh was PARC's Alto, watered down and reimagined as a consumer product for the computing masses. That fact pretty much everyone can agree on, but in Silicon Valley the ultimate significance of this technology transfer is still a matter of debate. Was Apple's visit to PARC the Promethean moment, where Steve Jobs stole the graphical user interface from Mount Olympus in order to give it to us mere mortals? Or is that oft-told story just Valley hokum and mythmaking—and furthermore, fundamentally backward? To hear the story as told by some who were actually there, Jobs wasn't astute enough to recognize what PARC had actually invented. And thus he stole the wrong thing. The Macintosh copied the Alto's look and feel, but none of its deep structure. In this telling, it was Steve Jobs himself who fumbled the future, because if Jobs had truly understood PARC's fundamental breakthrough—and had kidnapped that, too—then we really, truly, could have had the power of a Prometheus.*

Alvy Ray Smith: Okay, so what did Xerox do?

Chuck Thacker: Xerox was doing pretty well in the copier business—they had this big fat cash cow that they could just milk—and these upstarts come in . . .

Alvy Ray Smith: They just could not believe that people who looked like us could have business ideas.

Dick Shoup: I looked like a complete homeless hippie person.

Chuck Thacker: They had no idea what they had.

Bob Taylor: Most of the stuff that people use today when they use a computer—windows, a bitmap display, Ethernet, electronic mail, servers of all kinds, distributed computing architecture...

Alvy Ray Smith:...and color graphics and the windows-based UI and all of that, and then the laser printer—I mean it was just all happening: the personal computer was being invented!

Bob Taylor: All that stuff was invented by PARC.

Alvy Ray Smith: Here they had the gift. The computer was theirs and they could have been the computer company of the world.

Steve Jobs: They could have owned the computer industry. It could have been the IBM of the nineties. It could have been the Microsoft of the nineties!

Trip Hawkins: It's one of the greatest examples of disruption in history: how badly they blew it, how far ahead they were, what they could have had, how easily they could have had it ten years before anybody else, and how they kept blowing it over and over, even when they had plenty of time to catch up. It was pretty bad.

Alvy Ray Smith: They had everything and they kissed it off and everybody cried. All the supposedly hard-core scientists like Alan Kay and Dick Shoup? These guys literally cried when they heard that Xerox corporate was kissing off the computer.

Alan Kay: Businesspeople should be shot. They always say, "We are in business to make money." And I say, "Well, not really, you just want to make a few million or billion." But the return from PARC is about thirty-five trillion! Count those extra zeros and tell me what they are really doing. They are just trying to be comfortable.

Alvy Ray Smith: And when I was a Young Turk that's what I thought, too. They really blew it.

Alan Kay: The whole thing was just a complete bust, but the urban legend about Xerox not making money is completely false. Xerox made billions on the laser printer. They paid for PARC several hundred times over. They just missed everything else.

Alvy Ray Smith: They kissed off everything except the laser printer. The only thing they kept out of all of that was what they absolutely knew really well, and that was the laser printer.

Chuck Thacker: Because Xerox wasn't a computer company!

Alvy Ray Smith: What you say in business these days is they "stuck to their core competency." Their core competency was black-and-white, ink on paper, right? That's what they did.

Alan Kay: They just were so brain-dead and so unable to understand anything about any process that was not simple business.

Alvy Ray Smith: Did they make a mistake? From a corporate point of view probably not. After having started two companies myself and having had all the Young Turks come to me and say, "Alvy, we should do this, this, this, and this," it becomes quickly evident that you can't do every bright idea that comes down the pike. You just can't. You have to prune the tree. And your job as management is to prune the tree and stick to what you know. I still think they blew it, though.

Bob Taylor: It's pretty clear that Xerox fumbled the future.

Butler Lampson: But the "fumbled the future" idea is quite misleading. Obviously it's true that Xerox developed all this technology and the only real benefit of that was the computer printer. But in fact they made a huge effort to commercialize it in the form of the Star system. They set up an independent division, a so-called System Development Division. They had a very good engineering team, and this engineering team had a vision of what it was that they wanted to build.

Chuck Thacker: The Star was the beginning of the era of what I call scientific workstations because they were built for performance and much less so for cost.

Butler Lampson: And they built it, much to my astonishment. They shipped the Star system in 1981 and it was better than anything you could buy from anybody else for at least a decade. The irony of this is that back at PARC there were people like Chuck and me who were saying to these guys, "This is too ambitious. It's going to cost too much. You should build something much more like the Alto." But they didn't do that.

Charles Simonyi: At the time, I was working at Star basically out of necessity, and I was very frustrated, I saw biggerism.

Chuck Thacker: Biggerism—I think I coined that word. It grew out of an observation that had been made by Fred Brooks in his book *The Mythical Man-Month*. One of the things Fred pointed out is that the second system is the hardest one. When you're building your first system you're very careful because you've never done it before, and you use well-understood engineering principles, and if you get it right and you are successful and then you feel great and you're willing to tackle something that's much bigger, and that's where the danger comes. This goes under the mantle of "the second system syndrome"—that's what Fred called it. But it's also exemplified in something called Myhrvold's First Law of Software which is that "software expands to fill the available space—it's like a gas."

Charles Simonyi: Star was biggered to an unbelievable extent.

Chuck Thacker: It was a biggered version of the Alto.

Bob Taylor: The Star was designed to be not just a computer but a complete system, and it was very expensive because it had many components—printers, Ethernet, computers.

Adele Goldberg: Minimum buy was four Star workstations on a network with a printer and separate storage—you shared storage.

Bob Taylor: And it sort of out-designed itself in a way. Its abilities could not live up to the ambitions, so it fell short. It was too slow, it cost too much.

Charles Simonyi: The Star was not going to work. We were not going to win.

Chuck Thacker: So Simonyi went to Microsoft.

Charles Simonyi: In February of '81 I drove my car up to Seattle and started working right away.

Alvy Ray Smith: Years later I was at Microsoft and I met up with Simonyi. He took me over to his mansion there on Billionaires' Row on Lake Washington. I said, "How did you happen to leave?" He says, "You guys thought that you were at the center of the universe, didn't you?" I went, "Yeah." He said, "If you had met this kid Bill Gates and talked to him and heard his vision like I did...I realized you guys are not the center of the universe. This kid is." He left and went with Bill.

Adele Goldberg: Back at PARC we wrote a detailed proposal for a machine we called the Twinkle, the "little Star." And we told them that they could have product out in a year if they were willing to go with a 68000-based implementation of Smalltalk running a simple text editor and painting program with the mouse interface and some tutorials for how to do all that and hooked it up to printing. We pretty much had the prototypes done. It needed to be productized. And it could have been out in '83. We worked hard on that. And we sent that to office systems group and we never heard anything back, not a thing.

Bob Taylor: Xerox could have reengineered the Alto to make a better machine than the Macintosh, I believe, but they chose not to do that. They did not make another run at the fence.

Adele Goldberg: And then in '84 out comes the Twinkle but they called it a Macintosh. We all had the same idea. So we looked at it and we thought, *Too bad*. We could have done that. We would've had that out.

Bob Taylor: Remember Apple failed with the Lisa. Lisa was their first attempt.

Butler Lampson: Lisa had exactly the same properties as the Star system, and it failed for exactly the same reason.

Bob Taylor: And so Xerox looked at that and said, "We're done."

Butler Lampson: But Apple had a second go-around with the Mac. With a lot of pruning as well as the advantage of three-years-later technology, they were able to make something that was in many ways quite similar to the Alto.

Bob Taylor: Computers previous to the Mac I would regard as toys. But I don't regard the Mac as a toy. It was just a very weak manifestation of an Alto. Very weak. I mean it fell far short of the Alto in every respect, except cost.

Alan Kay: When I got to Apple I wrote a memo; the title was "Have I Got a Deal for You: A Honda with a One-Quart Gas Tank." The first line of it was, "Looks great in your driveway but don't try to go down to the corner store for a pack of cigarettes." And went on from there. It was short. It was like two or three pages. It was pointing out, "Hey, what the fuck are you guys doing?"

Andy Hertzfeld: Everyone knew what he meant. The main thing he was referring to was RAM. And we knew it didn't have enough RAM. It was just completely a technological compromise.

Dan Kottke: RAM was just very expensive in those days, and we're talking about an all-graphic screen—it's all RAM you're looking at there, so it's very expensive.

Bruce Horn: The Mac was simply not big enough to do any sort of object-oriented programming system. Basically if you look at a Mac and you look at Smalltalk, you have a mouse, you have overlapping windows. In the Mac you had pulldown menus and you have modeless text editing. Other than that there's no Smalltalk. It's not programmed in Smalltalk. It didn't use Smalltalk concepts under the hood. So there was basically no Smalltalk or any Smalltalk concepts whatsoever inside of the Mac.

Alan Kay: So you wound up with something that was skin-deep compared to what the PARC thing was but that had lost the soul of everything being accessible, everything being malleable and so forth, and we never got it back. And that messed things up forever. That messed things up to today.

Bruce Horn: Computing could have been this incredible tool for everybody. That was the vision of people like Doug Engelbart, Alan Kay...people like that. They were all basically trying to make the world a better place through computing. And personal computing is the most powerful way to extend an individual's human abilities. That's the bottom line. So for example, one of the things that Doug always said is "boosting mankind's capability for coping with complex urgent problems." That was the point of his work, his life's work.

Andy Hertzfeld: We couldn't be in the world we are living in today with Doug Engelbart's approach. Doug Engelbart did not care about making his software accessible. He really couldn't care less. Doug didn't care about it being easy to learn or really that easy to use. He cared about what you could do with it: "increasing human potential." It couldn't work for ordinary people, because ordinary people have very low tolerance for complexity. Steve saw that as much as anyone. He had such a passion for simplicity. And he's right. If you want it to be mainstream it's got to be as simple as possible. You never could have gotten to a billion people using it if you had a learning curve that took weeks. So you know both ideas are right. It's a balance.

Dan Ingalls: I look at it as a great collaboration. It may not have been planned that way, but that's how it happened.

Alvy Ray Smith: Everybody says, "Steve Jobs ripped off Xerox PARC." He didn't. Bill Gates did. He took Simonyi, who didn't just know the look of the Alto, he knew how it worked. And Simonyi came up there and wrote Word and Excel based on stuff he had done at Xerox PARC. That was the end. That was the real taking of the guts of the place.

Jim Clark: SUN Microsystems were really the ones who copied Xerox PARC. They made a workstation that would do what the Xerox Alto would do—a bitmapped graphical workstation.

Chuck Thacker: There aren't a whole lot of good ideas in the world, really. I mean the fact that the graphical user interface that we got together at PARC has lasted for as long as it has is to me quite amazing. Why haven't we been able to do better?

Bruce Horn: I think that there is another computer revolution to be had and we haven't had it yet, and a bunch of us are very sad about that and trying to figure out ways to make it happen. But there's just so much momentum and so much money and so much investment. If you think about Windows and all the people who have to do Windows support and so on, there's this huge financial infrastructure to keep it the way it is. And you know society doesn't like to tolerate risk. And anything that allows you to have access to the internals of the computers is risky. So it's very difficult to give people a canvas, a computing canvas on which to paint.

Alan Kay: Computing is terrible. People think—falsely—that there's been something like Darwinian processes generating the present. So therefore what we have now must be better than anything that had been done before. And they don't realize that Darwinian processes, as any biologist will tell you, have nothing to do with optimization. They have to do with fitness. If you have a stupid environment you are going to get a stupid fit.

Bruce Horn: I thought that computers would be hugely flexible and we could be able to do everything and it would be the most mind-blowing experience ever. And instead we froze all of our thinking. We froze all the software and made it kind of industrial and mass-marketed. Computing went in the wrong direction: Computing went to the direction of commercialism and cookie-cutter.

Jaron Lanier: My whole field has created shit. And it's like we've thrust all of humanity into this endless life of tedium, and it's not how it was supposed to be. The way we've designed the tools requires that people comply totally with an infinite number of arbitrary actions. We really have turned humanity into lab rats that are trained to run mazes. I really think on just the most fundamental level we are approaching digital technology in the wrong way.

Andy van Dam: Ask yourself, what have we got today? We've got Microsoft Word and we've got PowerPoint and we've got Illustrator and we've got Photoshop. There's more functionality and, for my taste, an easier-to-understand user interface than what we had before. But they don't work together. They don't play nice together. And most of the time, what you've got is an import/export capability, based on bitmaps: the lowest common denominator—dead bits, in effect. What I'm still looking for is a reintegration of these various components so that we can go back to the future and have that broad vision at our fingertips. I don't see how we are going to get there, frankly. Live bits—where everything interoperates—we've lost that.

Bruce Horn: We're waiting for the right thing to happen to have the same type of mind-blowing experience that we were able to show the Apple people at PARC. There's some work being done, but it's very tough. And, yeah, I feel somewhat responsible. On the other hand, if somebody like Alan Kay couldn't make it happen, how can I make it happen?

BOOK TWO

THE HACKER ETHIC

We are as gods and might as well get good at it.
—STEWART BRAND

What Information Wants

Heroes of the computer revolution

*B*y the mideighties, the technologists who were creating the future in Silicon Valley started to see themselves as more than simply engineers. Alvy Ray Smith and his midnight crew at Xerox PARC, the game makers at Atari, and the Macintosh team at Apple all became convinced that they were pioneers in a new expressive medium. They felt like designers, authors, even artists—and they wanted to be recognized (and compensated!) as such. But while the engineers fought their individual battles for money and credit, it took a writer from New York City to realize that this new class of creatives added up to a bona fide culture complete with its own lore, jokes, and ethic. Steven Levy made the argument in a popular ethnography entitled Hackers: Heroes of the Computer Revolution, *and the weekend-long book party for its release was the first* Hackers Conference. *At the confab, the hackers of Silicon Valley (and beyond) met each other for the first time and awoke to the fact that they had nearly everything in common.*

Steven Levy: When I first started writing about technology, I did a story for *Rolling Stone* about hackers at Stanford, but it turned out to be totally different than what I thought it was going to be. I thought I was writing about kids in college that were computer addicts. But I picked up on all the excitement involved in the personal computer revolution. So after that *Rolling Stone* story I wanted to write more about it and got a book contract for *Hackers*. Originally, it was with one publisher, but my editor went to Doubleday and I went with him. Doubleday was also doing the *Whole Earth Software Catalog*.

Kevin Kelly: Nobody was reviewing software then. It was considered completely nerdy, insignificant, hard to review: "Software" was floppy disks being sold in little baggies produced in people's bedrooms. What was good? What wasn't good? Nobody had any idea. So Stewart's idea was *Well, this is going to be big: Let's start reviewing this and make a guide to it. We'll have a* Whole Earth Software Catalog.

Steven Levy: The publisher spent $1.3 million for it. It was the most that was ever spent at the time for a softbound book.

Fred Davis: The *Whole Earth Software Catalog* would be the digital follow-on to the *Whole Earth Catalog* and this would be a mega-blockbuster. That was what we all hoped.

Kevin Kelly: When they started to hire for the *Whole Earth Software Catalog*, Stewart wrote me an e-mail and asked me if I'd come out and work for them. I like to say that I was the first person hired online. This was early '84. A lot of people were hired.

Fred Davis: I was young and wet behind the ears, just excited to have a chance to work for one of my heroes reviewing products.

Fabrice Florin: I edited the video section. I was a budding television producer at the time.

Steven Levy: I became the games editor of the *Whole Earth Software Catalog* and, you know, played volleyball at Gate 5 Road and got to know Stewart, who's like one of the most amazing people on the planet. That was great. So I was connected with those people. And when I finished my book I showed it to Kevin and Stewart—and they really liked it.

Kevin Kelly: I read *Hackers* and I was blown away, because I knew nothing of this world. It was all new. I mean new for me, because of course Stewart had written about the hackers in '72—he had been tracking this a whole decade earlier.

Stewart Brand: When I submitted that article to *Rolling Stone* in '72, Jann Wenner said, "This is going to set in motion a whole lot of reporting about these computer hackers and all of this," and I thought he was right about that. But it turned out there was a ten-year hiatus between when I did that article and when *Hackers* came out, so the lifetimes that I experienced around computers made it seem always very slow and kind of boring and frustrating.

Kevin Kelly: For me there was this new interest in this culture that was emerging around the programmers. I was talking to Stewart about the fact that there were three generations of hackers and they hadn't met— they themselves didn't know about each other.

Stewart Brand: And Kevin said, "What if we can get those three genera-tions together?"

Kevin Kelly: And the one genius of Stewart's is that he will often take an idea and, unlike me, he'll actually try and do something about it. He said, "We're going to do this." And he just kicked into his make-stuff-happen mode.

Steven Levy: They had a series of meetings with an advisory committee on who to invite. I was on the East Coast, so I didn't go to a lot of them, but I went to some. Andy Hertzfeld was on the committee, I remember.

Andy Hertzfeld: So I got a phone call from Stewart Brand, and once a week for seven weeks we drove up to Sausalito. There were like seven hackers he got to design the Hackers Conference.

Stewart Brand: We got a pretty good influx of folks. There was Ted Nel-son, obviously. Lee Felsenstein, the sort-of master of ceremonies for the Homebrew meetings, which I had never gone to. But his reputation was good.

Lee Felsenstein: By that time, Homebrew had ossified. It wasn't new people coming anymore. There was the same old faces. I called it "the old farts society." We had the meetings for the Hackers Conference at the tugboat that Stewart Brand lived on in Sausalito. The meetings were mostly where we threw out names of who else ought to be invited.

Andy Hertzfeld: My main contribution was just corralling the Apple crowd. I made sure all the key Mac people were there.

Lee Felsenstein: The launch of the Mac had been in January. And the con-ference was in November of 1984.

Steven Levy: It was on the weekend that was literally the publication date of my book, so it was like a giant book party. At that point you could some-times get a copy of the *New York Times Book Review* a little early. I just spotted one for the first time in the airport bookstore while I was flying out there, and so I picked it up and there was a tiny little capsule review of *Hackers*. It wasn't a full review. They trashed the book and said that it wasn't a real book, just an overblown magazine article. I was super-depressed on the flight. It was just awful. I just wrote my first book, and it's dead. *I'm ruined*, I thought.

Kevin Kelly: We had the conference at Fort Cronkhite in southern Marin. It was barracks, really primitive barracks.

Steven Levy: It was at this old army camp in the Headlands. A beautiful place, Fort Cronkhite.

Kevin Kelly: And so some of the people who we invited came. The number 114 sticks in my mind. I think that was maybe the number of people that finally were there. They were all hackers to varying degrees. I think there was only one or two women. Maybe three or four. All the rest were guys.

David Levitt: The Hackers Conference tried not to be a boys' club, but they did not try that hard.

Steven Levy: When I got out there my spirits soared. All those people were there and what they shared was that personality which I wrote about in the book. So it was like a living proof that what I wrote about was really there—no matter what happened to the book. It was the most amazing book party.

Captain Crunch: Wozniak was there. Stewart Brand was there; Ted Nelson was there. A lot of the BASIC people from Homebrew were there. A lot of the founding fathers were there, definitely: a lot of the old-school hackers.

John Markoff: I was there, hanging out.

Fabrice Florin: You really had all the players all in one place. It was a big deal.

Michael Naimark: I remember thinking to myself that this was a moment, a really significant event.

Lee Felsenstein: It really was a gathering of the illuminati. The important thing was that all the illuminati had never before gotten in a room at the same time.

Ted Nelson: It was the Woodstock of the computer elite!

Steven Levy: It was like a secret culture until then. So now you would say, "Boy, what would the ideal computer conference be? Who would be there?" And you would say, "Oh, you'd have Mark Zuckerberg and Larry Page…" Well, this was like that—but they were people you never heard of. But they were the people who really were driving things: the secret inventors of our culture. They were all there. None of them were really *famous*. Even Steve Wozniak, everyone knows him now, but he wasn't so famous then. He was just getting to be known a little bit, maybe through the US Festival and some other things. But basically back then no one knew or cared to know who these people were. We just introduced ourselves the first night.

Fabrice Florin: It started in the evening, everybody coming in. It rained a lot. It was gray skies, which was good, because we just needed to be inside.

Lee Felsenstein: That's a hacker thing, too.

Fabrice Florin: We all gathered in one large room and we formed a circle. Everybody started talking about who they were and what they were doing. So there was a lot of palpable excitement in the room that these people were meeting for the first time.

Steven Levy: This was the moment where the consciousness of that time really became something that you could pick up on. Just like in the gay movement where there was a time when people just felt incredibly alone, that there was no one else like them. There was this consciousness that came out there: *Oh, this is who I am!*

Kevin Kelly: Back then you had hippies and you had preppies. The nerds were just so off the radar that encountering them was exhilarating because you realized, *Oh, these are my people.*

Steven Levy: I think it's fantastic that now in high schools people can be themselves whether they are gay or a hacker, right?

Lee Felsenstein: It was really a very inspiring kind of occasion—that sort of general inspiration: *Hey, we are here. We can get together. We can do things together. We are interesting people!*

Steven Levy: The other thing was the laughter. Every session was joyous. People were really funny. There was so much laughter throughout. I don't think I've ever been to a conference where people laughed more.

Captain Crunch: Wozniak was pulling pranks on me. He'd go into my room and hide my stash behind some computer somewhere and I'd have to go looking for it.

Kevin Kelly: There was also that nerd humor. Now we all recognize it: Reddit, *xkcd*, *The Big Bang Theory*. That was it. Now this is sort of mainstream but at that time, it wasn't.

Steven Levy: It was a shared humor out of a shared experience there. And even though some people had never met each other, it was like you were part of a crew that had worked together for years and years. It was like inside jokes with people you'd never met before. They were also just genuinely funny.

Fabrice Florin: So, a lot of discussions that first night.

Steven Levy: It was sort of a surprise to me how amazingly it was set up. They knew because they had hackers on that committee that we were going to go all night and that we were going to need to have the machines on. Though the power did go out.

Captain Crunch: People were kind of speculating that Woz had something to do with the power being cut off.

Steven Levy: It went dark. But they kept going. They just kept talking. They were all in the same head. It was like this incredible energy. They had so much to talk about. They were just like talking to each other, they hardly noticed. Then the power came back and there was all this excitement again.

Fabrice Florin: I remember having basically a lot of interesting conversations, including the one with Bill Atkinson. Bill was one of the designers of the Macintosh. He had an idea at the time which he hadn't coded yet, but he was thinking about it. It was quite revolutionary.

Kevin Kelly: Bill was noodling with the idea of hypertexting but didn't have anything to demo at that point. He was talking about his ideas.

Fabrice Florin: Bill described this vision for enabling people to browse through nodes of information and jump from one node of information to another so they could learn by association, through serendipity, and be able to click on hyperlinks and just jump from one piece of information to another. It was based on the same concepts of hypertext that Ted Nelson and Doug Engelbart had been talking about for decades before. We all sit on the shoulders of giants. We're all inspired by these ideas.

Bill Atkinson: I viewed what was later called HyperCard as being a software erector set—where you could plug together prefab pieces and make your own software.

Kristina Woolsey: It was basically a way to take your personal documents and link them together. Just remember, when the Macintosh came out, it had an object on it that no one had ever seen. Nineteen eighty-four is when the mouse was something you could buy on a machine. And the mouse and linking go together. If you have a pointing device, linking is a natural piece. You can click on this, click on that. Fundamentally the mouse and linking are the same.

Bill Atkinson: Looking back, I sort of see that HyperCard was kind of like the first glimmer of a web browser—but chained to a hard drive.

Fabrice Florin: Then immediately after the group discussion people went downstairs. We had lots of tables and computers. It started looking more like a hackathon, which back then was a novel concept. It probably came out of this Hackers Conference. But essentially you had all of the tables and you had the Mac designers showing off their latest stuff. Andy was showing his latest hack.

Andy Hertzfeld: Software, groundbreaking software, is at its most dramatic at the very beginning.

Kevin Kelly: Andy was jumping up and down, he was so excited. He was literally jumping up and down. And he was calling his hacker friends over and he wanted to show them something he had hacked up the night before.

Andy Hertzfeld: Switcher was written the day before.

Fabrice Florin: He had done a hack just for the Hackers Conference.

Andy Hertzfeld: I said that, but that was a lie. I wrote Switcher because there was a 512K Mac, not because there was a Hackers Conference.

Steven Levy: At that point the "Fat Macs" were just coming out.

Kevin Kelly: He was working on a Fat Mac, and something he calls Switcher allowed you to have more than one program open at once, and you could switch between them.

Fabrice Florin: Because before Switcher you had to basically open an application and then quit it and open another one.

Steven Levy: Basically Switcher was a way to hack multitasking on the Macintosh. Which you couldn't do at all, no computer did that. No personal computer allowed you to do multitasking. So it was incredibly valuable.

Kevin Kelly: He'd just done it the night before: just pulled it out of his own brain! But the cool thing was seeing him explain it and show it off to other hackers and see that this was actually the currency, this was the dynamic. He was doing stuff and giving it out to everybody: "Yeah, here's a copy of it." And they were going to go and hack on it some more. It was just the very thesis that Steven was talking about in his book, this hacker ethic of sharing and building upon in an open-source sense and there it was. It was right there.

Steven Levy: It was just like endless demos and showing people things and cool stuff. They hooked up the games.

David Levitt: Amazing games no one had ever seen before.

Fabrice Florin: They were up until the wee hours just showing off stuff, discussing, hacking, teaching each other how to do stuff.

Lee Felsenstein: A lot of people had stayed up pretty late. And it was a hostel with these sort of pro forma beds that they were using with waterproof mattresses like in prisons.

Steven Levy: We were in bunks. It was cold at night.

Stewart Brand: What we knew was that hackers would be perfectly comfortable in low-rent circumstances—and they were!

Kevin Kelly: This was still early days, so they weren't like the billionaires that they later would become. They were people who had real jobs and had some money. And they were camping out basically for a weekend.

Fabrice Florin: Then in the morning Stewart Brand went through the dorms and started kind of kicking the beds and saying, "Anybody who wants breakfast, now's the time!" So you had all these droopy figures heading out to the little cafeteria.

That next day, the conference portion of the Hackers Conference started in earnest. There were no featured speakers: The whole point was to get the hackers talking to each other.

Fabrice Florin: We had different topics. I forget exactly what the topics were, but there's one that's an evergreen topic that's still being debated today.

Steven Levy: The "hacker ethic" was a term I coined to describe what I noticed was a shared set of values that hackers of any generation all have. You know, the way they saw the world and the way they operated and what drove them. The way the hackers saw it was that information should be free. That's what I tried to grapple with in *Hackers*: what that meant.

Stewart Brand: It was a discussion when the whole group was gathered. There was talk of free software versus commercial software.

Steven Levy: It was one of the first sessions in the conference—the session that launched a million other sessions. Every conference in the last twenty years has had at least one to ten sessions on that very subject. And here I was discussing it for the first time and probably with the best audience, the best participants. It was amazing to moderate, if you want to call it that, a session like that. Because it was so participatory. People were chafed and they all wanted to talk. Everyone wanted to get in on it there.

Doug Carlson (at the Hackers Conference): The dissemination of information as a free object is a worthy goal. It's the way most of us learned in the first place. But the truth of the matter is, what people are doing has more and more commercial value, and if there's any way for people to make money off it, somebody's going to try to get an angle on it. So I think that it ought to be up to the people who design the product whether or not they want to give it away or sell it. It's their product and it should be a personal decision.

Robert Woodhead (at the Hackers Conference): Tools I will give away to anybody. But the product? That's my soul that's in the product. I don't want anyone fooling with that. I don't want anyone hacking into that product and changing it—because then it won't be mine.

Steve Wozniak (at the Hackers Conference): Hackers frequently want to look at code, like operating systems, listings, and the like, to learn how it was done before them. Source should be made available reasonably to those sort of people. Not to copy, not to sell, but to learn from.

Stewart Brand: Wozniak made the point that there's a whole bunch of work creating a piece of software that does something useful and actually works well.

Steve Wozniak (at the Hackers Conference): Information should be free— but your time should not.

Stewart Brand: So, putting these things out for free is kind of nuts. And, I said, "Well, software—this kind of information—wants to be expensive, but it also wants to be free because it's so easy to copy."

John Markoff: It was a dialectic, right? Stewart is not a Marxist, but it was a very Marxist view of the information economy.

Steven Levy: It was a conversation, they were engaging. The whole thing was almost like a jazz improvisation. Just like building up in one of those long Coltrane songs or something like that.

Stewart Brand: I was really just restating something that was written down in Levy's book as "the hacker ethic."

Steven Levy: Information *should* be free.

Stewart Brand: My only addition to that was to take away the "should" and turn it into a "want."

Steven Levy: He hacked me! That's the way I put it.

Stewart Brand: "Information *wants* to be free" was the meme that got loose and went viral from that discussion.

Kevin Kelly: It was just another throwaway line at the time. There was no sense at all that this would become Stewart's most famous saying. He's said so many brilliant things over the years. I'm still surprised that this is what wants to be on his tombstone.

Stewart Brand: It's giving information its own desires, I think, that makes people jump.

Steven Levy: It sort of sang in its own way and described something that was happening—and would happen much more on the internet, in a way that just made people grasp it.

Kevin Kelly: Maybe the reason why it was launched into prominence is because of what I call the zero price point option, the idea that free would become this huge economic model and would come to dominate much of the wealth that was being generated. There was maybe a glimpse of that there.

Steven Levy: Information wants to be free on the internet in particular.

Stewart Brand: The quote in full is "Information wants to be expensive and information wants to free, and that's a paradox which will never go away." And the quote keeps reviving, because that paradox keeps driving people mad. Ha-ha!

The Whole Earth 'Lectronic Link

Welcome to the restaurant at the end of the universe

*T*he Well started as an experiment in community, in publishing, in the future—all things that Stewart Brand was expert in. From a technology standpoint, it was little more than a glorified BBS (bulletin board system) hacked together from a few telephone lines, a couple of early modems, and a nearly obsolete computer from the seventies. Still, a broad swath of notables from Silicon Valley and the greater San Francisco Bay Area joined Brand's primitive social network seeking, like him, to understand the new online world. What they discovered when they logged on was a bohemia and a frontier. That the Valley circa 1985 was also a bohemian frontier was no coincidence: The Well was Silicon Valley online. The Well attracted hackers from the Hackers Conference as well as Brand's many high-flying friends from the disparate worlds of journalism, ecology, philanthropy, and business. Almost immediately The Well became the place to be in cyberspace and, in fact, is the place where the word cyberspace first acquired its contemporary meaning. Excited by the success of this new and very social medium, Brand was sucked in—and burned. He discovered that The Well had a darker side, too. Flame wars, trolling, cyberbullying: All the social media dysfunction that is so familiar today came as a disturbing shock in the mideighties.

Larry Brilliant: I knew Stewart Brand from the Merry Prankster days and the Hog Farm days.

Stewart Brand: It was mainly through the Hog Farm. Larry Brilliant had been there as a kind of a resident physician.

Ram Dass: He used to be known as Dr. America. The Hog Farm was a hippie commune in Berkeley, run by a fellow named Wavy Gravy.

Stewart Brand: And I also knew Larry Brilliant by reputation: what he had done suppressing smallpox in India, and eliminating it, eventually.

Ram Dass: It was very exciting stuff: You've got maps with pins of each case and you go with vaccination needles in jeeps and boats and you raid villages. It's like a war!

Larry Brilliant: By then my Rolodex was pretty weird—Ram Dass and Wavy Gravy and the head of the smallpox program, UN diplomats, a bunch of rabbis and Catholic priests with eclectic backgrounds…And Steve Jobs.

Ram Dass: When it was over they all got depressed. They didn't know what to do next.

Larry Brilliant: So Steve comes back from India, he starts Apple with Woz. I went back and finally got a degree in public health.

Ram Dass: Larry decided he'd start a foundation and get all of his buddies together—and he drew his buddies from all the different parts of his life.

Larry Brilliant: And often Steve would call me, I remember once I was in the middle of painting my fence in Michigan, and my wife said, "Steve's on the line." I rushed to the phone covered with paint, and he said, "Larry, I understand you're thinking of starting a new organization?"

Ram Dass: The organization he called Seva, which meant "Society for Epidemiological Voluntary Assistance"—only he knew that it was also the Sanskrit word for "service."

Ram Dass: And I said, "Yes, we decided we would tackle blindness, we've got money from the Dutch government to do it in Nepal, and we want to start it off with a survey—we want to do it the right way." Hippies we might have been, but we were scientific hippies.

Lee Felsenstein: You could say you were a hippie and didn't like technology, but what *Whole Earth* was telling us all and we all believed was that you are going to be using technology if you are a human. And so, as Stewart said, "We'd best get good at it."

Larry Brilliant: And so Steve said, "Larry, I have just the thing for you. And—are you ready for this? I know the software that you can use to run your survey: It's called VisiCalc, it's an electronic spreadsheet." To which I said, "What's a spreadsheet?" And so we talked for about two hours, and finally this package arrives and we smuggled it into Nepal. This was the Apple II—serial number 12. They said, "What is it?" And I said, "It's a typewriter!" And they said, "No problem—go in." So I had that, I had one of the first acoustic modems, and I had all sorts of crazy software that Steve had sent me. And on the last day of the survey, after we'd finished the last case, the helicopter crashed, and it crashed in a town which was just impossible to get to. Nobody was hurt, they say it came down like an oak leaf, but the engine was totally destroyed—it had digested itself. So now I'm running a UN program, I'm a UN officer, my helicopter is down. So how do you get it out? Who pays for it? It had all the hallmarks of the dozens of other planes and helicopters that have crashed in Nepal and that are still there, thirty or forty years later. But I was determined to get it out, and so I thumbed through all the pieces of software that Steve had given me, and one of them was "Satellite Access." So I put it into the Apple II, and managed to access the University of Michigan computer from Kathmandu, and I set up a computer-conference call—this became known as computer conferencing—and it included Senator Hatfield's office, the Evergreen Helicopter Office in Oregon, the Seva office in Ann Arbor, Michigan, the ambassador to India in New Delhi, the UN office in New York, and Aerospace Seattle's office in Paris.

Kevin Kelly: This was the dawn of bulletin board systems.

Larry Brilliant: There was no e-mail, remember.

Howard Rheingold: On a BBS, there was one thread, and you had to wait for the person who was online to get off-line so that you could log in.

Larry Brilliant: And using store-and-forward asynchronous communications, we were able to get the engine—which weighed maybe a couple of tons— flown in. We put it on the back of a jeep and rode it through the jungle to the downed helicopter, where we had three or four men acting as a hoist. We changed the engine and flew the helicopter out. That was 1979 or 1980. It was pretty cool, and I'd never seen anything like that before. I'd never seen anything happen like that in the UN where a broken helicopter could be repaired and flown out seventy-two hours later! So a reporter from *Byte* magazine, who heard about it, called me and did a story in *Byte* about computer conferencing. And I then used that, when I got back to Michigan, to create something we used to call DEO—Distributed Electronic Office. This is 1980, 1981.

Kevin Kelly: When I got a modem in the early eighties it was a big "a-ha" moment for me. Because I started to explore bulletin boards in the evening and there was something happening, something really important happening. It was different than anything else I encountered. Something odd was happening there. There was this tone of voice. I mean there was this sensibility and people were talking in an informed way about stuff that they were interested in. There was this sort of underground aspect to it. There was this community aspect to it. I was an old hippie and was very suspicious of technology. I owned a bicycle. I didn't have a car. I didn't really own anything. I was sort of not so interested in what we called technology, which was like cars and factories. That's what technology was. What I saw when I plugged the computer into the modem was a much more organic version of technology. It was much more Amish-like. It seemed more appropriate. It seemed humane. There was something very robust and biological about it. And Stewart noticed that, too.

Larry Brilliant: And then Steve Jobs happened to come to Ann Arbor. He stayed with me, and he had given me the first $5,000 to start Seva, and I started to hit him up for some money to analyze the data from the survey, and he said, "Well, show me what you've done." And I showed him Seva-Talk, and I showed him how we could talk to each other all over the country, and then when I asked him for the money he said, "Look, Larry, instead of asking me for money, why don't you take your own fucking software, make your own fucking company, make your own fucking money, endow your own fucking NGO, and get rid of blindness yourself! I'll help you. I'll help you build what you've got into a company, and find you investors to take it public." And he did, and that's what happened.

Ram Dass: The company he started was a software computer company. What it does is set up conferencing through computers. The first beta-test site was the Seva Foundation.

Kevin Kelly: Larry is a very entrepreneurial guy.

Ram Dass: The stock of the NETI Corporation endowed the Seva Foundation.

Larry Brilliant: Then I said, "Stewart, I took this company public, I have some money, how about we do a joint venture?" This is how The Well got started.

Stewart Brand: Larry wanted a *Whole Earth Catalog* teleconference of some sort, basically an online version of the *Whole Earth Catalog*. He was a lover of the *Whole Earth Catalog*, I guess.

Fabrice Florin: The counterculture movement idolized the *Whole Earth Catalog*, which symbolized this kind of holistic view of the world.

Stewart Brand: I said, "I'd be interested in working on that. What do we get?"

Larry Brilliant: I said, "I'll give you some technology and a couple hundred grand, and you provide the customer base, the family, the community, you run it, and we split it fifty-fifty?"

Kevin Kelly: So we started The Well in 1985.

Fabrice Florin: *The Well* stood for "Whole Earth 'Lectronic Link."

Stewart Brand: I had seen online teleconferencing. And I had seen things that had made it wonderful and things that had made it terrible, and there were also bulletin board systems around at that time which had also established a certain amount of behaviors that were wonderful and not so wonderful. So, based on those experiences, we designed what became called The Well to reflect what had been learned about online discussions at that point.

Kevin Kelly: What *Whole Earth* wanted to do was to try out the experiment of what happens if you open this up to ordinary people. How does it all work? No one had really opened this up to the public beyond what CompuServe and Prodigy were doing—very controlled, minimalist systems.

Fabrice Florin: Stewart Brand recognized very early on the importance of computers for information sharing, and so very early on he wanted to be able to serve his community with the right tools that would allow people to communicate and share ideas effectively.

Kevin Kelly: Whether it was ever articulated or not, what *Whole Earth* was doing was trying to make it work in terms of governance, best practices, cultural nudges, pricing…

Howard Rheingold: Part of it was *not* trying to shape it, enabling the shape to emerge. Letting the people there make whatever it was.

Kevin Kelly: One of the design principles for The Well was to try to get as close to free as we could get. The idea was to see how cheap can we possibly make this, cheap for the user.

Stewart Brand: I priced it to be very inviting.

Kevin Kelly: We would bring the power to the people by having really cheap telecommunications. And that I think in some ways, that idea came from the hacker community, the Hackers Conference.

Stewart Brand: We made it easy to make conferences, and people would invent conferences. Anybody could start a conference.

Kevin Kelly: It was kind of a hack: The software was buggy, but sort of open and easy to modify.

Lee Felsenstein: All the people who had been at the Hackers Conference were offered memberships.

Stewart Brand: I wanted hackers inside the system, so we invited them in and writers, journalists, all got free accounts and that was our marketing. Initially it was just sort of friends of the *Whole Earth Catalog*.

Larry Brilliant: Ram Dass was on The Well. Wavy Gravy was on The Well. All the Seva people started using The Well. It was pretty cute.

Howard Rheingold: Stewart Brand brought together the environmental network, the people who were interested in community, self-sufficiency, communes, personal computers, tools. So it was really the meeting of those communities and this kind of ongoing party.

Larry Brilliant: And then the Deadheads started sharing audio tapes. They were digitized and available online, but the only way you could find the location online was to do it through The Well.

Kevin Kelly: And once they would come into the Deadhead conferences they would spill over. They would say, "Oh, you have a thread on gardening. Wow!" Or a thread on parenting, whatever. So they would come on to swap playlists and then they would go on to discover the rest of the world and become some of the most active contributors.

Fred Davis: It was like a hip pre-Craigslist thing where you'd hear about where a secret rave would be or who had tickets to the Dead concert or whatever. There was a real community, a cyberculture community.

Michael Naimark: You couldn't even send a picture on The Well. It was text-only, and of course it was asynchronous: There was no live chat. But nevertheless a lot of us were on The Well if not hourly, then multiple times a day. After Atari collapsed, Brenda and I were putting together some grand plan for virtual reality—we were striving to make virtual reality experiences that felt just like being there—but the irony was that this little trickle of text that wasn't even live gave us the greatest feeling of live-ness! It felt more live than any VR stuff we could possibly do.

Howard Rheingold: And it was all just words on a screen!

R. U. Sirius: These were just text-based bulletin boards, but in many ways they were superior to social media today. You had really great conversations with extraordinary people.

Larry Brilliant: Because it was Stewart, he attracted people who had these incredibly eclectic minds, and they were phenomenal writers, people who think in paragraphs. And the writing was fantastic!

Kevin Kelly: That made for a very literate salon-like environment where people who could write were writing—and writing well and writing very directly. So some of the best writing I think of that decade was happening on The Well.

Larry Brilliant: So just the opposite of Twitter.

Lee Felsenstein: The Well, for its first five years at least, was the San Francisco bohemian scene online, where you could join the roundtable of whatever-it-was. There was a whole bunch of roundtables. And in there were the people who were the ones you had read about and so forth. Or had firsthand connections with the people you read about. San Francisco had had such a scene since the nineteenth century. And here it came direct to your home at your fingertips.

Howard Rheingold: It was like there was a party happening in the walls of your house. You know, people were talking about serious things and exchanging knowledge, but they were also having fun.

Kevin Kelly: Stewart and I, we were living on The Well. This was the beauty of it.

Lee Felsenstein: It was an experiment. An attempt to infuse their culture into an online system. They were feeling their way, as anybody would have to, since it really hadn't been done before.

Kevin Kelly: I can't remember what the first conference was, but very quickly there was a conference about The Well itself, because we quickly learned that talking about The Well and the policies of The Well would kind of creep up in every single discussion. "Why can't we do this? Why can't we do that?" Because nobody knew anything. We didn't even know we needed moderators! That was not even clear. That was not something we thought about before. Why do you need moderators?

Lee Felsenstein: They assigned me to moderate the hackers' section of it.

Kevin Kelly: That became one of the biggest jobs that we had: people having to moderate the conversation.

Lee Felsenstein: And so I had to bust up some kind of paranoid discussion threads that mostly had to do with other personalities in the conference. A couple of times I had to jump in and say, "Now you are all making far too much out of this. There's nothing there." And, you know, "Calm down, for God's sake." It was worth the effort. But it took effort.

Kevin Kelly: These systems are natural amplifiers, and negative things are somehow easier to amplify or become much louder than positive things, there's something about a negative amplification that just powers up. And so we saw these phenomena where small slights would be amplified into huge harm and pilings on, and people who were normally very civil would get sucked up into battles. And they would have what we call flame wars. It was sort of like a flame in the sense that the hotter it got, the more that would be sucked into it and burn.

Lee Felsenstein: We discovered early on about the tendency to flame. Hackers do that, of course. But we thought that was just a hacker thing and it turned out not to be.

Kevin Kelly: And we began to see trolls, although we didn't use that term at the time, where there were people who were getting satisfaction out of starting fires or nudging people. They would do that over and over again just because they liked to see what would happen.

Stewart Brand: People learned how to deal with trolls. If you respond to them, they will make the flame even brighter.

Kevin Kelly: So we had to deal with that. And issues about people wanting to remove what they had said and whether that was okay. And so there were all these things that are now very familiar dynamics that were completely new to us. And each one we had to address, and we were spending days and nights and evenings trying to manage these things.

Howard Rheingold: The Well had a policy that people should be who they are. And so you had to use a credit card, or otherwise go to the office and show some ID, to prove who you were. That was a good design decision.

Stewart Brand: I had seen a situation online where people behaved very, very badly, and I knew that even famous intellectuals would behave badly to each other if they were able to post anonymously. Based on that, I made it impossible to be anonymous on The Well. However, you could put on a handle, which would be sort of pseudoanonymous.

Kevin Kelly: We did have an anonymous conference. That was a true experiment, and that was driven much more from the users who said, "Can we have an anonymous conference?" And we said, "Okay." We had no idea what would happen. Sometimes completely amazing stuff would occur in the anonymous conference, but in other cases this could be kind of scary and weird and creepy. People would confess to things or do stuff.

Stewart Brand: The anonymous conference was easy to set up with that software—and it lasted less than a week, because people immediately behaved absolutely viciously to each other. They pretended to be each other. They thought they were just spoofing, but actually it was mortally insulting stuff they were doing.

Kevin Kelly: We were asking ourselves, "What do we do about this? Are we legally liable?" And again that's the kind of stuff that, at the board level, I was concerned with. I didn't have an opinion about anonymity before the anonymous conference. But after the anonymous conference I had an opinion: It was not good.

Stewart Brand: So, The Well kind of grew and got a life of its own and established a certain amount of online practice.

Howard Rheingold: There was kind of a social policy: "You own your own words" was mostly about people had to get permission if they were going to quote you, but it was also about taking responsibility for your words.

Larry Brilliant: And the reason that The Well succeeded was because of those things—not because of the software, not because of the money.

Kevin Kelly: It was all new territory and it was very formative, because there's very little that's happened since that wasn't present in the decade that The Well was running at its height. Almost all the things that we've now come to see as a marked characteristic of this time emerged then, and we were dealing with it for the first time.

Howard Rheingold: And in some ways it was a forecast of not just the best of what online community could offer but also the worst. You know, people who just like to snipe at you.

Kevin Kelly: This was years into it but at one point Stewart basically quit The Well, because as the leader of this enterprise—he really wasn't the leader but as the figurehead—he was just getting trolled. He was getting pounded and harassed. And it was no fun and so he thought, *No fun? I'm out of here.*

Howard Rheingold: Stewart got irritated, and who can blame him? He created the place. A lot of antiauthoritarianism was projected on Stewart because he was Stewart.

Stewart Brand: There was a classic kind of gang-up thing that one continues to see, occasionally. And I just bailed at that point.

Howard Rheingold: It's very difficult to get thrown out of The Well, especially for being an asshole. They have to argue about it for months. A lot of people, including myself, just sort of got sick of it.

Stewart Brand: In any community, new people show up, and they want to participate, and the old hands typically close ranks and sneer at the newbies. I should have known that would happen at The Well. We should have made it the case where part of your job as a member of The Well was to make new people feel welcome. We never did that. It was a part of what kept The Well from growing.

Kevin Kelly: After Stewart left, everything started to kind of get really big. This was the era of ISPs, and you had Pipeline and Echo and AOL, and it was clear that this was going to stick around. Some of them were growing fast. And so why can't we grow fast? The problem was we were a nonprofit. Who's going to invest into this nonprofit? And so that was the issue. I looked at it in different ways. *Do we want to sell it? Do we want to turn commercial? What's the point of that?* So in the end it was like, *No, I think we can be more useful being who we are. We could grow and we could make a lot of money, but a lot of people are going to do that.* And that might have been the wrong decision or the right decision to make, but it was my decision to keep it sort of experimental.

Stewart Brand: It was never a commercial success. It may have paid its own way, just barely. What could be tried with this medium? That was the thing.

Kevin Kelly: Eventually it was sold to *Salon*, but it was really too late at that point.

Stewart Brand: It had great continuity and those people stayed in touch online for decades and all that. But it ossified...

Fabrice Florin: And a lot of the intellectuals that were sharing ideas on The Well went on to branch out into different areas. But you can really trace back a lot of the origins of this new movement to The Well. A lot of the folks were there.

Howard Rheingold: I remember I got a friend request on Facebook early from Steve Case and I said, "I know who you are. But why do you want to friend me?" And he said, "Oh, I lurked on The Well from the beginning." So I think, yes, it did influence things.

Larry Brilliant: Steve Jobs was on it—Steve had a fake name and he lurked.

Howard Rheingold: Steve Jobs, Steve Case, Craig Newmark: They would all say that they were informed by their experiences on The Well.

Fabrice Florin: The Well was the birthplace of the online community.

Larry Brilliant: All that goes back to Steve giving me the computer, letting me use it in Nepal, the experience I had with his software to access the satellite, and then coming back and Steve seeing what Seva-Talk could be.

We showed it to hundreds of people and nobody saw anything in it. Steve got it immediately.

Fabrice Florin: And Stewart basically gave the technology a set of values and ethics that all the developers could share. They already had their own hacker ethic, but he helped to amplify it and bring people together. And then it became big business, and it was hard for intellectuals to be the primary driving force anymore. It became the businesspeople who started driving it. Which is understandable given the scale and scope of what happened. It just became too large for intellectuals to hold.

Reality Check

The new new thing—that wasn't

*V*irtual reality is Silicon Valley's next new thing. Facebook and Google (not to mention Microsoft) are making huge bets on the technology and battling to control its future. Hollywood has jumped on the bandwagon, too: The best VR content is now showcased every year at the Sundance Film Festival. But VR is not as new as it seems: It germinated inside Alan Kay's research lab at Atari. The lab collapsed when Atari did, scattering Kay's people across the Valley, but a young, dreadlocked, programming prodigy, Jaron Lanier, continued the research on his own dime. His original goal was to revive an old dream. Like Doug Engelbart and Alan Kay, Lanier wanted to create a computing environment that was immersive, flexible, and empowering. The difference was the interface. Engelbart invented the mouse. Alan Kay added the desktop metaphor. And in Lanier's iteration, one donned goggles and gloves and stepped into virtual reality. Lanier actually coined the phrase. And the whole point of this new, all-enveloping interface was to be able to program the computer from the inside.

There was just one problem: Once people got inside the computer, virtually no one wanted to code. There was a whole world in there, a cyberdelic Disneyland just waiting to be explored. Lanier thought he was building a next-generation programming language with the corresponding next-generation graphical user interface, but what people experienced was something a lot more fun. VR was The Well's cyberspace made real. Taking advantage of the ensuing limelight, Lanier swiftly assumed a more Jobs-like role and marketed the heck out of his virtual reality machine, but in the end, the cost of an E ticket was just too high.

Jaron Lanier: I had already been working on virtual reality stuff in garages in Palo Alto before I went to do the Atari Cambridge thing. We did all kinds of weird experiments, all very funky and all short-lived. Tom came up to me after one of my talks in the early eighties.

Tom Zimmerman: I met Jaron at a Stanford electronic music concert at night in the outdoors. And I told him I had this glove.

Jaron Lanier: I said, "Oh, what does it do?" And he says, "Well, it's got continuous sensors." "Oh my God! We have to talk."

Tom Zimmerman: I had invented the dataglove between graduating MIT and joining Atari Labs.

Jaron Lanier: There were a lot of people who had done sensor gloves of one sort or another for, like, hearing-impaired applications or different things, but nobody had actually made a glove to my knowledge that had continuous sensing before. And so lightning fast I introduced him to a bunch of the other people who were kind of in the circle—including Young Harvill and his wife, Ann Lasko.

Young Harvill: Jaron had a house with Daniel Kottke, and we rented a house across the street from Jaron.

Dan Kottke: Jaron was a lean and trim game programmer when I met him.

Young Harvill: He was finishing a game called *Moondust* that was for the Commodore 64. It was a success and he did well with it.

Tom Zimmerman: What was cool about *Moondust* was it was probably one of the first nonviolent games. And the goal was just to create a visual-acoustical orgy of eye candy and sound, which was kind of a nice goal. It was this game where you were a little spacecraft and there's like dust and you have to surround the dust and corral it and then it just, like, explodes! So I think that spoke well of Jaron, and his life-affirming nature.

David Levitt: Zimmerman may have introduced Jaron and I. We had all this overlap in our interests with music. We resonated. We had a lot of the same ideas. People called us "the twins" because I had dreadlocks then, too.

Jaron Lanier: So about my hair. I gave up combing it at a certain point because it was taking over my life, and so I basically had three options. I either would spend all my life on hair care, I'd shave it off, or I'd just let it do what it wanted. And I just felt that of those options the laziest was the one I chose.

Young Harvill: Anyway, he was just coming off finishing this game.

Howard Rheingold: There was the personal computer, and video games were a big cultural splash. And so at the time people were beginning to ask, "What's the next thing?" Because things at that time seemed to be happening quickly. Although at the time I don't think there was any talk of virtual reality.

Jaron Lanier: Dan Kottke would kind of bring prerelease Macs home wrapped in towels on the back of his motorcycle and things would happen. The first MIDI program that I'm aware of was made on a prerelease Mac that Dan had snuck out of Apple.

Dan Kottke: I did that MIDI interface. It seemed like an exciting thing.

Tom Zimmerman: And then Jaron started getting into this programming language idea.

Jaron Lanier: I had a lot of money come in in '83 because this one game of mine called *Moondust* was quite successful.

Tom Zimmerman: And Atari blew up, so at some point Jaron said, "I'm forming this new company. Why don't you join me?" This was a lot more exciting to me, because it centered around making the glove that I had been developing for a while. And so I joined him, and that was the start of VPL.

Jaron Lanier: I believe VPL was incorporated in '83.

Young Harvill: VPL originally stood for "Visual Programming Language."

Tom Zimmerman: Jaron was developing a programming language. He really wanted to go for it.

Jaron Lanier: Back then we had the freedom to think in the big picture, which very few people have now. So the way people thought is they started with a general philosophy of life and tried to turn that into a computational paradigm and then thought about software architecture and hardware interfaces as part of the same unified concept. So that's exactly what Doug Engelbart did with his Augment system, and it's exactly what Alan Kay was doing with his systems: Smalltalk and the familiar interface and computer came about as one thing. And there were many other examples. So this is the way we all did it back then.

Young Harvill: Jaron's idea was that you could have a visual language. Kids could choose a series of actions that were represented by pictures, watch what they did by individually triggering them, and then collect them in a little order. And then that would become essentially a subroutine represented by this new picture that you made. Then that would become another one of these items that you could then demonstrate how it worked.

Tom Zimmerman: Jaron saw the glove as an input device for his programming language.

Young Harvill: What Jaron was pushing for, and Tom I think also, was for gestural interface rather than just click and drag and place. The visual language that you get with a mouse and a keyboard is limited. If you wanted to have a visual-type input, or say you wanted to have more analog-like input to trigger music or bend notes, then the gestural interface was really important. And so that was his whole point with the glove and gestures.

David Levitt: It's a direct manipulation interface.

Young Harvill: Jaron was very interested in Koko the guerrilla and the whole gestural thing. We were interested in sign language as just a beautiful gestural language. So I think at some point Jaron felt that this was a way to go after the whole Engelbart thing in terms of just pure expression. And everyone knew what the guys at Xerox PARC did and what they put together and how far they pushed that system—it was a touchstone.

Jaron Lanier: I was really interested in systems where you could change them while you're using them without limitation.

Young Harvill: So the first thing I ended up doing for VPL was the output for the optical glove that Tom was working on at the time.

Jaron Lanier: We very quickly added a tracker to it so we knew where it was in space, and integrated it with some of the software we already had so you could pick up stuff. And it clicked beautifully. I know it ran, I mean I have a picture of it running on a really early Macintosh, so the glove was already running and interfaced in '84. Glove-based interactions were very hard, because the machines were so slow back then. So it required heroic programming, which was in those days mostly supplied by Andy Hertzfeld on the sly.

Andy Hertzfeld: I have zero interest in talking about VPL or Jaron.

Alan Kay: Jaron's first version of VPL was certainly the best demo of anything I'd ever seen on a micro. It was just fantastic in every way, a programming tour de force with four or five really great ideas in it about how you might think about programming—particularly for younger kids.

Dan Ingalls: It definitely had this feeling of being able to do drag-and-drop in a 3-D virtual world. There is a lot that can be done in programming using essentially drag-and-drop: You can drag properties out of a parts bin and drop them on objects, and then the objects take on those properties.

David Levitt: The glory of programming this way was that by linking the graphic objects together and saying, "This one has that strength, this one has this weight," you could make gravity collisions. You could make a whole game. And if you drew it in front of a kid they would get every step. It was just miraculous: They understood this as programming inside a video game, rather than about a video game in some external text language.

Alan Kay: And it was just great. And of course it used a very, very simple glove he had. And I loved it.

So, as a by-product of building this novel programming language, Lanier had the beginnings of a virtual world. It was, essentially, a 3-D version of the graphical user interface pioneered at PARC. Instead of a mouse, there was a glove. Instead of a two-dimensional desktop, there was a three-dimensional workspace. Yet Lanier's virtual world was still trapped on the other side of a screen. It was another Atari Research alumnus, Scott Fisher, who took the final step—the leap into cyberspace itself.

Brenda Laurel: Scott Fisher, who was part of the proto-VR scene at Atari, went on to NASA Ames and really put the first systems together.

Scott Fisher: God, after I was fired from Atari it must have taken like nine months to do the background checks, but by the beginning of '85 I'm at NASA. I was also really passionate and obsessed with this idea of getting these head-mounted displays put together. And within the first week I discover that they have already put together this really great display.

Tom Zimmerman: So Scott had a head-mounted display at NASA.

Jaron Lanier: It was a very, very low-res display. So basically each pixel was as big as like a kitchen tile or something. But it had a head tracker and it was reasonably fast. And you could see fairly detailed things in it.

Scott Fisher: It had like, I think, one-hundred-by-one-hundred resolution.

Brenda Laurel: It was just green vector graphics on a black background.

Tom Zimmerman: Scott's vision was an astronaut could manipulate things outside of the capsule while being inside the capsule. Sort of like what they now have with the shuttle arm.

Scott Fisher: So I tracked down Tom and he by then had hooked up with Jaron to do VPL, and they were trying to figure out how to use the glove for this visual programming language that Jaron wanted to do. And I'm like, "Well, can you guys build a glove for us?" Because I needed a way to reach into that space and touch these virtual objects.

Tom Zimmerman: So Scott Fisher did telerobotics, while Jaron was talking about synthesized worlds. And he came to us with that vision and then contracted us to make a glove to do that.

Scott Fisher: We spent a lot of time with Tom and Chuck Blanchard, who did most of the software. He was from VPL. He helped us interface it with our system and wrote some great code to make that work. We ended up getting another glove so we could have two hands. And Tom continued to work on the glove technology. So it was getting better and better.

Tom Zimmerman: It was kind of the industrial use of virtual reality.

Brenda Laurel: They were doing training. I can remember laughing because one of their demos had a pull-down menu and it was like, "Why the fuck would you go to so much trouble to model a real world and then put a pull-down menu in it?" It's just like using a speech interface to say, "Move arrow on my screen." It was dumb. But I could immediately see the possibilities.

Jaron Lanier: It taught me how VR worked. It made a huge impact on me.

Tom Zimmerman: So Scott gave us entrée into matching the gloves with the head-mounted display. So that was a crucial element.

Jaron Lanier: Then we developed our own head-mounted display.

David Levitt: Our hardware engineer took apart a Sharper Image TV, a little Sony TV that was relatively uncommon in those days, and reverse-engineered until we found the video signal and said, "Yes, we can make a Silicon Graphics machine drive that."

Jim Clark: Silicon Graphics was a company that primarily made what would now be called GPUs—special purpose graphics processing units—to accelerate graphics, so that people who used our equipment could visualize the models that they made.

David Levitt: We used two Silicon Graphics machines, one for each eye, of course. So that became the first EyePhone.

Jaron Lanier: EyePhone, E-Y-E, obviously.

Mitch Altman: Our resolution was superlow: 480 by 680.

Jaron Lanier: And the EyePhone was the first commercially available head-mounted display. We sold a lot to labs all over the world.

Scott Fisher: I made it very clear in talks that I gave and talking with our collaborators like VPL that our next big push at NASA was for a multiuser system—having multiple people in the shared virtual space.

Jaron Lanier: The whole thing of being in there with other people where each person becomes an avatar was superimportant. That was the whole point.

Tom Zimmerman: And so what Jaron did was he took the head-mounted display, which was done by Scott Fisher at NASA, and he now created this whole interactive virtual world where he was really emphasizing interaction between humans in this synthetic world.

Scott Fisher: And next thing I know VPL has put out press releases saying that they've done it! It was competitive, and I hadn't been thinking of it as competitive. So again I guess I was kind of naïve about that.

Tom Zimmerman: There might be some jealousy. Because Jaron definitely got 95 percent of the media attention.

David Levitt: Our lead product at VPL was called "Reality Built for Two." RB2, that was an SKU at VPL.

Jaron Lanier: This is actually a bit of an inside joke and it goes back to Alan Kay. So Alan Kay had once called a computer "a bicycle for the mind." And in fact Steve Jobs put that slogan on some of the early Apple literature. So I thought the idea of it being a "reality built for two" recalled bicycles.

Alan Kay: A lot of this stuff is just modern engineering versions of old ideas. Ivan Sutherland did it first.

Jim Clark: Ivan had created a head-mounted display in 1968. I had the idea that one ought to be able to work in a three-dimensional environment doing three-dimensional design. Making all that work and putting it together as a system, and doing the mathematics to make the curves and all of that, was essentially my PhD thesis. It was the first early beginnings of being in 3-D.

Tom Zimmerman: Jaron really took the idea and ran with it and did a great job of marketing. Jaron used to say we were "the haberdashers of virtual reality."

Jim Clark: I kind of resented Jaron Lanier for coining that term "virtual reality," but he is a very smart guy. He is not a technology guy particularly.

Jaron Lanier: See, we didn't make the first head-mounted display, but we did make the first commercial one. And the requirements for a reproducible one were totally different. So we had to be able to make them on an assembly line. That was all us.

Tom Zimmerman: So when we started Jaron saw the glove as an input device for his programming language. And the message we got pretty quickly was no one wants to program but the glove is cool. So we had a glove. We had a pretty powerful new paradigm of being in the simulation. So we were a hammer looking for nails. We were capabilities looking for applications. And it turns out entertainment was the killer app and probably still is.

Kevin Kelly: I don't remember how I connected with Jaron—maybe I was the first—but I somehow wound up at VPL with Jaron in '88. I came down to try out his goggles and gloves setup. This was pretty early on, so there wasn't a lot of other people interested in this. I got to try on this thing and entered in a world that Jaron had just made that afternoon that he hadn't even explored. And it was pretty amazing. It was just like, *Wow, this is amazing!* I thought that it was the beginning of something really big immediately. I could see it.

Jamie Zawinski: I was working at Berkeley in the AI group there, and somehow we got an invite. Yeah, waited in line, put the helmet on, got to play around, it was running on SGI Indigos or something. It's like *Oh, I can reach out and I can grab the diamond, and then it turns into a rose! Wow, that's so cool!* I probably got to wear the thing for maybe a couple of minutes, but it was amazing! It was so cool.

Jamis MacNiven: You'd put the headgear on and a glove, and then you sort of walk out into virtual space and you just drift through the room. It was all very geometric, but you felt as if you went out of your body, drifting, and the glove would be the way you would direct it. I think you saw a little icon of a glove that was your hand.

David Levitt: To move in the world you needed navigation. The standard way was to let the dataglove be your controller and to point in the direction that you wanted to go and click your thumb to accelerate.

Jamis MacNiven: And you'd drift around corners and down streets, and you felt like you had gone into the Matrix!

David Levitt: I wound up being one of the big people demoing people in it and expanding and improving the demos over time. I experimented with gravity, and I made it sideways. I set it up so that any time I closed my hand, a ball would appear, and when I released it would be pulled by gravity, but horizontally so it would bounce off a wall and then come back to me.

Jaron Lanier: David Levitt was the first person to fall asleep in VR and then woke up inside VR which is this amazing thing.

David Levitt: And waking up to that could not have felt more alien. "Oh my God, gravity is sideways!" It was just too much.

Kevin Kelly: And it was very fun, kind of surprisingly fun to just explore a fantastical world with nonsense entities and creatures and stuff. And so the thrill is probably closest to visiting a really weird temple in Kathmandu. It's sort of exotic in that sense. It's strange. And you are also aware that this is all happening in your living room or whatever and that's kind of exciting, too. And of course Jaron has got his whole philosophy and he's going on and on.

Jaron Lanier: Postsymbolic communication, yeah. I used to go on and on about that stuff.

Kevin Kelly: I didn't understand anything he was saying.

Tom Zimmerman: Then Young Harvill and Young's wife, Ann, started working on these bodysuits instrumenting the whole body.

David Levitt: People, of course, like new technology and they quickly said, "What are the sexual possibilities?"

Howard Rheingold: A lot of it was, "Okay, we're going to have sex at a distance through computers somehow."

David Levitt: The term *teledildonics* quickly became common.

Howard Rheingold: I got so much weird attention for that one word. There were, I'll tell you, journalists from all over the world wanted to talk about teledildonics for a little while. It was part of this cyberculture vision, the *Mondo 2000* vision.

Jaron Lanier: There was a tremendous pressure from the cultural underground: "Oh, I am supercool. I publish the underground magazine from Amsterdam and I know Tim Leary." It was, "I did this and that and I'm the coolest and you have to give me a demo." There were a lot of people like that.

Scott Fisher: When VR got into popular culture and the media went crazy, I felt that it was kind of over at that point.

Jaron Lanier: They were really wild times back then. I mean there was a flow of celebrities into the place. My personal assistant would be like, "The Dalai Lama is stuck in traffic. Go ahead with Leonard Bernstein, it's okay."

David Levitt: We demoed for Spinal Tap.

Jaron Lanier: I remember Spinal Tap showing up. There was this big fuss. I was like, "Why do we want to show it to Spinal Tap?" Then some of the engineers said, "Oh, come on! You have to let Spinal Tap in."

David Levitt: We made a tiny virtual Stonehenge in honor of their movie!

Jaron Lanier: I said, "You know, if you use the wigs they'll just get messed up by the head mounts." And they said, "Fine."

Jaron Lanier: And that was like a day at work at VPL—just kind of completely insane.

Young Harvill: I remember being worried about the hype. You know that inevitably in pretty much every technology it reaches this point where people really get what it could possibly do and then they just kind of spin creatively on. Then they kind of convince themselves it's kind of already there. That's a really dangerous phase for any technology.

Scott Fisher: People started saying that VR was dead or it was never going to go anywhere.

Kevin Kelly: Why did it not take off? I certainly expected it to take off. I think it's not just that it was the collapse of the hype.

Jaron Lanier: My worst mistake with VPL was the business model. The problem was that there's not that many labs ultimately: After maybe a couple thousand customers in the world you run out. And it gets hard to come up with something so dramatic for the next-generation machine that you get them to spend the money again. And so at that point all you can do is either try to go down in price point to sell something to larger numbers of customers—or turn into a military contractor.

David Levitt: So the RB2 system, the Reality Built for Two, included a Mac as the main control, and one or more Silicon Graphics machines. In those days it was hard to get the performance you needed without a separate Silicon Graphics machine for each eye, so that got expensive.

Mitch Altman: RB2 was a quarter-million-dollar machine, not many people could afford it.

Jaron Lanier: For the full-on thing, it was about a million bucks a person. And there were only a couple thousand customers in the world who could pay a million dollars for a VR system.

Kevin Kelly: Without Moore's law you couldn't take the next step. You could say the expectation ran ahead of VR and that was the fault of the expectation, but actually I think the expectation was valid. It's just that there was a gap. There was some evolution needed to take place. We went to the moon and then didn't follow up.

Jaron Lanier: VPL started to hit a ceiling around '92, which forced the question of what to do next. I wanted to put all available resources into selling Microcosm, which was this amazing product that we developed that never shipped. It was this beautifully sculpted one-piece virtual reality unit that had all the tracking equipment and the computers and everything. There was a really innovative head-mounted display—sort of like opera glasses where there was a handle so you can get in and out of it really fast. And then you could wear a glove on the other hand—and it worked beautifully. I thought that we might be able to get that to a price point where we could sell enough of them to start to follow Moore's law down and sell it to larger and larger volumes and make it to the big time. So that's the path I wanted. The majority of the board wanted to turn into a military contractor, patent as much as possible, hold on to the patents until they were valuable, and then sell them. It's a very reasonable business dispute. So I left, and the company continued on trying to do the other plan.

Tom Zimmerman: The problem was that we started spending more money on protecting our patents than on innovating, and that consumed more and more of our resources. It's sad.

David Levitt: It was terrible timing, because within a year of VPL closing its doors, the PlayStation came out. For $300 you could have replaced our $50,000-per-eye Silicon Graphics machines—in 1994. What is so interesting is that no one picked up on that. Had they not noticed that the leader in the field had left it open to them and that the hardware was now affordable? Who was going to pick it up? No one did.

Michael Naimark: What happened? The spotlight shifted to the internet and the web.

Scott Fisher: So my lab at NASA fell apart and out of that came lots of little companies that really wanted to build pieces of the technology and figure out what the content was and where we could go with it. And for the next decade some of us did that, while the internet stuff went through a similar kind of crazy roller-coaster ride. I thank God for the internet, because that became the next big meme that took the attention away from VR. Which was fine because then we could get back to work.

Clive Thompson: The weird irony or poetic strangeness about VR is that it was the production of a billion mobile phones with their need for screens and little sensors that could figure out whether you were tilting them or whatnot that created this new generation of hardware: You've got Facebook's Oculus, Google's Cardboard, Microsoft's HoloLens, HTC's Vive, and they're all trying to figure out what this environment is good for.

David Levitt: These VR fads come and go. Before VPL ended, I used to think that there is a technological inevitability to VR—that it was unstoppable! After VPL was gone I realized that it takes a visionary: a Steve Jobs or a Jaron Lanier or an Engelbart determines whether these things are overlooked or not. A true visionary just makes the future out of what is available, and does not wait for the technology to catch up with him.

Tom Zimmerman: There are three archetypes that are essential for invention. The dreamer, the first person who thinks of an idea. The engineer who reduces the idea to a prototype. And then the entrepreneur who takes that one prototype and works out the marketing, financing, and business model to make thousands or millions of them. Jaron was the entrepreneur.

Jaron Lanier: If all a lab needed was a head-mounted display, we would sell them that. If all they needed was a glove, like Scott, we could sell them that. We had all these modules: a spherical video system so we could gather and replay spherical video; we did 3-D sound. We really had the complete thing—all in sort of early versions that would be considered somewhat crude today, but we had the whole package. We demonstrated what virtual reality could be as a product. Nobody else had done that.

Young Harvill: And I think we also touched a place in terms of the cultural dream where it kind of flourished in people's imagination and literature and movies. So I think we did well.

Jaron Lanier: Nobody had put the whole thing together, and nobody had just shown the joy of it before. We showed the joy of it.

From Insanely Great to Greatly Insane

General Magic mentors a generation

*C*hances are that you've never heard of General Magic, but in Silicon Valley the company is the stuff of legend. Magic spun out of Apple in 1990 with much of the original Mac team on board and a bold new product idea: a handheld gadget that they called a "personal communicator." Plugged into a telephone jack, it could handle e-mail, dial phone numbers, and even send SMS-like instant messages—complete with emoji and stickers. It had an app store stocked with downloadable games, music, and programs that could do things like check stock prices and track your expenses. It could take photos with an (optional) camera attachment. There was even a prototype with a touch screen that could make cellular calls and wirelessly surf the then-embryonic web. In other words, General Magic pulled the technological equivalent of a working iPhone out of its proverbial hat—a decade before Apple started working on the real thing. Shortly thereafter, General Magic itself vanished.

Andy Hertzfeld: The Macintosh had a great launch; it was really successful at first. Steve laid down a challenge at the introduction, which was to sell fifty thousand machines in the first hundred days, and it exceeded that. But then starting in the fall, sales started dropping off.

Steve Wozniak: It just didn't have any software at first.

Fred Davis: MacPaint and MacWrite were like demo programs. They weren't real tools that you could use to do stuff.

Steve Wozniak: The Macintosh wasn't a computer—it was a program to make things move in front of Steve's eyes, the way a real computer would move them, but it didn't have the underpinnings of a general operating system that allocates resources and keeps track of them and things like that. It didn't have the elements of a full computer. It had just enough to make it look like a computer so he could sell it, but it didn't sell well.

Andy Hertzfeld: By December of 1984 the forecast was to sell eighty thousand Macs, and in fact they sold like eight thousand. So something had to be done. If the Mac wasn't going to catch on, Apple didn't have a future. So what to do about it?

John Sculley: In 1985 Steve introduced the Macintosh Office, which was a laser printer—the LaserWriter with PostScript fonts from Adobe—and a Macintosh. One problem: The product just didn't work. It wasn't until at least a year later that the microprocessors were powerful enough that you could actually do the kind of things we were actually promoting back in 1985. So people weren't buying it. Steve got depressed. And then he turned on me.

Andy Hertzfeld: It got to the point where Steve was openly sabotaging Sculley. Something had to be done. Sculley and the board didn't fire him, but they removed his responsibility from running the Mac. And so he had to leave.

Ralph Guggenheim: Steve left in 1985. There was a rumor that he was starting a new company, NeXT. The NeXT machine was Steve's effort to build his vision of a desktop workstation with lots of computing power and a CD-ROM. It was his attempt to show what he could have done at Apple.

Steve Perlman: It was big, it was black-and-white, it was clunky.

Andy Hertzfeld: NeXT was a revenge plot. That's the reason I didn't work at NeXT. Steve denied it and we'd argue till he was blue in the face. But it was true: The purpose of NeXT was to eclipse Apple. And I loved the Mac. I didn't want to work against the Mac. But Steve wanted the Mac to fail, and so he started NeXT.

Meanwhile, back at Apple:

Larry Tesler: Steve was gone. Where are the new ideas going to come from now that we got rid of Steve? We're going to run out of Xerox's ideas. Where are we going to get ideas from? The management wanted new ideas. And so they decided they needed what they were calling the Advanced Technology Group. It was really an R&D group, a lab.

Steve Perlman: We were building a color Macintosh.

Steve Wozniak: The Mac wasn't ever really selling until we introduced the Macintosh II in '87. It had color.

Larry Tesler: One of the things that grew out of it was the Apple Fellows Program. The first three Apple Fellows were Steve Wozniak, Bill Atkinson, and Rich Page. The initial definition of a fellow was someone who had made a big impact on the industry. Al Alcorn was recruited—he had done *Pong*. And they also wanted to recruit Alan Kay. So we brought them both in.

Alan Kay: Steve never forgot where his ideas came from.

Andy Hertzfeld: Alan Kay was my hero. I was like *Damn!* But it was still time for me to quit.

Larry Tesler: And then as the program expanded and it even included a person who was not an engineer—Kristina Hooper Woolsey.

Kristina Woolsey: I came in '85 when the HyperCard stuff started. Bill Atkinson spontaneously decided to do the product. He and a small team just popped this thing out.

Al Alcorn: HyperCard was developed by Bill and two or three guys in Advanced Technology and—it wasn't supposed to be released. Products are supposed to be released by Product Development, not by Advanced Technology, understand? But Gassée, who succeeded Jobs as head of Macintosh development, wasn't going to have it. So Bill and his team just conspired. And while Jean-Louis Gassée did his French thing and went off for a two-week summer vacation on the beach in the south of France, Bill just released it. And when Gassée got back he was furious, but he couldn't stop it because HyperCard was really well received. It got great press. There are a lot of stories like that.

Andy Hertzfeld: But the beginning of General Magic is really Marc Porat. He's a very clever nontechnical person—an impresario businessman who in the thesis he did at Stanford came up with the first use of the term "information economy." Marc was a college friend of Larry Tesler's. So that was his connection to Apple. And he was hired on in the fall of 1988 to work in the Advanced Technology Group at Apple to try to help figure out what's the next thing beyond personal computers.

Megan Smith: Marc was very tied into Nicholas Negroponte and is part of that whole MIT Media Lab conversation.

Andy Hertzfeld: Marc started by interviewing everyone at Apple trying to come up with a consensus—what's in the wind. Eventually he decided the next thing beyond the personal computer combined two things. One is communication. The other thing is instead of being on the desk it was in your pocket.

Megan Smith: And Marc came up with this idea that he called a Personal Intelligent Communicator—a smartphone, basically. The whole idea was there.

Marc Porat: Remember something called the Sharp Wizard? It was a screen with some chiclet keys: one of those silly little organizer things, but useful. I took that and a Motorola analog cell phone and I duct-taped them together. That was the concept: "Imagine this screen was beautiful, imagine the applications were amazing, imagine that it was linked wirelessly with you everywhere." And that little concept went around, and then we reduced it to something that was smaller and more beautiful.

Andy Hertzfeld: And the smartest thing he ever did to make his vision real was make little models out of plaster.

Steve Perlman: He was showing what looked like a little wallet—a block of plastic wrapped in leather.

Marc Porat: Everything was going to become very small and in your hand, and intimate and jewelry-like. You'd wear it all the time, and if there was a fire in your house you'd think, *Family goes first, and then I grab my Pocket Crystal.*

Megan Smith: Pocket Crystal—that was his nickname for it.

Andy Hertzfeld: The first thing I said when I saw his models was, "Well, they're not realistic." And he said, "I know—but they will be."

Michael Stern: Marc Porat was an incredible visionary. His Pocket Crystal project at Apple was really a previsioning of everything that we now take for granted.

Marc Porat: I wrote a book on it, all the things that this thing is supposed to do and why this computing-communication object would change the world.

Michael Stern: Devices in your pocket, social networking, social media, a notion of an electronic community, anytime-anywhere communication, handheld devices that could enable you to do just about anything from shopping to research to talking to your mother. It's all there, in that book he wrote. He was at Apple to help springboard the project and get it funded.

Al Alcorn: So Marc was pushing this thing, and he infected Bill Atkinson and some of the other guys with this idea.

Andy Hertzfeld: Marc met with Bill Atkinson, who had just finished Hyper-Card and was kind of looking around for what to do next. And he got Bill really excited about it. So one day right after my friend Burrell Smith went insane and I was dealing with that emotionally shattering experience and all of the fallout there, I get a phone call from Bill, incredibly excited, "You've got to see this new thing at Apple! It's the next thing!" And my first thought was, *Oh my God, Bill went insane too!?* But he didn't. And I called Bill and he said, "You've got to meet with this guy Marc." And so Marc called me up. And I said, "Yeah, this is pretty interesting but I don't want to work at Apple again." And Marc said, "I understand you don't want to work there. Why don't we just pay you to be a consultant, and you know, just get started?"

Marc Porat: Bill loved it. Andy saw it, was impressed that Bill was there. Bill was impressed that Andy was there. They immediately started seeing how to do it, and they signed up.

Andy Hertzfeld: Bill created a prototype user interface in HyperCard. I wrote a server that allowed you to send little electronic graphical messages.

Michael Stern: There were these beautiful little "tele-cards" with animations and handwritten notations and sound embedded in them.

Andy Hertzfeld: All kinds of whimsical wacky animations—like we had a walking lemon. As well as these little graphical decorations also had meanings associated with them. Some of them you could even interact with, like one that was lips with a speech bubble. When you tapped on it a microphone appeared and you could talk into it and it would send your voice as part of the tele-card.

Michael Stern: These beautiful little things you could exchange with each other. That was Bill's thing.

Andy Hertzfeld: We had what are now called stickers, which is part of instant messaging. You know how you could use little stickers in your messages? We had that working twenty-some years before the iPhone, in a better way than most people have it now.

Steve Jarrett: You've got to remember, this was before even digital cellular existed.

Marc Porat: Many of us had analog cell phones, which were bricks, and some of us carried batteries in a little, tiny briefcase to go with the brick.

Steve Jarrett: People were just using mobile phones to talk. So, it's very, very early.

Marc Porat: The theory and the strategy, right from the beginning, was to create a global standard. Apple and Microsoft would have the personal computer, IBM and Digital would have the bigger hardware, and we would do everything else. We would do telephones, we would do television set-top boxes, we would do kiosks, we would do absolutely everything else that had an operating system.

Al Alcorn: But it clearly wasn't going to get supported by management at Apple. It wasn't going to happen.

Marc Porat: It would need a lot of communication networking, including wireless, and it would need something that did not exist, which was a network to run on, which Apple did not have.

Andy Hertzfeld: One company can make a computer, but it couldn't make a communicator all by itself, because it wouldn't be able to establish the communication standards. So it couldn't be an Apple product.

Marc Porat: But it was also clear that politically, it was going to be okay to spin this thing out.

John Giannandrea: I think a big part of spinning General Magic out of Apple was this idea that it was too big even for Apple, right? Apple couldn't deal with this thing.

Marc Porat: So there were two projects established: General Magic on the outside, and Newton on the inside.

Andy Hertzfeld: The Newton project was an 8½-by-11 tablet that was supposed to cost $5,000.

Marc Porat: The Newton was a hedge in a big strategy that we had described to Sculley.

Andy Hertzfeld: Basically Marc convinced John Sculley to not only let us spin out from Apple but also to help us convince Sony and Motorola that they should join Apple in trying to create this new standard. We called it "the Alliance."

Marc Porat: Apple, Motorola, Sony: So now we had the three board members as licensees, who were the core of the founding partners.

Michael Stern: This is before the web; this is 1990–1991. And so after we'd already formed a company and spun out of Apple...

Andy Hertzfeld: We decided to bring AT&T into the fold. AT&T became the fourth investor in General Magic on par with Sony, Motorola, and Apple. And that kind of worked, I was surprised. We were off and running.

John Giannandrea: It was a ridiculously ambitious company.

Marc Porat: People of enormous quality began to come over and see what this crazy team in downtown Palo Alto was doing—and joined.

Andy Hertzfeld: We had a lot of the original Mac people.

Steve Perlman: It was a great group of people. To have a chance to work side by side with Andy and Bill? Come on! They're the real deal. These guys are just geniuses. I didn't care if they called it General Pickle. I was going to work side by side with Andy Hertzfeld and Bill Atkinson; what else could anyone ever dream for, right? I think I was employee number thirteen.

Michael Stern: Andy and Bill were the gods of the universe, and then these kids came to work for them. Andy in particular was the mentor for and helped train a cohort of brilliant kids. Many of whom had their first job at Magic, like Tony Fadell. Tony hung around and would just talk to people until we finally hired him.

Tony Fadell: I had my own start-up in Michigan doing educational software and getting frustrated being a big fish in a little pond. There was no internet then, so I would religiously read *MacWEEK* and *Macworld*, and there was always the last page of each of those rags which were like the murmurs, the rumors, the goings-on, and this company called General Magic kept popping up in it. I was like, *Whatever that is, I have got to learn more about it.*

Marc Porat: General Magic had all the buzz. All the buzz. It was where the pixie dust was. It was where you wanted to work if you were cool and smart.

Tony Fadell: I am looking all over for General Magic, and I find out that their office is in Mountain View in this high-rise, and so I show up about eight thirty in the morning with a tie and a jacket on, with my résumé in hand, and walk in the door. There was no one around. Or at least I did not think so. So I walked down the hall, and I saw a couple people who looked like they had been there all night. I was like, "Hey! I wanted to bring my résumé by, wanted to see if you guys were hiring?" They looked up at me with these bloodshot eyes, like, "Leave us alone, kid."

Michael Stern: But Tony just keeps pestering people until they hire him.

Tony Fadell: I was humbled in the first ten minutes of being there, I was like, *Oh my God, this is not like Michigan, these are the smartest people ever, I have got to be working here.* Whereas Apple was still making computers, these people were making the next form of a Mac! It was the iPhone, fourteen years too soon. It had a touch screen, it had an LCD, it was not pocketable but it was portable—the size of a book. It had e-mail. It had downloadable apps. It had shopping. It had animations and graphics and games. It had telecommunications—a phone, a built-in modem. I was like, *I want to go with them, where they are making this stuff from scratch. I can be part of it.*

Marc Porat: He just walked in the door and impressed Bill and Andy and got his first gig to do stuff. General Magic was Broadway, and Andy and Bill gave this kid from the Midwest somewhere a break.

Tony Fadell: Ultimately, they gave me the job, and I was the lowest guy on the totem pole. I came in and I was working with my heroes.

Michael Stern: We had a building right on the Mountain View–Palo Alto border. That building had been empty for ten years, and it had a pack of feral dogs living in the basement. So we were the first tenant. We had the top floor. It was a jungle. I mean they had built like a model railroad that ran up the top of the building.

Tony Fadell: That was not a train set, that was a remote-controlled car track, but I had a Lego train set in my cube and put that up, and a bunch of other toys in my cube.

Michael Stern: We had Bowser, the pet rabbit which lived there and would shit everywhere.

Steve Perlman: Bowser was never trained to use kitty litter, so he would leave little gifts around the office. I typically walk around barefooted, and I would collect these little gifts between my toes. But Bowser was so cute and adorable—we all loved him.

Michael Stern: People brought their dogs and the parrot flew around and the trains rattled around and everybody lived there kind of 24/7.

John Giannandrea: Zarko Draganic famously slept under his desk for months at a time. Zarko shared a cube with Andy Rubin.

Megan Smith: You'd say "Hey Zarko, let's meet at three o'clock." And then he would say, "A.M. or P.M.?"

Michael Stern: The kids, they just worked nonstop. They'd work in bouts.

Megan Smith: That's what the kids in Silicon Valley do in their twenties. We work like crazy and have an amazing time. Maybe some other folks are busy partying or doing whatever they're doing, but the Silicon Valley kids are inventing together and that's the community.

John Giannandrea: It was fun times. The place was kind of crazy, too.

Marc Porat: The culture and the environment was Apple done even more fun. We had meals, we had spontaneous music things going on. We had Bowser running around. We had gunfights—Tony was the chief antagonist or protagonist with water weapons. He would start battles with water guns and pistols, and it was really fun.

Steve Perlman: So much fun! But too much fun.

Tony Fadell: Remember Gak, or Slime, the green slime? We would leave it on stuff, and we had all kinds of practical jokes on each other.

Amy Lindburg: I hid a hot dog in Tony's computer, and he retaliated by putting a pickle inside my computer and it shorted out a whole bunch of stuff. There was just always some little joke going on. It was just, we spent all of our time there. Every night was late night.

Tony Fadell: We were known to work really late. Eleven or twelve, even four in the morning. It just depended. So this was about ten or eleven o'clock at night, and we were all getting punchy. We were like, "Let's get out the Gak and the slingshot!"

Steve Jarrett: Somebody, late one night, had built a funnelator. It's like a slingshot, it was surgical tubing that could shoot anything. And so, Tony and a bunch of other guys decided that they would put a bunch of this gel inside of the funnelator...

Steve Perlman: People were having a meeting in an upper-story conference room and we thought it would be really funny if we went and shot Gak so it splattered on the outside window. We'd watch it drool down.

Tony Fadell: So who was it? I think Perlman and Andy were holding it on either end, and because earlier in the day we had lined it up, I was like, *I am really going to pull this thing back.*

Steve Jarrett: Tony's really strong.

Tony Fadell: Everyone is cheering me on: "Go, go, Tony!" So I am pulling it back and pulling it back, and they are like, "Go!" and I am like laying down on this thing, and the thing goes "whoosh" and then "pop!"

John Giannandrea: The green slime went straight at the window and then the window just shattered—fell out of the building. And we were like, "Ooops."

Tony Fadell: I just looked, and my whole life flashed before my eyes. I am like, *Oh fuck, I am going to get fired, what am I going to do!?* And we have an all-hands meeting the next morning. So I go, and I sneak in the back just as it was starting, and then of course everyone knows what I did, all of a sudden everyone is just like, "There's Tony!" and they just start clapping, and they are laughing and cheering, and I am like, *I broke a six-thousand-dollar window, and everyone is laughing and cheering?*

Marc Porat: I said all was forgiven. Do not even worry about it. There were no rules about having fun.

There were other sorts of breakthroughs, as well—technological ones.

Steve Jarrett: So Tony and I were two of the youngest employees and visibly so. At the time, we were probably—I don't know, I think we were maybe twenty-three and twenty-four. We have a trip to Japan which, because I'm the Japanese business development guy, I helped organize. We go to see all our partners there and we're going to describe the design of the new hardware—the new chips and the entire layout of the next generation of the hardware that we're creating.

Tony Fadell: Japan was *it* for electronics in the early nineties. Even Apple could not make small stuff back then. And Mitsubishi Electric was a partner of General Magic's. Our job was to work with them on building new chips for our devices. So we presented the architecture for the next-generation General Magic device—showing it off.

Steve Jarrett: We go into this meeting and Tony gets up and on the overhead projector, puts down a slide, and he starts to walk through the diagram.

Tony Fadell: I am going through the block diagram, explaining it, and they were all, "Oh, yes, that is great, wonderful."

Steve Jarrett: And so, Tony gets to the part where he's describing the modem. He says, "Well, then here's where you can connect the telephone cable."

Tony Fadell: And they go, "But Tony-san, where is the modem?"

Michael Stern: In the block diagram there was no modem chip.

Tony Fadell: I said, "In software."

Steve Jarrett: They all look at each other and one of them says, "*Honto desu ka?*" which is "Can it be true?" The other guy just kind of shrugs his shoulders. Then they immediately just start yelling at each other in Japanese. One of them jumps up and grabs the telephone that's in the room and yells down the phone. They ask us to stop and wait.

Tony Fadell: Just then the boss, Nagasawa-san, comes into the room and sees this activity. And as this happens, the loudspeaker goes, "Beep beep beep," and we are like, "What is that?" And somebody says, "Oh, there is a typhoon warning. We all need to leave before noon, because we need to get home because the typhoon is coming." I go, "Really?" and they go, "Yeah, but we are going to keep working."

Steve Jarrett: Five minutes later, guys in all these multicolor factory jumpsuits show up and fill the room. And then they say, "Tony, again, please. Please describe your architecture." So Tony gets to the part and he says, "Well, this is where there's no modem because we've done it entirely in software." The one guy grabs his head in his hand and just hits his head on the table. I said, "Is everything okay?" The guy who organized the meeting says, "You have just obsoleted Kato-san's division."

Tony Fadell: So while the whole of Mitsubishi Electric is being evacuated, the CEO and that team and us stay there the whole afternoon and we sat in this room, and I think even at some point the power went out and we just kept talking. How do you do this modem in software? How do you do this device? They were blown away by the system-on-a-chip low-voltage all-in-one design and the software modem. The CEO had never seen anything like it.

Michael Stern: They were selling hundreds of millions of dollars in modem chips. And we made it so you don't need a chip anymore. So that was the kind of stuff that Magic accomplished. Again, Magic's DNA was Apple. We were Apple all the way. I mean here is a little start-up with limited working capital, and we are designing our own chips instead of going and buying them? Absolutely insane.

Amy Lindburg: The OS, the hardware, the soft-modem: All those achievements were amazing at the time. It was like the Apollo Project.

Steve Jarrett: The user interface was really visionary, too. It was groundbreaking. It was based on the real world.

Tony Fadell: Each thing was a mock-up of the real world, and you would interact with those things like you would interact with the real world.

Steve Jarrett: You started at a picture of a desk and you could drag things around on the desk and you tap on things to open them.

Tony Fadell: It was skeuomorphic—like file cabinets to put your things in, a desk to write on...

Steve Jarrett: And then you could leave the desk and go into the hallway. And in the hallway there'd be doors.

Michael Stern: You'd click and behind the doors were various rooms—like the game room, the media room, the library. In the library there would be all your books, all of your electronic books. And we'd written some manuals so there would be something to populate the library. And in the game room we had the first networked games. You could play online with opponents. No one had ever done it before! Again, this was something Tony just kind of whipped out in a day.

Tony Fadell: And you would go out the door and you could go down the street and go to a shop which would be a virtual shop.

Michael Stern: This is where the third-party developer community was connected. Because remember it was all about being a platform, not a closed system, so that third-party developers could develop applications. Productivity applications, games, you name it: They had them. We actually had wonderful games in the game room.

Megan Smith: Pierre Omidyar was running all the developer services stuff.

Steve Jarrett: Pierre was doing really early, really interesting work of thinking up online services that would work on these mobile devices that we were creating. He created a turn-based chess game that was easily the most popular game on our platform.

Tony Fadell: It was pretty neat, you know?

Michael Stern: So we were demonstrating all this stuff. Meanwhile Marc is talking up the vision of the global community enabled by anytime-anywhere communications. So by 1992 we had all the ideas and we were trying to build the thing. That was really hard. And we kept slipping. The schedule kept slipping. And the partners kept getting more and more anxious.

Andy Hertzfeld: At the same time, we found out that our leading benefactor, our father, decided to kill us.

Marc Porat: I remember a board meeting where we were laying out all the secrets and all the plans. In that meeting John Sculley was taking copious notes, word for word almost, just huge amounts of what was being said, and Andy turned to me and said, "What's up?"

Andy Hertzfeld: The Newton project had run aground. It wasn't going to work and it was going to cost $10,000 instead of $5,000. You couldn't get the display and it was too complicated and it was failing. And with the

Dynabook vision of the Newton failing, he decided that Newton should instead copy Marc's vision and make something based on Marc's prototype. It should fit in your pocket at a low price point.

Marc Porat: That was really a problem. Why did Apple not just put all their weight behind Magic? Why did they have to hedge their bets? Sculley set up the competition for essentially the same IP and the same market space.

Andy Hertzfeld: The Newton was supposed to cost $5,000. They remade it to cost $500. It was supposed to be the size of a notebook and it became the size of a postcard. Obviously, it was an existential threat.

Marc Porat: The design center of the Newton was handwriting recognition. Our design center was personal communication. Personally, I thought there should have been enough blue sky between them so they could coexist, but it was widely viewed by both the teams and by outsiders, for example our licensees and the media, that they were set up to compete with each other. That was the perception. And there was some bitterness about that.

Andy Hertzfeld: Sculley was on our board! It was a real betrayal.

The Newton debuted in August 1993, and it flopped. The main problem was the handwriting recognition software: It was not very good. But the fact that Garry Trudeau made the Newton into a recurring punch line in his nationally syndicated comic strip, Doonesbury, *didn't help much, either.*

Mike Doonesbury: "I am writing a test sentence."
Apple Newton: Siam fighting atomic sentry.
Mike Doonesbury: "I am writing a test sentence."
Apple Newton: Ian is riding a taste sensation.
Mike Doonesbury: "I am writing a test sentence!!"
Apple Newton: I am writing a test sentence!
Mike Doonesbury: "Catching on?"
Apple Newton: Egg freckles?

Apple had rushed to market and paid the price. General Magic, meanwhile, wasn't going to make the same mistake that Sculley did. Instead, they made the opposite mistake. Pursuing perfection, Magic fell several years behind.

Marc Porat: AT&T was late; they could not get their network done. Sony was late, they could not get the consumer electronics piece done. We were late, because we were perfectionists. We were late because we pushed ourselves to the limit. We felt a lot of pressure internally because we were late, and we had to synchronize our releases with how ready our partners would be. Everyone was late. Two years late.

Michael Stern: We missed Christmas of '94, so we really thought, *Okay, next year.*

Amy Lindburg: The first thing that we shipped was the Sony Magic Link.

Andy Hertzfeld: And as soon as the Magic Link came out I gave one to Steve.

Marc Porat: Because it was inspired by the same kinds of things that the MacOS was inspired by, it was built by the same people, it was designed by the same graphic artist.

Michael Stern: And then as soon as we shipped we went on the road show for the public offering.

Marc Porat: I spent 80 percent of my energy managing the founding partners. We needed to be clear of them, and an IPO would get us enough money where we did not actually have to be so beholden to them. The IPO hinged on getting everything organized, and of course AT&T was the last one in, to deliver what they needed to deliver, and then we were off to the races.

Steve Jarrett: We'd already sold a significant amount of stock to these large partner companies, and we had a bunch of licensing revenue coming in, so the company looked superstrong on paper. And so when we went public in 1995, it was one of the first internet IPOs in the sense that we had a particular price and then we opened way, way above the S1 price.

Marc Porat: We were priced at $14, opened at $32, and took in tons of money. So the IPO as an event was very successful. But then the Magic Link sales were not good, and we said to ourselves, "Oh man, this is not going to go well."

Andy Hertzfeld: We were hoping to sell a hundred thousand of the first Sony devices, but they only sold like fifteen thousand.

Steve Jarrett: The hardware was too large. The battery life wasn't great. They were very expensive. The first Sonys were $1,000. Remember, these weren't mobile phones. These were devices that, to communicate, you had to plug in a telephone cable.

Megan Smith: The devices were doing so many new things that nobody knew what we were talking about. Why would they want something like this?

Amy Lindburg: We were designing a product for Joe Sixpack—literally they would call the customer Joe Sixpack—and the trouble is that at that time Joe Sixpack didn't have e-mail! It was too early.

John Giannandrea: One of the great Silicon Valley failure modes is "right idea—way too early."

Amy Lindburg: And then the internet started to emerge.

Marc Porat: One day, as I recall, someone brought in something called Mosaic, and they said, "This is the future." I said, "Okay, what is it?" We loaded it up and it crashed immediately. "Well, why is this the future?" "Just watch, just watch." He put it back up.

John Giannandrea: We downloaded this thing and people were crowded around the workstation—I think it was a Silicon Graphics Indigo—and we were like, "Look at that!"

Marc Porat: We were looking at a really early browser, and there was already awareness inside the team that we had to go internet. We knew that release 2.0 would have to be an internet machine. The question was, can we ship the first one and still have enough time, money, energy, and stamina to get to the second one?

Michael Stern: We have no revenue. Things aren't working. The partners start bailing. Things are getting pretty ugly.

Steve Jarrett: We kept taking the first-generation hardware to trade shows and kept getting asked, "Oh, does it have a phone inside of it?" It became really clear that the next product that we should try to build should actually be a cellular phone. And so Andy and the other engineers started to make a smartphone—a handheld cellular phone that was running our software. Again, remember, like this is before digital cellular existed, so SMS didn't exist and you couldn't send data over cellular. And so, they built this very early prototype which had wireless web browsing.

John Giannandrea: So there were two devices. There was the device being built for Sony. And everybody knew that it was a dog. I was working on the second product, which was going to be cheaper, faster, have better battery life…

Marc Porat: Andy and Bill demoed a Magic machine, which was in a format like an iPhone, and they said, "So this is what we are going to ship next. It is a phone, and you can see all the Magic Link icons on it, and let me show you how it works." They go through contact the manager, the telephone, sending e-mails...the works. That was iPhone form factor and that was iOS functionality in 1995. We were very excited about it, and they said that this is the next thing that we are going to do. That was for 1997.

Steve Jarrett: But was this actually physically possible for us to do? It was clear that was going to take years to build a supercompelling product like that. For me, that was the sign. That's when I left the company.

John Giannandrea: I quit. Part of it was realizing that we had a better product and we weren't going to ship it.

Marc Porat: We were one year away from it. We kind of ran out of gas, we ran out of steam, and ran out of the will to take the next step. Most of us were exhausted, for our own reasons. Andy had probably worked eighty hours a week forever. We were just plain tired.

Michael Stern: After his first big surge of creativity, Bill kind of checked out. Andy was running engineering for three years. He had just had enough by '95.

Marc Porat: It was a physical fatigue, when you keep something alive for five years on a vision and on just the raw energy of passion to do something amazing, five years is a long time to keep that going, and you want some validation at the end of it.

Michael Stern: So everybody leaves. I stayed on because I felt some loyalty to Marc. It was really sad and ugly. Everyone who touched the project at their respective companies—at Motorola, at Sony, and AT&T—their careers were wrecked. Because these companies had gone public saying, *We're doing the next big thing. We're creating the future here!* And then it crashed. It was a train wreck.

Marc Porat: Engineers do things because they want millions of people to touch it. That is the ultimate reward for a top-level engineer. And when millions of people do not touch it, where is your source? Where is your juice? Where is the passion coming from? Where does the juice come from to take it to the next level? You need some affirmation; companies need some affirmation to keep going.

Michael Stern: The company failed—we didn't ship a product that people wanted to buy. But this cadre of brilliant kids in this incredible pressure cooker went on to do amazing things and create the world that we now take for granted.

Megan Smith: There was this group at Apple that apprenticed with Steve Jobs, Woz, Mike Markkula, and all that crew, and learned. And then at

General Magic they became the wizards and then we're the apprentices. Phil Goldman and Zarko and Tony, Amy—we were the junior group to this senior group.

Amy Lindburg: Tony went on to do the iPod and the iPhone.

Steve Perlman: Phil Goldman and Bruce Leak both founded WebTV with me. Andy Rubin joined later.

Amy Lindburg: And then Andy went on to do Android. And Zarko spun out the software modem—which was the first modem ever in software in the world.

Michael Stern: Megan left Magic and went to San Francisco and founded PlanetOut, the first online community for lesbians. It became very successful and of course she ended up as the CTO of the United States of America. Not bad!

Amy Lindburg: General Magic was the kind of company where a guy who is going to be a billionaire in a couple years didn't even rate a window cube.

Chris MacAskill: Pierre Omidyar was running AuctionWeb on his Mac IIci in his cubicle.

Pete Helme: Supposedly the idea came at a local Chinese restaurant while he and a bunch of General Magic guys were talking, you know, "Wouldn't it be great to sell stuff on the internet—and what about an auction format?"

Michael Stern: So Pierre came to me when I was the general counsel in 1994 and said, "I've created this little electronic community. I'm getting people to talk to each other about trading tchotchkes. We're creating traffic on the network and getting people into a community. That's kind of in our sweet spot, isn't it?" That was General Magic's thing: the whole notion of electronic community. I said, "That's the stupidest idea I've ever heard! If you want to do it, bye, see you later." That was eBay.

Amy Lindburg: Everything came out of General Magic like just a big explosion.

Michael Stern: iPhones, social media, electronic commerce, it all came out of Magic. That's the story.

The Bengali Typhoon

Wired's revolution of the month

*B*y *the early nineties, Silicon Valley had racked up an impressive array of fascinating failures—The Well never scaled up, VR never took off, General Magic was in the process of imploding. The Valley's breakthroughs were not breaking through. Meanwhile, up in Washington State, Microsoft was having great success in incrementally leveraging its humdrum operating system into de facto control of the personal computer industry. The money and power was shifting decisively north. Yet the Valley was still the center of an emerging nerd culture, and that culture had a powerful new tool. The Mac II, when loaded up with desktop-publishing software and paired with a laser printer, was a new kind of printing press. It allowed small, tech-savvy editorial teams to create magazines that looked like they were produced by publishing empires. Suddenly the bohemian world of West Coast publishing (of which the* Whole Earth Catalog *was the prime example) had an opportunity to disrupt the "real" publishing world. San Francisco's techno-underground launched a raft of new titles, including* Future Sex, Mondo 2000, *and most importantly,* Wired.

John Plunkett: Where did *Wired* come from? You have to go back to Nicholas Negroponte and his insight that led to the creation of the Media Lab.

Nicholas Negroponte: The original idea for the Media Lab was very simple. We foresaw a coming together of three industries which were previously completely distinct. You would pull the audiovisual richness out of the broadcast-entertainment ring. You would pull the depth of knowledge and information out of the publishing ring. And you would take the intrinsic interactivity of computers and put these three things together to get the sensory, rich, deep, interactive systems that today we call multimedia.

John Plunkett: Stewart Brand wrote a book, *The Media Lab: Inventing the Future at MIT*, which Louis and I read. It was a big factor in Louis's realization that, *Hey, there could be a magazine about this!*

Louis Rossetto: So previous to *Wired* we were in Amsterdam and we produced a magazine called *Language Technology*, which became *Electric Word*. We changed the name in the middle of its life.

Jane Metcalfe: We sent a copy to Kevin Kelly.

Louis Rossetto: We had a piece on Anthony Burgess about how word processing affects writing. And a piece on Nicholson Baker, who was a word worker, a novelist. This is like the seed for everything that came out ultimately in *Wired*.

Kevin Kelly: And I thought, *Wow, this is interesting!* So I reviewed it for *Whole Earth*. I called it "the least boring computer magazine there was."

Louis Rossetto: He had written this gracious thing. And so we said, "We want to go meet Kevin Kelly." And so we went to visit him in his chaotic office. He had billions of books all over the place.

Jane Metcalfe: Stacks and stacks and stacks of books everywhere.

Kevin Kelly: I was interested in the culture of technology. I felt that there was something big moving there, and I was interested in trying to move *Whole Earth* in that direction. My thing at *Whole Earth* was I did the special issues and I did this special issue called *Signal* in 1988 as a kind of a trial, because to me the hippie colors of *Whole Earth* were beginning to wear a little thin. This was the same territory as *Wired* would be, but it was mostly about ideas and concepts. There was not a lot of profiles or stuff like that.

Louis Rossetto: We had an incredible high-bandwidth conversation. We really enjoyed each other for the hour, hour and a half that we met.

Kevin Kelly: All I could say was, "Well, if you're going to do the magazine like this, you can't possibly do it in Amsterdam. You have to do it in San Francisco." That's all I could say. That and, "Get on The Well."

Louis Rossetto: Then we went back to Amsterdam. We tried to keep *Electric Word* going and we couldn't do that. Then we came up with the *Wired* idea and we came to California.

Kevin Kelly: Two years later, maybe, they reappeared with the news that they were moving to San Francisco and they were going to start the magazine that they were trying to do in Amsterdam but couldn't. My thought was, *Good luck!* Starting a magazine is an assignment for dreamers.

Louis Rossetto: There was a lot of trying to meet people who ended up becoming subjects of what we were doing.

Bruce Sterling: They called me up and said, "Hi, we're publishers, and we'd like to show up on your doorstep because we hear you're a science fiction writer and we need more science fiction writers to be reporters for our new magazine." I'm like, "Yeah, come by the spread." They were on their way to San Francisco to start the magazine. They were kitted out in full Amsterdam euro-techno. Jane was wearing some very handsome shoes, the kind you would never see on a tech chick. I assumed that they were complete cranks. They didn't look like any of the *Mondo 2000* cyberhippies I was hanging out with at the time.

R. U. Sirius: *Mondo 2000* was really the first technoculture magazine starting in 1989. Previous to that, computer magazines were sort of like car magazines. They were entirely designed and oriented toward the mechanics of it, if you will. *Mondo 2000* took technology and treated it as an element of counterculture.

Fred Davis: I had been working for Ziff-Davis as editor of *A+*, *PC Magazine*, *PC Week*, *MacUser*. Louis and Jane showed up at my house in Berkeley, literally broke, with nothing but this idea to do *Wired* as a high-end version of *Mondo 2000*.

Mark Pauline: The *Mondo 2000* people were like, "Well, we'll see about that!"

Dan Kottke: Before it was *Mondo* it was called *High Frontiers*, and then it changed to *Reality Hackers*. It was a mixture of technology and culture and literature and music and film with drugs and consciousness studies.

R. U. Sirius: So it was this merger of the psychedelic influences, technological influences, and strange ideas that were going around in the area of science and pseudoscience and quantum physics and pop culture.

Fred Davis: *Wired* was closer to what I would consider potential to be a mainstream mag.

Jane Metcalfe: Before we launched, we had a private conference on The Well where potential contributors, anybody who was a friend or whatever, could come and comment or participate.

Fred Davis: Louis had a good concept, but to raise money they needed a good story about the audience. I helped put that together.

Louis Rossetto: But it became apparent that we still needed something else to sell it. So we said let's bear down and create something that looks like a real magazine—a real prototype that shows the edit and advertising, the attitude that the magazine was supposed to represent. We began to gather editorial material, advertising, and put the thing together as best we could, and then got together with Eugene.

Eugene Mosier: I had just left a job at *MacWEEK* that didn't work, and I was trying to decide what I should do. Fred Davis said, "Come to a party at my house, you'll meet some interesting people."

Fred Davis: I had a fantastic big house in the Berkeley Hills. It was a great party pad. People like Todd Rundgren would come, Jaron Lanier. Something about programming and music go together, and there's a high coincidence of programmer-musicians. But what I was trying to do was get the tech intelligentsia together with some of the literary intelligentsia.

Eugene Mosier: The most interesting people I met were Louis and Jane. They told me their idea, I said, "Wow! I'd like to get involved." Afterward I called them asking if they needed anyone.

Louis Rossetto: I said, "We have no money." He said, "That doesn't matter, I'll just help you."

Eugene Mosier: I had no idea what I was getting myself into.

Jane Metcalfe: We were in the building right on the corner of Second and South Park.

Jack Boulware: It was this building at the end of an alley in a pretty funky neighborhood where we were doing our satirical magazine, the *Nose*. Then *Future Sex* moved there. Dave Eggers was doing *Cups*, which was a coffee shop magazine, then he launched *Might* magazine—but they all had an office in the same building.

Coco Jones: It was an open postindustrial space that a lot of magazines shared. I was working for *Mondo* at the time. To my left were Jane and Lewis.

Louis Rossetto: It was a really sketchy neighborhood. We got there around mid-June, and within two weeks somebody put a bullet through the front window, a big plate-glass window in the front, and it landed behind Jane's desk.

Jane Metcalfe: And then on the Fourth of July somebody shot a bullet through the front door of the place.

Louis Rossetto: Two incidents. Two bullets.

Eugene Mosier: They said, "We don't have a computer." I said, "That's okay." They were borrowing office space near where I lived. It was a matter of walking to the office with my stuff. I put my Mac IIci and Apple LaserWriter II on a cart and rolled them down the street.

Louis Rossetto: Eugene brought his computer, got another computer and another screen for me to work on while Jane got a hard drive, an incredible gift from the universe to help us finish. For the last two weeks Eugene called in sick to his day job and then worked to produce the prototype and then helped us actually print it.

Jane Metcalfe: He'd worked for *MacWEEK* when they converted to desktop publishing and just didn't want to work for the Man or whatever.

Fred Davis: Desktop publishing was the killer app for the Macintosh. It fueled an explosion of people buying Macintoshes just to have the LaserWriter—which cost $6,000, by the way. That is the most anyone in the personal computer industry had ever charged for a printer. People bitched about the price. But the Apple LaserWriter could do real typography and allow you to create camera-ready copy from a device that cost a tiny fraction of what a big printing press cost. This was a whole new thing in publishing. It was a whole new underground culture, which was the Mac culture.

Erik Davis: The counterculture of the sixties and seventies had split into a wide variety of subcultures, and one manifestation of this was the explosion of "zines," which were a pre-web, pre-Listserv way for small communities of obsessives to get together. The eighties really was the decade for zine culture. And pretty soon people realized that, by bringing in these new technologies, you could sort of make a whole cultural space that was partly fantasized, and then almost instantly became partly real because when you have the publishing tools that let you be slick, you can make your own pop universe. You can make something that feels pop because it looks pop, and then you draw people to it, and it becomes pop. So it was kind of like a subcultural hack that just did very very well.

Jack Boulware: I don't remember anybody ever buying a copy of any of the software, never. It was a sign of weakness if you bought software. People were trading floppy disks all the time. You would just ask somebody. "Oh, I'll make you a disk!"

Jane Metcalfe: Eugene was a whiz at figuring out digital production issues.

Eugene Mosier: The challenge, early in the desktop publishing days, was how to get the stuff out of the computers and onto paper so it could be shown to someone. The first PostScript interfaces for color copiers were just coming out. We found this copy shop in Berkeley that had one. Jane convinced them that it would be good publicity if we ran out the dummy of our mag on their machine. So we brought our files and started printing. We quickly realized it was taking way longer to reproduce a page than we had imagined. It was going to be an afternoon. That stretched into evening, and then became obvious it would run overnight. The people at the copy shop gave us the keys and went home. We proceeded to spend the entire weekend pressing the Print button and taking catnaps. We managed to print an entire magazine over the course of three days.

Jane Metcalfe: Oh, we did not sleep. And of course they opened up the next morning and people started coming in so we had to go over to a little corner and share the printer. This was when I had started having hallucinations because of the spray mount.

R. U. Sirius: The guys from *Wired* were upstairs. I remember Louis coming down and showing me a mock-up of the sample issue they had put together to get funding. I definitely had a visceral reaction, the suspicion that some people within the tech culture, establishment people, had looked at *Mondo 2000* and said, "Well, this won't do."

Louis Rossetto: All this time we'd been trying to get in touch with Nicholas Negroponte. His secretary said, "He's going to speak at TED. He'll look forward to talking to you three."

Richard Saul Wurman: When I first did TED every single conference was white guys in suits that were either politicians or CEOs sitting on a panel, or a guy who stood at a lectern and put down a speech and read it. Every conference was about one subject—siloed. TED was not siloed. It was about a convergence between disciplines. I had captured this audience at the convergence of technology, entertainment, and design. These two people came, the beautiful Jane Metcalfe and the somewhat acerbic and very bright Louis Rossetto, a handsome couple. They said, "We found an audience for everything that we've been thinking about," and TED was their audience. TED was *Wired*—what they had in their minds.

Jane Metcalfe: The conference started at eight a.m.; I think we met with Negroponte at seven thirty. He said looking at a business plan this early is like doing a shot of bourbon for breakfast. He flipped through it.

Louis Rossetto: The lights are down. Nicholas Negroponte sits there and looks at it page by page. He closes it and says, "I'm in."

Nicholas Negroponte: I asked them, "How much money do you need?" They gave me a number, and I said, "Fine." It was a handshake.

Louis Rossetto: It was like the sun came out and rainbows appeared.

Nicholas Negroponte: And to protect my investment, I told them I wanted to write a column.

Jane Metcalfe: *Oh my God! He's going to help us! And he's going to write a column! And help us find more money!*

Louis Rossetto: Warmth hit us in the face. He's the first guy who said, "I think it's a good idea, and I'm going to help with real support." It was like, "Finally! Yes!"

Jane Metcalfe: He was clearly on our side from that moment forward. It was phenomenal. He said, "Let's go find Bill." We ended up at some party that night, and Bill Gates was perched on the buffet table with his smeared glasses, swinging his legs back and forth the whole time. He turned through the magazine.

Louis Rossetto: He looked at me weirdly, like, *I don't think this is going to do it*. Sort of shrugged and gave it back to me.

Coco Jones: I was at *Mondo* the day the check arrived from Negroponte. It was not an easy conversation when I later quit to go work for *Wired*. They called me a traitor.

R. U. Sirius: They had gotten funding from a bunch of seriously well-known people in the tech-cultural world, Nicholas Negroponte being one of them, and I had the distinct sense that these would be the sorts of people who would not like us, and would be happy to see an alternative pop technoculture magazine come up, because it won't be so anarchic and druggy and somewhat disreputable.

Coco Jones: *Mondo* hated *Wired*. They felt it was completely derivative of what they were doing.

Gary Wolf: *Mondo 2000* was the drug side of this culture. *Wired* was the other side.

Jane Metcalfe: The world was changing. And so everybody's seeing it from their own worldview. Some of them thought it was changing because we were all doing the same drugs. Other people thought it was changing because we were using the same digital tools. And so there were people who could surf in between those worlds and so there were some common threads there in terms of contributors.

Louis Rossetto: At that point we started to get serious about putting the band together. I was the editor in chief and also the publisher and CEO. So I knew that we needed somebody to keep the trains running on time: a managing editor. So I called Kevin; he came down to our office. He listened to this whole idea.

Kevin Kelly: I was surprised. I thought I had successfully discouraged them from pursuing this idea. But they weren't giving up. Louis had a prototype that had been funded by Negroponte, a color-Xerox spiral-bound copy of articles that were taken from other sources and reimagined, re-curated into this prototype. It was a magazine about the culture of technology rather than just technology itself. I had never seen anything like that. I said, "Oh my gosh, this is going to work!"

Louis Rossetto: I said, "So we need to hire a managing editor. Who would you recommend to do that?" Never thinking that I was offering him a job because I didn't think he was the guy. He was already editor in chief of his own magazine.

Kevin Kelly: So the question I had for Louis was "What's the budget?" We paid almost nothing. *Mondo 2000* paid almost nothing. Louis mentioned a very adequate, professional editorial budget. I said, "I think this is so good that I'll help you launch the first issue."

Louis Rossetto: And so we created a position for him, which wasn't in the business plan, we had no money for it, but he became the executive editor of the magazine. And then we still needed a managing editor.

John Battelle: I was going to move to New York to be a freelance writer. I wanted to start a magazine about the impact of digital technology on culture. My catchphrase was "the *Rolling Stone* of the digital age." I was at a pay phone picking up my voice mail, and there was a message from Louis. I almost deleted it, which would have been the biggest mistake of my career. But I called him back. He explained what the magazine was, and that did it. I was completely hooked.

Louis Rossetto: John came up on Monday, and we were in the office at Second Street and the place was completely deserted, there's nobody there because we hadn't really started yet, and there was one chair and a sofa in this big empty place underneath the casement windows. John and I sat there and talked for a couple hours, and by the end I'd somehow convinced him that this is what he should be doing. So instead of him moving to New York he ended up convincing his fiancée to quit her job, she was a producer at CBS News, and move to San Francisco. So he became the managing editor, just a fucking demon, and she became our PR person. The two of them in lots of ways made the whole thing happen.

Kevin Kelly: John Battelle was a key hire. He was right out of grad school with a lot of experience, tremendous drive, very organized. The one question I had was whether he was too square. He dressed preppy. This was a renegade outfit. Louis was a born renegade. I was an old hippie. John didn't look like one, but it turned out he had an inner renegade in him.

John Battelle: We all brought different networks to bear. I had been a trade reporter covering Apple and technology generally for five years, so I had those writers. Louis knew fabulous freaks from all over the world—crazy artists, hackers from Japan. Kevin had the *Whole Earth* network, Stewart Brand and those guys.

Fred Davis: *Wired* had writers just hovering, the cream of the crop. We were cheering them on, forcing them to produce this magazine because we all wanted to write for it so badly! We were all so frustrated writing reviews of printers.

Louis Rossetto: We had raised $250,000, but now we were paying salaries, rent, electricity, and all the stuff it takes to make a business. We were going to run out of money in four months. I wanted to make sure that when we ran out, if we weren't on press, at least we'd have a magazine we could take to a major publisher and say, "Take a flyer on this."

Michelle Battelle: There were only ten or twelve people working at *Wired* at the time: Amy Critchett was there, and Kristin Spence, Eugene Mosier . . .

Amy Critchett: I was one of *Wired*'s first interns.

Jane Metcalfe: Amy came in and wanted a job, and this was back when we had Kristin working for no money and Eugene tapped to work for no money. And I said, "Sure, you can have a job. You can work eighty hours a week having the time of your life. The only problem is we're not going to pay you." She's like, "Great. That sounds great."

Michelle Battelle: I couldn't believe that they were putting something together that didn't require a whole bank of desks. They barely had any supplies.

Amy Critchett: Louis was very clear in his thinking that if you bought people a bunch of pens they would all disappear. So why buy them in the first place?

Kristin Spence: We put the office together with spit and tape. I was responsible for furniture procurement. South of Market was a wasteland. Louis and Jane were mystified that we didn't have an IKEA store nearby! We went to salvage yards where we got a bunch of crap-ass hollow-core doors and put them on top of old used filing cabinets or sawhorses for desks. Completely ad hoc.

Amy Critchett: There was no heat and certainly no air-conditioning. It was a freezing cold winter. So at one point Louis said, "Amy, go rent a heater." And I rented this industrial heater, it was like a jet engine that literally had flames coming out of it. And we sat it on a mail cart and it had this really loud kind of roar. And that was our heating system. Everyone wore cut-off gloves. It was freezing, it was a freezing winter.

Jane Metcalfe: We pushed it around to blow heat on people because it was cold and wet. People were getting sick; the water was literally coming through the roof. It's like, "You can't get sick! You can't get sick! The magazine has to be at Macworld."

Michelle Battelle: *Wired* was supposed to launch right around January 15, 1993, because they were aiming for Macworld.

Kristin Spence: That added to the whole vibe. It was a skunkworks operation in the extreme. We were so committed and passionate about what we were doing. We were living it 24/7. There would be times when we'd be working at one a.m.; there would be a handful of us working, we'd be cranking house music because the rave scene was blowing up and Eugene was deeply involved in it. The rave scene, the revolutionary aspect, we're guerrilla revolutionary journalists!

Fred Davis: It was this pre–Burning Man, Anon Salon, Survival Research Labs, Multimedia Gulch group of artsy, nerdy people.

Amy Critchett: We were just working so damn hard, and so every night had the potential of turning into a party. Nothing that scandalous. There was definitely drinking. There was definitely…I'm sure there was some coke. I'm sure definitely there was ecstasy around. After work. We were blowing off steam.

Jane Metcalfe: Everyone was sleeping with everyone at *Wired*. Louis and I were sleeping together; John and Barbara—the magazine's graphic designers—were sleeping together; John and Michelle were sleeping together, and that only left a few people who I'm sure had found other people to sleep with because we never left.

Amy Critchett: There was a couple couches in Jane and Louis's office. Eugene basically lived there.

Michelle Battelle: I was from a different world. I was from CBS Network news, where we still had to wear stockings with our skirts. We were much more buttoned up. Seeing Eugene emerge from one of the back offices where he had been sleeping was really interesting to me. It was the first time I'd seen a lip ring in my life.

Kristen Spence: For those of us who were superyoung, it was just this cool thing happening. But Louis, Kevin, and John Plunkett had the long view. They understood how much things were changing.

Amy Critchett: Louis was magnificent. His presence, force, and passion— that's what made *Wired* happen.

John Perry Barlow: Jane had the real juice. She got people really excited. There were a lot of people that gave them meetings because of her. She had a lot to do with creating the energy in the magazine and, to a fairly strong extent, its attitude as well.

Louis Rossetto: Battelle was the most focused, hardest-working manager I've ever met. He was intense yet calm, the foundation on which the editorial team was built, the guy who got it done, no matter what. He could sit at his computer for hours on end, his fingers jabbing at his keyboard like it was a karate exercise, just getting it done.

Kevin Kelly: For the last week Louis came up with these kamikaze headbands. He wanted everybody to wear these headbands in solidarity or something. I never got the allusion. We were going to kill ourselves?

Louis Rossetto: The payoff was seeing the magazine come off the press: a big six-color Heidelberg roll press, and seeing for the first time what that means. The rolls of paper are up to my shoulders. I can't put my arms around half of them. And gargantuan big black barrels, fifty-five-gallon drums filled with screaming orange Day-Glo ink. Rolls of paper the size of Volkswagens and ink we're buying by the barrel-full.

John Plunkett: It was a built-in contradiction. We were using an old medium to evoke a new medium that didn't quite exist yet. We knew it would be some sort of electronic networked future, but we didn't know what that would look like. How could we make it visible?

Kevin Kelly: Color on the computer was no extra cost. Therefore there should be an extravagant use of color, because it's free.

Eugene Mosier: Adding fluorescent colors was a way to evoke the screen, transmissive color rather than reflective color.

Fred Davis: You could use color on every page if you didn't have to pay for color separations. It was possible for *Wired* to have very low start-up costs because they were done with desktop publishing.

Louis Rossetto: Then we get on press. The first sheets come out; they rip them off the caddy, put it on this big table at the press control panel with the lights that are adjusted to get true color. The guy takes out his loupe, and he makes some adjustments; finally, he says "We're ready." John looks at the sheet and says, "I want more ink." The guy says, "It's perfect." This was a press that had never done a job before. Every piece was pristine. We were the first clients to go on this press. There wasn't anything that was sloppy or worn or out of adjustment. The guy says, "This is the perfect first sheet."

John Plunkett: What printers define as "good" is the least amount of ink that they can put down.

Louis Rossetto: John says, "I want more ink." The guy looks at him like he's got two heads. He does the same thing all over again. John says, "More ink." They do this two or three more times. John says, "All I want you to do is turn the ink up until it starts to smear. Then dial it back until it doesn't. That's what I want." The guy is disgusted. Out comes a sheet and it looks like *Wired*.

Eugene Mosier: *More!* That was the theme of the first issue in terms of getting it physically created.

John Plunkett: The experience of picking up the magazine and flipping through those first pages was intended to be disorienting and disruptive. To take Marshall McLuhan's advice and make the medium be the message.

Kevin Kelly: In the first issue you have Bruce Sterling, you've got Stewart Brand, you've got Van Der Leun. I was basically taking the same subjects and the same people that I was talking about and working with at *Whole Earth* and I was doing it in *Wired* in color. But it was like totally different. Suddenly the entire world was paying attention. Suddenly it was like the most amazing thing in the world. Suddenly it was like, *Oh my gosh, this is revolutionary!*

Justin Hall: I remember seeing an advertisement on the side of a bus rolling by with Bruce Sterling's face cropped and Day-Glo'ed staring out toward me.

Jane Metcalfe: You can't believe how little we paid for that.

Justin Hall: I was so excited to see a computer magazine with a human face on the cover. I thought, *Oh my God! Like these people get it. Technology is about this culture that's happening!* And I'm so excited that these people got it.

Louis Rossetto: We weren't making a magazine about feeds and speeds. It was about the people, companies, and ideas making the digital revolution.

Carl Steadman: There it was: this mix of business, technology, and culture, that was right. *Wired* got it, so I was entranced. I was both entranced and horrified, because it was clear that *Wired* was horribly awry when it came to its politics. I didn't know the term *libertarianism* back then, it hit me as very conservative.

Louis Rossetto: I was at Columbia in '67, ' 68, '69, and '71, and of those four years in college, two of them were wiped out by eruptions in the spring. It felt like the world was coming unglued. And if you had eyes you could tell that there were issues in society that weren't right. But the analysis from the left just seemed wrong. It was like refried Marxism and this colonial grievance bullshit and that was so inauthentic and their prescriptions— what they wanted to have happen—so wrong! So then I started looking for other stuff. And I think I'd read Ayn Rand and then from there you realize there's more than Objectivism. There's this libertarian strand and then you realize that libertarianism is really deep. It goes back to Pierre-Joseph Proudhon and beyond and has all sorts of other manifestations in American history. And I just got farther and farther into that and realized this is a kind of good way of thinking about things.

Carl Steadman: *Wired* had this ideology that had no sense of social justice. Instead, it was Social Darwinism. And I couldn't abide that because I knew *Wired* would succeed. It was clear that *Wired* was *it*. So that day I had it in my mind that I would have to go to *Wired* and change it from within.

Justin Hall: My immediate reaction was I want to join. I want to be a part of this. So I called and I left a voice message on Louis Rossetto's answering machine and I said, "Hi. I'm in Chicago and I'm connected to the hacker pirate underground of the bulletin board scene out here. I'd really love to write something for you or connect you with those people. Here's my phone number." Then the craziest thing happened. He didn't call me back!

John Battelle: It was overwhelming. Really overwhelming. All of a sudden we were a hot ticket. The company and the people behind it—the writers, photographers, illustrators—became rock stars.

Fred Davis: Having *Wired* on your coffee table, man. That was a status symbol. That said you're part of that digital visionary crowd. You're part of that cyberculture.

Kevin Kelly: And every day there were cameras and TV companies sitting in our offices. And for me it's like I'm doing exactly the same thing I've been doing for years at *Whole Earth*, with very few resources, for a very small audience. Suddenly I was doing it on a big stage with colored lights at the center of the universe. It had been hard to get anyone to pay attention before, and now everyone was saying it was brilliant. It was peculiar but wonderful. And yeah, that was a surprise I was not expecting at all.

Louis Rossetto: There's something about investing your humanity, your eccentricity, your exuberance in the things you do. That's what the first issue of *Wired* was about. It was about being naked with your craziness, your enthusiasm, your genius. It was pushing to be as great as you can. Be naked. Be human. Let people see who you are.

Amy Critchett: One thing for me postlaunch that was, just I'll never forget till the day I die, is all of a sudden I think, *Wait, we have to make another magazine? And then again and again and again?* Like the idea that it was a monthly occurrence was beyond mind-boggling to me. But we did it.

Louis Rossetto: Jane and I had been sprinting flat out for three years to get out the first issue, and then the marathon began.

Jane Metcalfe: We were exhausted. Everyone needed a vacation. Then we looked at each other and said, "Oh, shit! We're late for the second issue!"

Toy Stories

From PARC to Pixar

*T*echnically speaking, Pixar is not a Silicon Valley company, as its head-quarters have never been in the Valley proper—but its roots lead directly back to Xerox PARC. Back in the seventies Alvy Ray Smith was PARC's unofficial artist in residence. He'd come in at night to play with Dick Shoup's SuperPaint machine and ponder the potential of computer graphics. Like so many others before him, Smith believed that the computer could be much more than a productivity tool or an entertainment device: It was a new kind of paint-brush in a new kind of art form. Xerox explicitly rejected that idea when they kicked Smith out of PARC in 1975 for being what Stewart Brand called a "computer bum"—a hacker, essentially. Smith turned his fall from grace into the start of a twenty-year vision quest. The dream? To make the world's first full-length computer-animated film. The journey first took Smith to a research lab in Long Island, where he met his future partner in Pixar, the like-minded Ed Catmull. Then, in 1979, the two boomeranged back to the Bay Area where they found new patrons: The first was George Lucas, who was in the process of finishing The Empire Strikes Back. Smith and Catmull were the core of Lucasfilm's computer group—the proto-Pixar. Steve Jobs was their next patron. He took over from Lucas in 1986, after leaving Apple, and set it up as a technology company which made little computer-animated shorts on the side. Ostensibly Luxo Jr. and other early Pixar filmlets were created to show-case what could be done with Pixar technology. In reality, the technology was just a means to an end. Pixar limped along for almost a decade while Catmull and Smith waited for the inexorable advance of computer power to catch up with their movie-making ambitions. The evitable result: Toy Story—which, like the prior era's Pong, marks the emergence of a new American art form.

Alvy Ray Smith: So when David DiFrancesco and I got booted out of Xerox we had already cast our lots together for an NEA grant. And suddenly we didn't have the basic instrument which was the frame buffer, the picture memory. Dick had the first one in the world, so we had to find the next one. We knew there was one was being built in Utah, so we went there and they looked at us, a couple of hippie freaks. They said, "You know we're Department of Defense funded, but there was a rich guy who came through here and bought one of everything in sight—including the frame buffer." It turned out to be Alex Schure out on Long Island.

Ed Catmull: He had this school funded by government money for Vietnam vets returning from Vietnam.

Alvy Ray Smith: New York Tech, or the "New York Institute of Technology," as they are very careful to say these days.

Ed Catmull: And it was, at that time, not a particularly strong technical school except that he was willing to fund this lab. We weren't really connected with the school other than being on the same property. And then the equipment arrived and we set up a computer room there and started programming. Then two more people showed up: this is Alvy Ray Smith and David DiFrancesco. So Alvy then became my third hire.

Alvy Ray Smith: We got ourselves out there and he snapped us up. He didn't snap David up, he snapped me up. David again didn't have a job.

Ed Catmull: It was a while later before I even hired David. So we had this equipment, and Alex would come in every day because he was just so excited about it. He also had an animation studio.

Alvy Ray Smith: So then Jim Clark shows up, and we hire him to build a head-mounted display for us. Because we all thought it was such a cool idea and he'd built the one at Utah, and so we wanted one of those. So he agreed to come. And Jim picked it up immediately that he wasn't going to last.

Jim Clark: I was fired from NYIT, which led to my going to Stanford.

Alvy Ray Smith: He went off under this cloud—which was unfair in our opinion—to build a geometry engine and start Silicon Graphics and all that.

Butler Lampson: Jim Clark was at Xerox PARC, as an intern. He was obviously a hotshot, and had cool ideas about how to do graphics.

Jim Clark: I saw the Alto. I saw bitmap graphics the same year that Steve came through. Steve Jobs said, "I am going to put this into a personal computer," and they made that little quaint box with the little tiny screen with bitmap graphics. I looked at the Alto and thought, *This is boring. It's just moving bits around! Why aren't they doing 3-D graphics?* I left that summer with a project that I began to implement, which subsequently led to Silicon Graphics. And Ed was, at that very moment, in negotiations with George Lucas to be hired.

Steve Jobs: Now George Lucas, who produced the Star Wars film trilogy, was a smart guy, and at one point when he had a lot of money coming in from these films, he realized that he ought to start a technology group.

Ralph Guggenheim: You have to understand this was '79. So this was about a year or year and a half after the first Star Wars film came out in the summer of '77. Everyone in the world thinks that George Lucas has got to be the most advanced, the most technological, superlative filmmaker ever. The visual effects in *Star Wars* were incredible.

John Lasseter: When that movie started you were swept away. By the end I was just shaking.

Ralph Guggenheim: That little computer-animated scene at the end with Luke shooting down the Death Star? It was astounding!

Jim Clark: We all left New York Institute of Technology pretty much en masse.

John Lasseter: Ed Catmull was hired in 1979 and asked to do three projects, with a fourth added by Ed. They were insane ideas at the time: digital nonlinear film editing, digital sound editing, digital optical printing, and 3-D computer animation, which is something that Ed brought with him.

Ed Catmull: For the first year I wasn't allowed to bring anybody. They'd only had one successful film, and they didn't want to overreach.

Ralph Guggenheim: George didn't give Ed the go-ahead to hire anybody until *Return of the Jedi* came out in May of 1980, because if the film had opened poorly he didn't want to be committed to spending a lot of money on a research group.

Ed Catmull: Finally, we were allowed to hire people, and so I brought in Alvy Ray Smith to head up the graphics group. And I brought in Andy Moorer to head up digital audio, and then Ralph Guggenheim from New York Tech to head up video editing. Andy Moorer built a digital audio processor, and then with Ralph we built a digital editing system.

Ralph Guggenheim: Just after Labor Day 1980, I show up at our office, which was not much of an office for a computer research group. It was in a renovated Laundromat in San Anselmo that George Lucas owned. So here we are in this Laundromat and we are supposed to write these papers, and the great irony is, there's not a computer in sight anywhere in this building!

Alvy Ray Smith: So we decided to build this machine for George. It was basically a supercomputer for images.

Ed Catmull: The computer division was basically R&D. We were not part of Industrial Light and Magic. We had made some special effects. We did a minor special effect for the third Star Wars film, and then this major effect, which was for *Star Trek II*.

Tom Porter: The genesis effect for *Star Trek II*—that was a very technical piece of work. It was a bunch of camera movements. There were no characters walking around. There were no performances. There was no notion of character animation.

John Lasseter: The computer animation world at the time was primarily in university research labs. It was mostly TV commercials and mostly quite awful. Everything was made of stone and glass and very reflective and all that, because most of the stuff was being done by the people who created the software, and they loved mixing crazy imagery and then putting computer music to it.

Alvy Ray Smith: We decided we were going to show up at the next conference and show to the world that we did character animation—not flying logos and not special effects, but character animation. That's what we wanted to do: deliver the first character animation.

Ed Catmull: We weren't interested in flying logos. We wanted to have characters and objects come to life. And this is where John is the master.

John Lasseter: They brought me in to work on the very first 3-D computer animation of a character.

Alvy Ray Smith: And we knew what was going to happen. It was going to be an amazing event. And George Lucas was going to see it!

John Lasseter: It was called *The Adventures of André and Wally B.*

Alvy Ray Smith: The idea was that an android named André wakes up in this gorgeous landscape—and we knew we could do beautiful landscapes by then—and the sun is going to rise and he's going to rise and stretch and welcome the new world of full character animation. It's a terrible story. It's really corny.

John Lasseter: I was the very first traditionally trained animator in the world who worked with computer animation. And I just blossomed! I thought to myself, *You know what? I can make an object move around and give it personality and emotion through pure movement.* It was so exciting: Every day you would go to work and there would be something new on the monitor, on the screen. *Wow! That's great!*

Alvy Ray Smith: John really saved my bacon.

John Lasseter: *André and Wally B.* was a massive step forward. It was very cartoony, and people loved that. I remember one guy, who worked with a computer graphics company, coming up to me after a screening to ask what software I used to get the humor in!

Alvy Ray Smith: George hated it. And it just proved to him that we couldn't do animation. We were a bunch of techies. Also about that time it was falling apart at Lucasfilm. George is chasing Linda Ronstadt around. Clearly his wife had already left. And in fact, as soon as he got divorced I went to Ed and said, "You know, George never really got us. And we've got forty people here that we need to support. He doesn't have the money. He's lost half his fortune overnight. He can't afford us anymore. He's going to fire us."

Ralph Guggenheim: The management at Lucasfilm kept coming in and saying, "Look, we are going to shut this down because George can't afford to spend more money on you research guys."

Alvy Ray Smith: And one of the first questions to each other was, "Well, what are we going to do?" I knew we couldn't do a movie. You know I had just done the numbers. I had run them as tightly as I could. And I got the wrong answer.

John Lasseter: Ed's dream, even though he was doing all this other stuff, was always to do a computer-animated feature film.

Alvy Ray Smith: Everything was off by one crank of Moore's law—Moore's law being a factor of ten every five years. We needed another order of magnitude. We needed another five years. We couldn't do a movie for five years so what were we going to do for five years? We had a prototype called a Pixar Image Computer. Let's make a business out of that!

Ed Catmull: I thought, *Okay, there is some basis for doing a Silicon Valley sort of start-up.*

Alvy Ray Smith: Now this is two complete nerds saying this, we had no idea how to do a hardware manufacturing company. So we call up Jim Clark and he says, "It's easy, but it will take you about a year to get it."

Jim Clark: How do you tell a good friend: "Hey, you have got a shitty idea!" I was not going to tell them that.

Smith and Catmull had a prototype product—the Pixar Image Computer— and a plan that would allow them to spin out from Lucasfilm and continue on as an independent company, but they still needed a backer. The writing was on the wall at Lucasfilm. If they didn't find a funder quickly, Lucas would simply cut his losses and fire the entire group. They crossed their fingers and reached out to Alan Kay.

Alan Kay: Alvy Ray called me up on the phone and said, "We'd like to get out of here. Have you got any ideas?" And I said, "Yeah, I do. I'll go over and talk to Steve." He was at NeXT then.

Alvy Ray Smith: Alan Kay was the chief scientist at Apple, before Steve got kicked out.

Alan Kay: So I went over and talked to Steve, and then I guess the next week or two weeks later Steve and I took a limo together up to San Rafael, and basically I spent a long time explaining to Steve just exactly who these people were.

Alvy Ray Smith: Alan Kay brought Steve up to Lucasfilm to meet us.

Alan Kay: They weren't like any of the people he had working for him at Apple. This was like the first team. These are like first-class people of extreme talents: Take a good long look at what they are doing and see what you think and you know, don't fuck around with these guys.

Ralph Guggenheim: Steve came up and saw what we were doing with the image computer and was very impressed.

Steve Jobs: Apple had been doing stuff with graphics at some level for ten years, but it was all 2-D. The 3-D stuff that Ed and his team were doing was way beyond what anybody else was doing. It was very exciting.

Tom Porter: What Steve appreciated in us was that we were looking out ten years. We had a clear goal: photorealistic imagery.

Ralph Guggenheim: He was probably putting together ideas for his NeXT machine and he thought, *It can't be a $100,000 box with ten circuit boards like you guys have now, but if we can shrink it down to maybe one card in my new* *computer? This could be awesome! This could really give me a competitive edge over anyone else in graphics.* And he understood the value of being able to do graphics.

John Lasseter: Initially it was pure software and computer hardware that he was interested in.

Ralph Guggenheim: The animation thing? Pffft. He had no interest whatsoever.

Tom Porter: It wasn't reachable. It wasn't two or three years away.

Ralph Guggenheim: The story I was told was that there was a conference call between Steve and the execs at Lucasfilm, and you can imagine execs in a conference room with a speakerphone on the middle, and Steve's on one end of the phone and all of these execs are trying to negotiate with him at the other end. Steve says something like, "I'll give you five million dollars for the thing. I'm going to have to put in another five million dollars. That's my offer!" And one of the execs undoubtedly said, "What! Five million? That's not nearly enough." And I was told that Steve said, "Fine, go fuck yourself!" and instead of going and fucking themselves they took the offer and they closed the deal.

George Lucas: We lost a ton of money on the computer division.

Yet, thanks to Jobs's $10 million investment, the R&D group survived. It spun out from Lucasfilm to became an independent company: Pixar.

John Lasseter: Pixar had two lives: For the first ten years it was more of a technology company.

Tom Porter: At Pixar, in those early days, there were seventy people—that grew to be 120 people.

Ed Catmull: The Pixar Image Computer was the product when Pixar became a company.

Tom Porter: They were trying to figure out how to sell this thing for medical imaging, for satellite imaging.

Ralph Guggenheim: We were making short animated films on the side in the service of selling the box, and in service of building the reputation of the company, and in service of our own secret mission to make a feature-length film using computer animation.

Alvy Ray Smith: We sold a lot to three-letter agencies who did unknown things with them in DC, and to Disney.

Ralph Guggenheim: Disney were buying Image Computers to do their cel painting for them.

Ed Catmull: CAPS is what they called it.

Alvy Ray Smith: The problem was it was a bitch to program. The second problem was Moore's law was going so fast in the background that by the time we got everything out to market, the standard computers—like a SUN or Silicon Graphic workstation—were fast enough to do it.

Tom Porter: Then Photoshop was written by the Knoll brothers, one of whom was at Industrial Light and Magic, who saw what the Pixar Image Computer could do and thought, *This could be done on a Mac!*

Alvy Ray Smith: But for a short number of years in there it was the fastest machine on the planet but we didn't sell enough of them to carry the company.

Ed Catmull: Steve didn't know anything about how to run our business; and neither did we. None of us had ever done this before. And because we didn't have venture capitalists involved, nobody knew how to find the right people. We didn't know what it meant to find a good sales or marketing person. We were all completely ignorant; and in a high-stakes game. So we had to learn a lot. And we made a bunch of mistakes.

Alvy Ray Smith: We'd run out of money. We did it about four or five times. We should have been dead. Any other company would have shut the doors.

Ed Catmull: I learned a lot, a great deal, and Steve learned a lot. Each of us were learning. And with all these mistakes, we came close to failing.

Alvy Ray Smith: Steve could not withstand the embarrassment of failing on his first event out of Apple. He just couldn't. It was so embarrassing, so he would just write another check and eventually he had put $50 million in.

Ed Catmull: There were some really difficult times. And there were times that any one of us almost gave up.

Alvy Ray Smith: Steve tried again and again to sell the company. I wrote the business plans. If anybody had come in and made him whole on the $50 million he would have been out of there overnight. We should have gone broke.

Ed Catmull: And there were times it was, *Okay, I can't believe this is happening.*

Ralph Guggenheim: *Luxo Jr.* was nominated for an Academy Award.

John Lasseter: Our short film from 1986 called *Luxo Jr.* is only a couple minutes long, and we were so limited by the lack of computer power that we couldn't even give the characters a background. We just locked the camera down and had this wood floor that faded off to nothing.

Tom Porter: I doubt *Luxo Jr.*, an artistic piece, rang any bells for Steve Jobs. There wasn't any money in it. There wasn't computing in it. Frankly, *Luxo Jr.* is not very high in visual complexity. It was a nice badge of honor, no question. But that in and of itself wasn't the breakthrough.

Ralph Guggenheim: The next year *Tin Toy* won the Academy Award. *Luxo Jr.* and *Tin Toy* were huge hits and a big success for everybody except our computer hardware sales guys. They were pissed. We were getting all this credit for our animation but not for the hardware product. Even Steve was getting frustrated.

Ed Catmull: So *Beauty and the Beast* came out in 1991 and used all the software that we had written for CAPS and it was an incredible success. 'Ninety-one was also the year *Terminator 2* came out, which also used our technology. So all of a sudden the two big moneymakers of the year made heavy use of technology and the whole industry just like yanked their head around: "Holy cow! Something is changing here!"

Clive Thompson: It marked a very weird shift in filmmaking. When you shoot a regular movie, you shoot with a camera, and all you get is what the camera's eye sees. If you want to change things in postproduction, all you can really do is work with what light hit the camera lens. The thing that was revolutionary about Pixar is they were essentially generating an entire world—a virtual world—where they can move the camera anywhere they want, because it's a virtual camera which they can fly around anywhere they want. So, they can shoot a scene, and look at it, and then decide, "Hey, let's put the camera somewhere else." Thus the camera becomes just another postproduction element. That's a very jargon-y way of saying you create the scene—and *then* you figure out how you are going to look at it. That's revolutionary, that's a complete inversion of the way Hollywood movies had worked.

Andy Hertzfeld: But Steve came really close to going bust there in 1992 or 1993. NeXT and Pixar were failing at the same time. He was running out of money. He came pretty close to crashing, because NeXT was on a dead-end path and Pixar had a similar thing in that the technology wasn't quite there.

Alvy Ray Smith: Life with Steve was awful. There was this famous board meeting at NeXT. Steve comes in and he's busting Ed and me for being late on a circuit board for the image computer, which we were. And I said, "But Steve, you're late on one of your boards." Which was true. Now normally that would have just been okay. Not this time. He starts insulting me, making fun of my accent, playground bully stuff. This was not two intelligent people having a conversation or even a healthy debate. This was just sheer bullying. And what did I do? I just stood up and went right into him. Now I'm very proud of that, but it probably was an insane thing to do. I went right up into his face, screaming in rage. I still can't believe it happened. Just weird, just screaming at each other. And at that point I forced my way past him and wrote on the whiteboard. I wished I had written something clever. I was too insane. I just made a mark. It was a forbidden act, so he said, "You can't do that." And I said, "What? Write on the whiteboard?!" That was it. He stormed out of the room.

Ralph Guggenheim: Alvy and Steve, it was just oil and water.

Alvy Ray Smith: A lot of people jump to the conclusion I got fired. But Steve didn't have the right to fire me. Steve was the chairman of the board, so Ed would've had to fire me, and Ed wasn't about to fire me. And so even though I was there for another year I knew I had to get Steve Jobs out of my life, because he was a foul bullyboy down underneath it all. So what happened in there is that Disney finally came. They knocked on the door just at the right crank of Moore's law and they said, "Let's make the movie that you guys always wanted to make." This is the big dream, right?

Ed Catmull: They thought that computer animation looked like it might have a future. But it was "a boutique film." Those were their words. So they were willing to fund it at a very low level.

Alvy Ray Smith: So we go off and make the movie and, by the way, I left right in there. What finally freed me up to leave was the movie happened.

Tom Porter: *Toy Story* itself took shape in 1993 and 1994.

Ralph Guggenheim: Toys were something we could easily render realistically on a computer, so it seemed like something practical we could do.

Joss Whedon: They sent me the script and it was a shambles, but the story that Lasseter had come up with—that toys are alive and they conflict—the concept was gold. I went up to Pixar and stayed there and wrote for four months and completely overhauled the script before it got green-lit.

Ralph Guggenheim: But the Disney execs kept pushing us to make these characters more edgy: "We want them more hip, more edgy, more aggressive."

John Lasseter: "Edgy." That was the word that they kept using. We soon realized that this was not the movie that we wanted to make—the characters were so edgy that they had become unlikable. The characters were yelling, they were cynical, they were always making fun of everybody.

Ralph Guggenheim: And so we ran into story problems along the way. Mainly because it got to where Woody and Buzz were just horrible. And so we're now at like Thanksgiving of '93 and we show our latest story reels.

John Lasseter: And the movie was just horrible! The characters, especially Woody, were just repellent. Woody was just awful, awful, awful! And I was embarrassed because it wasn't the movie we set out to make.

Tom Porter: Disney actually went to shut down the production because they no longer believed that the story could work, and John and Andrew famously asked for two weeks.

Ralph Guggenheim: John, Pete Docter, Andrew Stanton, and Joe Ranft holed up in a room for a week or two and just rewrote the script. Joss Whedon came in again, too.

Joss Whedon: We sort of went back in the trenches and made sure we had everything we needed and nothing we didn't.

Ralph Guggenheim: And they rewrote the script top to bottom. The script got approved. We resumed production. It could have been a total disaster.

John Lasseter: From that point on, we trusted our instincts to make the movie we wanted to make. And that is when I started really giving our own people creative ownership over things, because I trusted their judgment more than the people at Disney.

Alvy Ray Smith: So Steve's busy running NeXT and Disney took the movie to New York and the critics saw it. They went nuts and they said, "This is going to be a huge success!" And as soon as Steve heard that he pushes Ed aside to be there when the cameras roll.

Ralph Guggenheim: You can see the lightbulb go off in Steve's head. This thing that he's not really been much a part of...

Alvy Ray Smith: He didn't have anything to do with it! He never had anything to do with any movies. In fact, they never let him in a story room.

Ralph Guggenheim: I would say between '86 and '95 when *Toy Story* came out Steve probably—I'm not exaggerating—I don't think he was in our building more than nine times in nine years.

Alvy Ray Smith: So Ed and I, the way we ran the company was to keep Steve out of the building, because he would come in and give one of his speeches, and it would take us a week or two to get it all back together again.

Ralph Guggenheim: Steve was a very generous benefactor, but kind of an absentee landlord, all this time. Within a couple of months we have a new CFO who knows how to take companies public. Steve is now going to be the CEO of Pixar, et cetera, et cetera.

Alvy Ray Smith: It was when I looked at the prospectus for the IPO when I first realized that Steve lies. He claimed to be the cofounder of Pixar and the CEO since its founding. In the prospectus! Cofounder and CEO forever! Bullshit. Both of those are wrong: lies. I don't like this "reality distortion field" idea. He lies. But you know his genius was to take the company public on nothing. The movie wasn't even a hit yet, and there was essentially no cash at all. But he saw his chance to make his $50 million back. And he did.

Alan Kay: Steve just hung in there and hung in there until they got into the sweet spot where everything that they knew suddenly was applicable in a way that made commercial sense.

Al Alcorn: Steve thought he was going to make a computer, and he wound up making one of the best movie companies in the world. Who would have thought?

Jobs timed the Pixar public offering to take place exactly one week after Toy Story*'s theatrical premiere, thereby insuring that the value of Pixar's public offering would be pegged to* Toy Story*'s box office. It was a bold bet on synergy, and Jobs hit the trifecta. His movie was the highest-grossing film of 1995, his IPO was a blockbuster, and his reputation was restored.*

Alvy Ray Smith: Steve made a billion dollars overnight. It may have made him a billionaire for the first time. So businesswise the guy was a genius, and he got everything he deserved on that front.

Alan Kay: I remember feeling extremely good when I saw that Pixar had become such a success, because any funder like that should be rewarded. Ten years! That was an incredible thing.

Ed Catmull: It was the biggest IPO of the year, it was bigger than Netscape. It was an incredible thing.

Jerry Garcia's Last Words

Netscape opened at what?!

*T*he internet is almost as old as Silicon Valley itself, but before the browser was invented, the Valley took the 'net for granted. There were online experiments like The Well, but for most the internet was simply the ether that carried e-mail hither and thither. In the eighties and early nineties, the conventional wisdom was that the real action online was inside the great walled gardens of AOL, CompuServe, and Prodigy. The world wide web was a latecomer that floated on top of the already-existing internet, providing an open-source alternative. Yet at the time few in the Valley took the web very seriously at all—if they had even heard of its existence. It was little more than an academic curiosity until Jim Clark, a Silicon Valley hardware mogul who had made a fortune selling specialized computer-graphics machines to Pixar and others, decided that the conventional wisdom was wrong. He founded Netscape in 1994 and in just over a year the company laid the foundation for virtually every technology that defines today's online experience.

Alan Kay: A lot of people think the internet appeared in the nineties. It started in 1969.

John Markoff: Today's internet started with the ARPANET, and the ARPANET started with two nodes, and one of them was in Southern California and the other one was at Menlo Park in Doug Engelbart's Augment project.

Doug Engelbart: Bob Taylor and Larry Roberts, the two guys running that office, told us all that they were going to go ahead and put together this network, and I volunteered to start a Network Information Center and that's sort of why they put me on early.

John Markoff: The NLS system was supposed to be the first killer app for ARPANET, which became the internet.

Bob Taylor: But the ARPANET was not an internet. An internet is two or more interactive networks. The first internet was put together by PARC. They put together three: the Ethernet—which they invented—with the ARPANET, with the SRI packet-radio net. That was in about 1975 or 1976. We had an internet without a world wide web or a browser. Those things we did not invent.

The first version of the world wide web was hacked together in 1990 by Tim Berners-Lee, an English computer scientist working in a French physics lab. The embryonic web was a geeky but efficient way to link a couple thousand physicists to a tiny number of supercomputers. Then Marc Andreessen, an American student working at NCSA, the National Center for Supercomputing Applications at the University of Illinois, built NCSA Mosaic—the first decent web browser.

Steven Johnson: You can't imagine how hard it was just to get on the internet in like 1991 or 1992. It was a colossal battle. You knew that there was this incredible thing out there, but it was just really difficult to get to. All the online spaces were discontinuous—there was The Well, CompuServe, AOL, but they were all separate dial-up universes. To go from one site to another you had to hang up the phone and make another call and listen to this crazy sound that the modem would make and then hopefully you would get a connection. It took so much work to move from one place to another. You really had to be committed to find anything useful on it.

Jim Clark: The web's original genesis was with Tim Berners-Lee, as a means for physicists to share publications and to pass around written documents of that sort. HTML and all of the Hypertext Transfer Protocol that is used is basically a format for passing around these documents, for shipping them electronically to other people.

Aleks Totić: Everything was academic on the internet. It was all .edu's. It was not that interesting.

Steven Johnson: And then starting in 1993 or 1994 you started to hear word about Mosaic, this new browser that was here—but we didn't really know what that meant.

Sergey Brin: I used Mosaic at the time. I tinkered with it, like a hobby. It was more of a fun thing to do than an *Oh! This is going to change the world.*

Brian Behlendorf: In the early days the internet was really small. Everybody knew everybody: researchers, programmers, people working in academia…

Stephen Johnson: And then the Mosaic browser becomes Netscape. Suddenly you had a unified front end for the internet.

Netscape was founded by two people: Jim Clark and Marc Andreessen. At the time Jim Clark was already a Silicon Valley legend. In grad school in the seventies, Clark had worked out the fundamental math behind what eventually came to be called virtual reality. In the eighties he put all that math into a silicon chip, which became the basis for Silicon Graphics Incorporated (SGI), one of the decade's highest-flying computer companies. Then in the nineties Clark abruptly quit SGI and teamed up with Marc Andreessen—a young Silicon Valley newbie—to form Netscape. The mission was to rebuild and reimagine the NCSA Mosaic browser—they were going to drag the web out of its ivory tower.

Howard Rheingold: Jim Clark is sort of like who Bob Taylor was to Doug Engelbart. He found this guy who had an idea and really made it happen.

Jim Clark: At the end of '93 I was just finishing my twelfth year after founding and starting Silicon Graphics with a group of my graduate students from Stanford. SGI was a company that primarily made what would now be called a graphics processing unit. The company grew to be quite large, $4 billion a year and ten thousand employees.

Marc Andreessen: Silicon Graphics at that time was what Google is today— the best technology company in the Valley. It was the one that everyone wanted to work at. They were just phenomenal technologists with phenomenal products.

Jim Clark: SGI was a hardware business, and it became a workstation business, and we ended up being the biggest high-end workstation company out there.

Marc Andreessen: It was just a great company.

John Giannandrea: Jim was like this macho kind of hardware guy. He would come into the lab late at night and say, "What the fuck is this!" and "Why is this not going?" That was how he ran things.

Jim Clark: I was sufficiently disruptive in that context, with that board of directors, with that management, that if I had pushed it much further I would have gotten fired like Steve Jobs got fired. And it was my company! I taught it everything it knew about graphics, literally.

John Giannandrea: He just became very frustrated. It was pretty clear that he didn't have control of the strategy of the company.

Jim Clark: I decided instead to call a board meeting and, on the phone, resigned. I told them I was not going to compete, I would not actively recruit from the company, and I was not going to stay on the board.

John Giannandrea: And so then he left. That was a big thing, when Jim left.

Jim Clark: The last day I was there, a fellow came by my office to say good-bye and I said, "I do not know what I am going to do. Every single high-quality engineer I have ever met, I recruited here. I made a commitment to not try to recruit them wherever I go. And I do not have any ideas about what I am going to do." He says, "You should call Marc Andreessen." I said, "Who is that?" He went over, he pulled a web browser down from University of Illinois, opened a search, and typed, "Marc Andreessen." It came up with this page and he says, "There you go. Just read that," and walked away.

Marc Andreessen: Mosaic was very successful and so I had gotten a whole bunch of job offers coming out of school. I had spent my whole life in the Midwest, so I knew I wanted to be on a coast. So it was either East Coast or West Coast. All the interesting job offers in my field came from the West Coast, which I guess is not surprising. And so I came out to Palo Alto and I went to work at a small networking company in Palo Alto called EIT. In fact, I was still getting job offers because in Mosaic there was an area in the Help menu that said "About the author" and it was my résumé. That was automatically generating job offers.

Jim Clark: I clicked around, I found Marc's e-mail, sent him an e-mail right there saying, "This is Jim Clark. You may not know who I am, but I am the founder of Silicon Graphics, and if you know me, you know that the news has come out that I am leaving Silicon Graphics. I would like to start a new company. I would like to know if you would like to get together and join me to talk about that." Ten minutes later, he responded.

Marc Andreessen: I said, "I know who you are."

Jim Clark: We thrashed around literally for months, meeting a couple times a week. My wife would cook dinner and we would drink some nice wine and talk about things. We became pretty well acquainted; he had respect for me and I developed an increasing respect for him.

Jim Barksdale: My wife, Sally, would work on his table manners. He dressed like a kid, and acted like a kid, but he had a great sense of humor. He was so engaged and engaging.

Jim Clark: After about March of '94, somewhere in that time frame, Marc says, "Well, I do not know, but we have to do something if we want to recruit my friends out of the University of Illinois because they are all interviewing for jobs right now." And I said, "Well, what do you want to do? Have you got any ideas?"

Marc Andreessen: Remember this is in early '94. The prevailing view, in the business world and in the world at large, was the internet is not a commercial medium and will never be a commercial medium. It's not for consumers, ordinary people won't use it, and there's no money to be made on it.

John Giannandrea: There was this sense at the time that it was a weird university thing, and that it wasn't serious.

Jim Clark: The world was basically saying, "You are crazy. You cannot make money on the internet, it is free."

Marc Andreessen: We were talking one day and we said, "Well, this internet thing—no one takes it seriously but it's growing vertically…"

Jim Clark: All I had to do was look at the growth of Mosaic for a year and see that it grew to a million users, and I figured there was a network effect of people getting online.

Marc Andreessen: The idea was to basically do a new version of Mosaic but from scratch, and in particular do it as a real product.

Jim Clark: I said, "Okay, let's hop on a plane and go out and recruit your buddies."

Aleks Totić: Then Marc calls: "Hey, we're flying in—I think we're doing it: We're going to do the Mosaic killer."

Jim Clark: The first night we were there, we got together for pizza with the entire group of guys that ended up founding the company. Even Lou Montulli came in from Kansas. All the guys that had built Mosaic were there.

Lou Montulli: Jim got us superexcited. He is a very energetic, charismatic person when he wants to be, and he was giving us the full sell. He changed our perspective from "We just want to keep working on the web" to "We are going to go to Silicon Valley, we are going to change the fabric of the universe, and we are going to basically rule the planet." To have somebody like that believe in us and want to work with us to go change the world is very exciting. We were excited. I was personally very excited.

Aleks Totić: We would have these fantasies. "Let's have a company called "com.com.com" and live on a boat somewhere and it will be awesome!" And then it happens!

Jim Clark: I said, "Great. I will put three million in and I will hire the whole team." I made myself the CEO and printed all of these stock agreements and offer letters, one with each guy's name, and everyone signed them and we were off to the races.

John Giannandrea: Jim Clark was very clever, because he took all of the young people that knew anything about the web—literally flew around in his jet, picking up these people—saying, "Hey, I have a job for you in California!"

Aleks Totić: A week later we're in California driving down 101. We saw Oracle, Sun, SGI, and I was like, "How come no one told me about this place before? This is awesome!" It was like the mecca.

John Giannandrea: And then he paired them with seasoned people from SGI. So the first twenty or twenty-two employees were a mixture of people right out of school who knew the leading-edge thing on what was going on with the web, and then also these seasoned engineers. The SGI DNA was there. And that was the magic that kind of worked.

Jim Clark: I did not have a financial plan, and there was no way I was going to take the time to write a financial plan. I was running on instinct about what the network effect could be. I thought, *If we can get a couple million people using our product pretty quickly, there is going to be money to be made.* The leap of faith that large numbers of people using your product is going to yield a profit does not seem to me like rocket science, but it did then.

Aleks Totić: We didn't know how to make money. Our moneymaking vision was not as strong as our engineering vision.

Jamie Zawinski: "Give it away, and make it up in volume!" Well, no, originally, we had two software products. We had a web browser, and we had a web server.

Lou Montulli: Our original plan was to give the browser away for free, or for near free, and make all of our money on enterprise server software. We were going to give away the razors, and the razor blades were going to be our web server and our commerce server and all the other servers we were going to create.

Netscape's give-away-the-razor-and-sell-the-blades strategy was almost immediately put into jeopardy. A rival team, Apache, was planning to make sure there would always be a free alternative to the server software that Netscape was developing—free razor blades for Netscape's free razors. Apache wasn't even a company. It was a group of part-time developers led by Wired's *webmaster, Brian Behlendorf, who took the hacker ethos—that information should be free—to heart. Behlendorf's motives were as high-minded as they come, but his arrows were aimed squarely at Netscape's business model.*

Brian Behlendorf: There was an e-mail sent to this list of users saying, "Unfortunately we have bad news. The entire NCSA Mosaic development team has been hired by this new company called Netscape. The good news is that now the web will be commercial. Now the web will have official support. Now there will be a company that you can buy a product from, and that will sell you support, and if you have any problems you will be able to call a number. And you will get that all for the low, low price of a couple of thousand dollars per server." It did not quite say that, but that was the implication.

Jim Clark: The instant response was, "You guys are crazy, and we hate you, because you are going to ruin the internet."

Brian Behlendorf: We did not want the web to be owned by anybody, and felt that at least at the layer of web protocols, getting pages to people, we felt like the web server is like the printing press. All of us were running our own printing operations—building interesting websites, building interesting stuff. We just did not want to wake up one morning and find out that we had to start paying a tax to do the things that we had been doing for free before. It was very much an idealism kind of thing. It was inherited from where a lot of internet technology had come from, which is this idea that tech should be public and distributed. It is not necessarily the same point of view as saying, "All software should be free."

Jim Clark: I said, "You do not get it. The only way the internet is going to be is if it is financed by businesses and operated by businesses, because the government cannot continue to put money into it ad nauseam. It has to be a commercial medium." The gnashing of teeth! The wailing! And I got so much hate mail and nasty e-mails that you just would not believe it.

Brian Behlendorf: We were happy to see that these college students were getting real jobs, I guess, but at the same time a little sad, like, *Well, okay. NCSA's server software continues to work for us, and if we continue to fix bugs and add features to it, do we need Netscape?* So I set up an e-mail list, and we started saying, "Should we give it a different name? It is obviously not the NCSA web server anymore. Maybe we should call it something else?" So we called ourselves Apache and figured out how to coordinate our efforts, we being a core team that was still ten or twelve people strong, even on day one, and grew to a couple hundred pretty quickly.

Meanwhile, back at Netscape . . .

Jim Clark: By June of '94, everyone was showing up. There was already a team of eight to ten people, administrators. We were in full swing. Over the course of the summer, I began to worry about running out of money, because we were building up people. I only had maybe $15 million to my name, and I had already put in three. And so I did not have a whole lot more money I could afford to put in it. What we needed was to get the product done.

Jamie Zawinski: When I decided to take the job at Netscape, I thought I'd been working really hard. Turns out, I didn't really know what "working hard" was yet. It was a lot of work.

Aleks Totić: It was great! Since we were all in the office all the time there was no calendar. You met several times a day and basically went over what's going on and where are we going next. I think I was drinking twelve Cokes a day at that time. Or more. You lived on Coke and not too much else. Every once in a while when your body broke down completely, then you would take a day off and you would say, "I'm not coming in today," and you would sleep all day. That would just go on and on and on.

John Giannandrea: General Magic was more intense, but the initial six months or year at Netscape were crazy. Jamie had a particularly hard job.

Jamie Zawinski: At the time, I was living in Berkeley and commuting to Mountain View, so I was only going home every other day. I'd either just stay awake for a day and a half, or while I was waiting for some code to compile, I'd take a nap in my chair.

John Giannandrea: He had an Aeron chair, an extrawide version. And he could sit cross-legged in his armchair, easily, and he would be programming with no shoes on. And he would have a pair of big headphones on playing thrash music of some kind, loud enough that you could hear it through the headphones. And when he was taking a break he would put a blanket over his head and sleep like that with the music going.

Lou Montulli: I have never had a more productive time in my entire life. A great deal of that was I could dedicate all of my energy to it, every waking moment, and not think about anything else because I had no distractions to pull me away at all. We were able to just get stuff done amazingly fast. I wrote about thirty thousand lines of code in three months. That is a ridiculous amount. I would not recommend it. It worked in that very specific circumstance.

Jim Clark: Toward the end of the summer, I wrapped up the financing, and we began to recruit some management to help. Then the University of Illinois began to rattle sabers, claiming we were violating their intellectual property, and implying that we were illicit and illegitimate. People at the University of Illinois had a lot of animosity toward us, because we were absconding with their crown jewels, and going to make money on them.

Lou Montulli: We were essentially doing a new implementation, a reimplementation, of something we already knew very well, Mosaic. Our goal was to create a brand-new web browser from the ground up. But we already knew what it should look like, and we basically knew how it should perform. We had all written the code once already.

Jim Clark: I had spent some energy trying to get everything smoothed out, and I wanted to pay them some kind of a fee just so they would be quiet. They would not take it. Well, they would, but they wanted it in the form of fifty cents a copy, and we were planning to give it away for free. I was not going to pay anyone fifty cents a copy when I was going to allow everyone to download it for free. That is a quick way to go out of business. Meanwhile, they gave Microsoft a paid-up copy for a couple of million dollars.

Aleks Totić: Microsoft gets the license; they have our old Mosaic code. We were worried. *If we had our old code we would be really farther ahead than we are right now. But the old code sucks, so maybe not? The new Netscape code was much better*... So it's like our own code versus our own code.

Jamie Zawinski: We all believed very strongly that shipping fast was really, really important, because there were other people who were trying to do the same thing we were. So we had to be first, and we had to be best, that was our goal. "First" being the higher priority.

Lou Montulli: The last part of the release process is the most stressful, because it is constant back-and-forth with bugs and just trying to get this thing right, and it is the most frustrating time, because bugs sometimes take an indefinite, unknown period of time to work out, and you are like pulling your hair out and trying to find this thing, and then you have the a-ha! moment, and you do not know if the a-ha! moment will be ten minutes from now or a day and a half from now. It is really that kind of *CSI* moment: You have to find this thing within a couple hundred thousand lines of code, which is difficult.

Jamie Zawinski: It's incredibly stressful. The days leading up to it were a lot of sleepless nights. The bug list is still long, but we have to decide if we are going to delay this release because of this bug, or not: If there's a bug, and someone has a fix, we have to sit there and argue about whether it's safe to put that fix in—is there a chance that this fix is going to destabilize other things and make it worse? So there's a lot of waiting and panicking.

Lou Montulli: Jamie and I were very close, then I started to notice this more annoying side. Especially in meetings, he would just be really abrasive about things. He was famous for sending out flaming e-mails: "How dare you do this to me, you are wasting my time, and you are a complete asshole for doing this, and do not ever fucking do it again or else I am just going to get really upset!" So those sorts of messages, and almost to that extent of meanness. I know he is a nice person inside, but he just could not help himself on that sort of stuff. I had gotten several of these, and more often than not, he was directing his flames at me.

Jamie Zawinski: All of us who worked at Netscape in the early days were terrible to each other. We were all really abrasive people. We did most of our negotiation by screaming and insulting. Not necessarily the most healthy environment, but that's how it was.

Lou Montulli: It was three a.m.; I was at the end of one of those double-day binges. Jamie had just done the same thing, and he had sent out a flaming e-mail and went to lunch or dinner or something, wherever you can get a hamburger at two a.m. I was just fed up with it. I virtually never do this, but I just sent back an e-mail that said, "Here is the problem, it was really your fault," and then I was really not nice, and I said something like, "I really feel bad for all the people who have to work with you," it was bad. I was just trying to give it back to him so he could understand how I felt, and this was right at the end of the release process so it had been boiling up.

John Giannandrea: Montulli and Jamie were needling each other. But Jamie was under so much stress, and is a very emotional person, and probably hadn't had enough sleep.

Aleks Totić: Jamie—he was temperamental. He coded well. He was different from us. He was more flamboyant. We were just pretty much corn. Straight corn. He had the funny look, he had an image, he had a sense of style which was fairly foreign to us.

Jim Clark: He had half of his head shaved. Some stylistic statement on his part. I completely ignored it. It did not matter to me; I did not give a hoot. He was a great programmer, brilliant young guy. I do not think anyone bothered. People have just got to realize that a computer geek is respected on the basis of how much code he can write and the quality of code he can write. People do not give a crap what he looks like. If you generate good code quickly, no one cares. Jamie made everyone a lot of money.

Jamie Zawinski: We were there a lot, tempers ran high. Like I said, we were not friendly people to each other.

Lou Montulli: So it was boiling up, so I sent this e-mail and I went over to the cube next to me, to my friend Garrett Blyth, and I said, "Read it, read it," and he reads it and he is giggling, and we hear Jamie walk in, and we are listening, and we hear him come in, log in to his computer, and then he starts typing really loudly, like really loud and vigorous, vigorous typing. Garrett and I know exactly what's happening. *Oh, this is going to be great! It's going to be amazing flame coming back. I just cannot wait to read this.* And we tried to take it in a good spirit, and right in the middle of the furious typing, we hear this sound. This very distinctive sound of a computer booting up. Or rebooting, in this particular case, for no particular reason at all. He is typing furiously and then his computer, for no reason at all, just reboots, which means everything that he has worked on just disappeared. So you hear the ding, you hear this slight pause, and then you hear, "What? What the fuck?!" and then he had this wall, this pyramid of Coke cans that was at least twenty layers deep, that was built up on this table next to his desk and it went up above the cube wall, and you hear this crash as all the Coke cans come off the desk in a sweep, and then this really loud sound to our left, and then Jamie storms off.

Jamie Zawinski: Lou was taunting me about something, and I lost it. I kicked a chair, and I walked out.

Lou Montulli: Garrett and I get up and we turn the corner, and there is Eric Bina with a chair like literally jutting out of the wall right next to his head. Apparently what happened is, Jamie got up, picked up his chair, and threw it over his head out of his cube. Eric had come around, because he heard the Coke cans, and the chair went right by his head and embedded itself into the wall. It was quite a scene.

John Giannandrea: He threw a chair and we sent him home.

Jim Barksdale: It was a big deal. We even had T-shirts printed up that showed the chair in the wall!

Jamie Zawinski: We made a T-shirt that had a coat of arms, and there was a chair in the middle of it. Yeah, because we were jerks and we taunted each other that way. That's it. That's the whole fucking story.

Lou Montulli: After that incident, he essentially went into his cube behind this thing—this large camo thing which you use to cover a tank or a plane—that was hung up over his cube, and he never talked to any of us ever again. It was kind of the end of the social world, at least between me and Jamie and the core team. It was a highly charged event; I do not fault him for any of it. I certainly provoked it as much as he did. It was right at the end of our release process—the beta release, the very first beta release. We had been working four or five or six months, seven days a week, 120 hours a week, and all of our core motivation was to get it out into people's hands because if nobody is using it, it is essentially mental masturbation, right? The end result has to be this thing that people can use.

The first Netscape browser was released into the wild—as a "beta" or experimental version—in October 1994.

Lou Montulli: So we are very very excited, and we all got in a conference room to celebrate the release. We were going to watch as people downloaded it live, and we kind of figured out how to do that.

Jamie Zawinski: There's a log file showing each download, and one of the guys rigged up a little script to parse that log file as it was scrolling by.

John Giannandrea: I don't know why, I just did it, and people thought it was kind of cool.

Aleks Totić: We had a different sound effect for different downloads—each different browser got its own sound bite. There was a frog, a cannon, and shattered glass. We're in a darkened room, kind of watching the monitor log, and there's nothing on the log. Then the first download came from Australia. It took him a long while. He got half of it then he had to restart. And then it finally went: Boom! The first download. Cheers, champagne, beer, and all of it. A few minutes later the next one comes, and then a few minutes later the next one and it gets faster and faster.

John Giannandrea: It started off as a trickle, and then it became a flood, and eventually it's just deafening.

Aleks Totić: Six hours later we're still there and it's like boom, crash, boom, crash! It's an avalanche, and it's never slowed down since.

Lou Montulli: I don't want to toot my own horn, but it was really good software. We were very proud of it. In the modern experience, going from any other browser to Netscape would be the equivalent of going from your flip phone to an iPhone kind of experience, like, *Wow, this finally works the way it is supposed to work!* We were ten times faster, we had a lot more features, it was a lot more polished. Images became a lot more practical on a web page. Before Netscape, you would wait minutes before you saw anything. We changed that completely. With Netscape images would appear ten times faster, because the content would all stream in dynamically.

Steven Johnson: So what had just been text-based pages suddenly had images. This was the point which the web stopped just feeling like an interesting new format for data, and really started to look like media. You could start to imagine that this was something that could become a magazine, that you could put advertising on, or maybe the kind of intimate media of your own personal pictures. I remember looking at Netscape and thinking, *Oh, I can learn this and publish my own magazine—with basically no distribution costs!* That was just mind-blowing.

Lou Montulli: The next step was to introduce those elements which allow for dynamically driven web content, allow for video, audio, a full programming language as a plug-in within the browser. Netscape started to lay the groundwork for the dynamic web that we know of today: We started to treat the web more like an application rather than a series of pages.

Marc Andreessen: Think of each website as an application, and every single click, every single interaction with that site, is an opportunity to be on the very latest version of that application.

Lou Montulli: The whole metaphor of moving through the web page to page is now disrupted, and you have a complete application written in HTML on the one page of your browser, which is really important. This allows you to deliver truly immersive apps in a web browser which changes, fundamentally, how software is delivered.

Steven Johnson: The internet was an asterisk that evolved into this browser-driven second platform. It kept Microsoft from being the dominant quasi-monopolistic force that we all thought it would become.

Lou Montulli: You can see very clearly, there is a correlation between the decline of Microsoft's power and the increase of the open web. We knew that would happen.

Steven Johnson: And because no one owned the open web, it unlocked this period of unbelievable creativity and innovation. It was not just variations on a theme. Whole new categories of software were created: user-created and -curated encyclopedias, social media, and a million other ideas. Suddenly a couple sitting in the proverbial garage could come up with something and change everything again—like Larry and Sergey starting Google or Biz and Jack starting Twitter. Or even Mark Zuckerberg and Facebook.

John Doerr: The greatest legal creation of wealth in the history of the planet!

Steven Johnson: It all happened remarkably quickly: All the exciting things and all the vast new fortunes began to be created on the open web.

Bob Taylor: The world wide web opened up the internet like nothing before, and the browser made all of that a lot easier.

The Netscape web browser was an unprecedented success—as was the Apache web server. And yet the two were like matter and antimatter: by all rights unable to exist in the same economic universe. Apache was giving away the server software that Netscape had planned to sell in order to keep the browser free. Paradoxically, it all worked out. The solution? An inspired bit of what Silicon Valley likes to call "social engineering."

John Giannandrea: It was a somewhat sneaky idea.

Lou Montulli: We had put a price tag on the browser but with kind of a wink-wink. We were really saying, "Well, if anyone wants to download it, just download it." It was freely available.

Marc Andreessen: There was a clause in the license that said if you use it for business, you have to pay for it.

John Giannandrea: The core engineers—the Jamies of the world—were a little bit ambivalent. They said, "This is not okay, we are going back on our word." Because Marc had said things like, "It would always be free." And now we were parsing that a little bit.

Lou Montulli: But what it meant was our salespeople could call up virtually any business in the world and say, "We noticed you are using our software, would you like to pay for it?" and they would say, "Oh yeah, sure sure." It was not very expensive; it was twenty dollars or something. It had a more expensive price but the discounting made it very cheap, as cheap as a dollar per person. People were like, "Yeah, absolutely, it is useful software, we love it! We will give you money." And our revenues just ballooned. It was ridiculous.

Marc Andreessen: The browser was intended to be a loss leader, but revenue shot up like a rocket because that split license worked! I mean, we never expected browser revenue to get that high. We always thought that would be free.

Jim Clark: Our revenue started going up exponentially and I started seeing that we were going to be profitable that first year. I thought, *Well, this is wild. We should just take this thing public!*

Jim Barksdale: We decided to go in in May. We hired the bankers. We did the necessary filings. We went on the road show. You could sense the momentum building as we went from city to city. The word was getting hot. The newspapers were covering it a lot. The *Wall Street Journal* was talking about it, and everybody else, wondering what this new IPO was going to do. We were a brand-new company, less than a year old. Back then, the conventional wisdom was that you had to have at least three years of solid financial data in order to go public. We didn't. We had a couple of quarters and some projections, which I thought were conservative. And they proved to be. It was getting exciting. In London, which was our last stop, we got to the end of it. "Mr. Barksdale, what is going to happen when Microsoft just bundles the browser into their product?" I said, "Well, sir, there are two ways to make money in this world. Bundling and unbundling. We have to catch a flight. Maybe another day, I'll go through that with you." We laughed at that one.

Aleks Totić: We had no idea what we were doing. All of this stuff was improv. The long working hours, the releasing of it on the internet, not having a business model to start off with, the IPO, all of this stuff, we were doing it because we didn't know any better.

Jim Clark: I remember the day we went public. I was worth $663 million at the close of market that day. So I put a tail number on my airplane of 663 Mike November. And I began to build that boat, *Hyperion*.

Jim Barksdale: Jim was interested in boats. He had this idea of building this computerized boat, the *Hyperion*. It was single-masted, the tallest mast in the world.

Jim Clark: It *had* the tallest mast in the world—until John Williams, this real estate guy in Atlanta, decided he was going to build a bigger one. He made his mast one meter taller. *Hyperion* had eight crew and three bedrooms. It was meant as a jewel box: It's a real work of art inside. It is 150 feet long, which sounds big, but it is not that big on a sailboat. To get to the size where you have another deck above on a sailboat, you are getting into the three-hundred-foot range. That is what I have now—a three-hundred-foot three-masted schooner, *Athena*.

Lou Montulli: It never even occurred to me that people could make this much money. I made probably $20 or $30 million. Something like that. But it did seem absurd to me and, in some ways, I felt a little numb. *This cannot be real; this money cannot actually be there.*

Jamie Zawinski: It was just a joke. Actually, my favorite joke from IPO day—it turns out Jerry Garcia died on the day of the Netscape IPO. So the very first joke I heard when I walked in that day was, "What were Jerry Garcia's last words?"

Jerry Garcia: Netscape opened at what?!

A Fish, a Barrel, and a Gun

Suck *perfects the art of snark*

*I*n a manifesto that accompanied the first issue of Wired, editor in chief
Louis Rossetto predicted a digital revolution that would whip "through our
lives like a Bengali Typhoon." Remarkably, the deluge showed up just as Rossetto
promised. Nineteen ninety-five was the year that toys started to talk; it was
the year that the web awoke. From his vantage point, Rossetto could clearly see
that the new medium was going to create a new crop of media moguls. And if
he plotted his course right, he could join them. It didn't happen. Instead, Rossetto was upstaged by two mutinous employees: Carl Steadman and Joey Anuff.
Disgusted with Wired and especially HotWired, Rossetto's attempt at an
online media empire, the two twenty-somethings secretly decided to launch a
"HotWired killer"—while still working for Rossetto. Suck was cynical, anonymous, gleefully Marxian: the anti-Wired. Yet it was produced in Wired's
offices and served out on Wired's bandwidth—and it established a format, set
a pace, and adopted a tone that came to define the world wide web.

Kevin Kelly: If *Wired* had been done in like 1989, it wouldn't have made it. It
would have been too early. *Wired* came just right at the dawn of the web,
and everybody kind of woke up and said, "Oh my gosh, now I get it! I need
this magazine to understand what's happening."

Louis Rossetto: We had no reference to the world wide web in the first issue
of *Wired*.

Kevin Kelly: We were accused of being completely clueless about the internet, because we had no mention of the internet in the first issue. But I was
so tired of talking about The Well and the internet that I didn't want to
talk about it in *Wired*. We actually didn't want it to feel like it was about

the internet. We wanted it to feel like it was about this broader culture, and so we avoided it. Maybe that omission seemed like an oversight, but it was on purpose.

R. U. Sirius: I remember when *Wired* started their conference on The Well. People didn't talk so much about how the internet was changing the world. They were already imagining virtual reality right around the corner and stuff like that.

Brian Behlendorf: At *Wired* in the summer of '93 I was setting up a website, but I was also fixing bugs, I was adding features, and sending them upstream over this e-mail list to the kids at the University of Illinois building, this along with everyone else, the other users, and we were trading these fixes like baseball cards. This was kind of like the Brownian motion of how technology improvement took place in the early days of the web.

Louis Rossetto: It was only in the second issue that we had a small news item in the front talking about Tim Berners-Lee in Geneva. That was the first mention of the web in our magazine. But not the last. Then it became something we were paying attention to.

Howard Rheingold: And so Louis thought, *Okay—let's create a web-based cultural publication that we'll make money on.* One of Louis's admirable traits was that he thought really, really, really big.

Louis Rossetto: The magazine people were having a ball. The advertising people were selling ads hand over fist. So the *HotWired* thing was a complete flyer. We had a meeting where we sat down and said, "This is what we're doing: We will create the first website that has original content and Fortune 500 advertising."

Brian Behlendorf: I was impressed with that, but none of them had deep experience, or any experience, really, with the internet.

Louis Rossetto: So let's find out what it is! The mandate for *HotWired* was "What is original content made for this medium supposed to be like?" We set a launch date for the fall of '94.

Jonathan Steuer: I was the nerd getting a virtual reality PhD at Stanford. What pulled me over to *Wired* to start *HotWired* was the data piece. To get interesting data about how people use media and new technologies, you have to pay college sophomores to do weird stuff for you. But on the internet people would leave logs.

Brian Behlendorf: I knew enough about the protocols and the technology to understand that you could start to use this new technology that a couple of us had proposed called a cookie.

Lou Montulli: The original design of cookies was specifically designed to be a tracking-free methodology, which is pretty ironic given how it is used today. The problem came about when cookies enabled tracking essentially by combining multiple technologies together to create a tracking technology.

Howard Rheingold: So one of the things that we did do at *HotWired* was to use forms, which had just become available, and cookies, which had just become available, to enable people to have forum-like conversations online.

Louis Rossetto: We invented web media. For the first six months we had 50 percent of the traffic on the web—some ridiculous number.

Brian Behlendorf: We wanted to cultivate an online community where people actually had identities and talked to each other. Everything was the beginning point of a conversation, almost more of a BBS than a newspaper.

Louis Rossetto: We separated it from *Wired* because we didn't want that idea that we're just repurposing content from *Wired*. We wanted these people to feel like they were doing something unique.

Brian Behlendorf: As we were putting together this *HotWired* division, Jonathan pulled together a lot of people he really respected from the digital world and from the BBS world, Howard Rheingold being one of those, Justin Hall being one of these. All of them were, like me, freaks.

John Battelle: Weird furry hacker freaks pushing the boundaries of the internet.

Justin Hall: In December 1993 there were like six hundred sites on the internet. They were all like, "I'm at this university and I study advanced mathematical modeling and here's a picture of my cat." I thought, *Boy, this can't be expensive or difficult. This guy's just put up a page about his cat!* So I figured out how to put up a home page and launched it in January of 1994. And I introduced myself like I'd seen the sort of professor-y people do it. I said, "Hi, I'm Justin. I'm going to college. I love rock 'n' roll music and LSD..." I started just writing and using the hypertext structure of the web to tell the story of my life.

Jane Metcalfe: Justin was a bona fide internet personality.

Joey Anuff: Justin's links.net was a bigger hit probably than anything *Hot-Wired* ever published.

Howard Rheingold: Julie Petersen was the one who found Justin. She said, "Hey, I know you wanted to have a nineteen-year-old around. There's this guy at Swarthmore whose site is getting more traffic than *Wired* does!"

Justin Hall: I went to visit the *Wired* magazine offices at 2 South Park and it was awesome. You know, just people with tattoos and long hair and ear-plugs. Like plugs in their ears, not like for sound but for aesthetics. I had been to Grateful Dead shows and I had been to rock concerts and I had been to all kinds of stuff, but walking into the *Wired* office you had the sense that all these different subcultures were converging.

Brian Behlendorf: There were other people that we hired eventually who I knew through my connections on the SF Raves mailing list that I administered. The list was a gathering point that a lot of young creative people seemed to be attracted to.

Jane Metcalfe: I called them the happy shiny people, and they attracted more happy shiny people. And they were good-looking and fun and totally on board with whatever needed to happen. They knew how to manage parties. They knew how to manage the list and the front door and all the rest of that thing.

Justin Hall: I met up with the online team for a while and we chatted. I can't really remember much of what we chatted about except that at the very end one of them said to me, "If you're going to be the intern, you're going to have to smoke the swag weed." That's when I realized, *Oh my God. I think it's going to be okay. I'm going to fit in with these people. These are my people. I get it!*

Howard Rheingold: When Justin joined, we went downstairs. There was an interesting guy who had a little magazine called *Might*, Dave Eggers.

Jack Boulware: South Park was a publishing hub. You might go to the burrito place for lunch and *Oh, there's the guys from* Might *magazine having a Gen-X rap session about something*, and then the *Wired* people would start floating in.

Howard Rheingold: He had this little magazine, this little storefront down there. So we smoked a joint down there and went back into the office and all eyes turned to us, because we had smoked it at the base of the elevator shaft and it just sucked all the smoke into the office.

Justin Hall: The online people were in a space called the grotto, in the back of 2 South Park. It was called the grotto because it was hot and it was far from the door and it was packed. I loved it, of course, because I was nineteen. *I'm elbow to elbow with people who love the internet!*

Amy Critchett: Justin ended up sleeping in the grotto for a summer. I gave him as a present once to a friend of mine for her birthday. A big bow, you know. So yeah, there was definitely some partying going on.

John Battelle: Justin was in the *HotWired* team, but whenever I interacted with him, I was struck by two things: that he was very smart, and that he was like ten years old. He was so thin and slight.

Louis Rossetto: Justin was cool. He was this young guy who was full of ideas and energy, and I was happy to have him around. He had his own thing going, and he brought a lot of insight into the medium and to help launch *HotWired*. We ultimately had a disagreement about a couple things.

Justin Hall: I started arguing with Louis Rossetto about how the web should take shape, and whether people should enjoy the web or whether the web was more exciting as a place where brands and curators should have the loudest voices. I was so worked up about the idea that individual publishing would be the ticket to happiness and a truly new medium. Louis was already feeling like the web was too unpolished for it to be popular, and that the web needed a firm hand to take shape as a truly great space for publications.

Louis Rossetto: Justin had just a sort of unwillingness to understand that the site needed to make money. For him to have a job the site needed to raise funds and you need an advertiser for that. There was some kind of a disconnect between the business reality of it and what his ideals were about what the site should be.

Gary Wolf: Howard and Justin, they had a vision of a future of the web in which anybody could say anything to anybody else. And in which platforms for enabling communities would be the new valuable form of media. And Louis had a vision of the web in which the web would provide a mechanism by which popular voices could reach even larger audiences due to the erasure of distribution costs.

Justin Hall: Louis was very different than the people in the online department. It was fun to be able to sort of go hang out with the people that were all sort of psychedelic and communitarian and then go to the other side of the office and argue with the guy who was—I'm trying to summon my best Louis impression here—like, "We're on a rocket ship to design expertise and content curation!"

Howard Rheingold: Louis absolutely wanted to be Hugh Hefner, Jann Wenner, Helen Gurley Brown, the dictator of what culture is. Not, as he put it, "the bozo filter for the web." It came down to lots of everyday decisions: Can you see any of the content before you sign up and give us your e-mail address? Or is there just a banner ad and you don't see anything more until you sign up?

Louis Rossetto: We *invented* the banner ad. The stupidest thing that we did was not patent that. We would have been centimillionaires for having done that.

Justin Hall: Anyway, we had some philosophical differences.

Brian Behlendorf: And it became personal for a lot of people.

Gary Wolf: And a kind of war broke out between him and his staff.

Howard Rheingold: So I quit. And Jonathan Steuer quit. And a few other people did as well.

Justin Hall: I said, "Well, I disagree, but whatever. I'm leaving in a month to go back to school."

Louis Rossetto: One thing about media is that it is software that ships on time, and so at a certain moment I just got completely frustrated with them. We were burning money and no website is happening, and so I ended up replacing them. I just kicked them out.

Gary Wolf: When I first walked in there, I felt this sense of a place where there already was some kind of hysterical traumatic history. All the answers that you got when you asked a question were inflected with this kind of emotional sense of betrayal. And that's because, I gathered later, that there had already been some kind of vision for what an online voice of *Wired* should be. And it was related probably somehow to both the utopian language that was accumulating around the early publicly accessible internet and to Louis's really over-the-top revolutionary proclamations.

Justin Hall: You know—the "Bengali Typhoon"? His ability to spool out hyperbole as a way to describe the oncoming rush of digital technology was historic.

Louis Rossetto: If you read the Bengali Typhoon editorial, it's completely spot-on. It's probably not even a big enough metaphor. It's probably a comet hitting the planet.

Gary Wolf: So you already had kind of an atmosphere of factionalism. And this is where the story starts, really. The story starts with attempting to create and launch the first commercial web publication. How would it support itself? Who would it be for? What would it look like? And that's really where I came in. Carl came in as a kind of webmaster. Production, I think we called him.

Carl Steadman: I end up getting a job at *HotWired*. They put me on the mailing list, and here I see this epic battle between employees on one side, and someone called Louis Rossetto, about T-shirts. Like who should get a free T-shirt? If each employee deserves a free T-shirt! I could not believe that there was no sense of organizational structure. I couldn't believe that there were all these little brats!

Gary Wolf: In this kind of crazy, hysterical microclimate, people who had six weeks of experience on the web comported themselves like captains of industry. You know like, "Okay, I can make a paragraph break on the web, therefore I am king." And in a way it was true, because if you couldn't do that you were sitting there saying, like, "Shoot, how do I get the paragraph to break? Can anybody help me?" Keep in mind, you can't just look it up on Google.

Carl Steadman: There were hordes of young people who had found their first employer, who had no notion of how lucky they were, and were completely undisciplined, and couldn't get anything done. And I make no friends because things have to change. We have to get things done.

Joey Anuff: The thing about *HotWired* is, there is the original *HotWired* and there is late-stage *HotWired*, and I was late-stage *HotWired*. Because the original *HotWired*, as far as I can understand it, were a group of people that sat around, smoked a lot of weed, and never did a damn thing.

Carl Steadman: I think *HotWired* before I got there took over a year to launch, and it was just ridiculous how nothing happened at that place.

Gary Wolf: In the two years after I got to *HotWired* it went from about a dozen people to about two hundred people.

Jack Boulware: *HotWired* was a massive website, it was far larger than anything else that had launched in terms of an online magazine publishing presence. They had so much money and it was online only, and they were just scooping up people from these zines right and left to contribute to it.

Gary Wolf: Louis's goal in expanding so fast was to do clearly branded media voices across all the major categories that would soon become important on the web. He wanted entertainment and the arts. He wanted tech. He wanted political coverage.

Louis Rossetto: So we set out to basically plant flags in a bunch of different areas, as many different areas as we possibly could. So we did travel sites. We did extreme sports sites. We did political sites. We did how-to sites.

Jane Metcalfe: How to mix cocktails!

Gary Wolf: There was an art section, a music section...

Louis Rossetto: The Beta Lounge was the first streaming music site.

Jack Boulware: They were doing book reviews, which seems sort of ironic. I did a bunch of stories for them. They sent me to the porn Academy Awards in Vegas.

Gary Wolf: You never knew what the new assignment was going to be. Literally any idea that somebody had might catch fire, and the next thing you know we were publishing that stuff. A disciplined organization we were not.

Louis Rossetto: The *HotWired* site just grew like crabgrass. Every time you turned around and there were another twenty people over there. A large part of the crew that's being assembled is coders. Literally people writing code for all the things that needed to be done: the editorial systems, the ad management systems, the billing systems. None of them existed, and there was no place to go for them so we had to write them ourselves.

Brian Behlendorf: Before we launched, I had a direct line in with Louis and everyone else building the thing, because I was the one kind of taking their ideas and six hours later having a running prototype. That was a very rapid, highly iterative kind of thing, and that was a key part of why that effort felt creative. After the launch, after there was a lot of attention and a lot of traffic, and suddenly advertisers and such, and understandably it skewed to be more of a business-driven kind of operation. But I think too quickly they moved into a mode that dropped off a lot of the engineers from the decision making and the C suite. I did not stick around and fight for engineering to be better represented at the top.

Carl Steadman: So I'm at *HotWired*. I ended up running engineering when engineering left. But keep in mind that I don't just have a technical background, I also have an editorial background, and I've got a design background as well. There's no clear reason why I am locked out of meetings which could determine the direction of editorial. Certainly editorial needed me. Certainly design needed me.

Gary Wolf: Carl had a combination of skills that almost nobody else had. He was very culturally savvy, he was a big reader, and he also knew his way around a web server. So he could do experiments in web publishing that were inaccessible to anybody else in the room.

Carl Steadman: So my frustration with *HotWired* is building, and my frustration with the magazine as well. Newt Gingrich ends up on the cover. They end up putting Zippies on the cover.

Joey Anuff: Burning Man was sort of the respectable face of what at one point was called the Zippies.

Carl Steadman: *Wired* for "getting it" really didn't really have much of a clue when it came to the web.

Gary Wolf: One of the things that we didn't have, for instance, was any way to easily look at traffic on our website. There were no reports. There was not even a way to kind of fake the reports. You could access the data directly from the server, but then you had to have access to the server.

Carl Steadman: *HotWired* was the biggest website on the web, based on my analysis of the traffic numbers. But we're going down the wrong path, and I'm not fulfilling my original ambition of changing the course of the development of this new medium. So I determine that I would have to launch something. I would have to launch what I called at the time a *HotWired* killer.

Joey Anuff: *Wired* had these occasional Friday lunch sessions where six people from across the different departments would go out and just have lunch and talk about whatever with one another. And I ended up on one of these with Kevin Kelly and John Plunkett. And I brought up the idea of the web needing its own version of *Mad* magazine. And I got a really thoughtful look, like, "Hmm, that could be interesting," from Kevin Kelly. In my life I would never say that I would think that Kevin's seal of approval is a great sign of either commercial, artistic, or even intellectual success. But at that moment I thought he was of peerless taste and insight: *A goddamn observational genius with his ear to the ground!* And I went back and when I told Carl the idea he liked it, too. And that became the logline of the *Suck* launch. It was like, "Okay, let's try to do something that's somehow vaguely inspired by *Mad* magazine." Or at least took itself as serious as the *Mad* magazine editorial voice took itself. In the gang-of-idiots sense.

Gary Wolf: *HotWired* was traditional journalism. *Suck* came along and had the good sense and recklessness to shake its hands free from all that pretension and the courage to state what was starting to become obvious: that the internet was going to be about novelty, sarcasm, and microcelebrity.

Joey Anuff: A big part of the premise of *Suck* was "What if somebody published something new every day?" Nobody was doing that.

Kevin Kelly: Every day is essentially what a blog is about. Because at the time we had home pages. "Build your own home page!" That was the refrain. But these home pages would go up and then they would never change. They were very static. And so the idea of changing things was seen as a big thing.

Joey Anuff: I thought, *Oh, we can do that and that would be really hard-core.* But that meant that since I was pretty much writing it all, I had to write it every day, and I never got to even midnight with knowing what I was going to write about the next day.

Carl Steadman: So I go and get a bunk bed and just throw it up in the damn office so at least we've got this place to sleep.

Joey Anuff: So I would go to sleep there, because I would only finally in desperation write six hundred words.

The Duke of URL (writing in *Suck*): Shit makes great fertilizer, but it takes a farmer to turn it into a meal. With that thought in mind, we present *Suck*, an experiment in provocation, mordant deconstructionism, and buzz-saw journalism. Cathode-addled net-surfers flock to shallow waters—*Suck* is the dirty syringe@hidden in the sand.

Joey Anuff: And Carl would have already gone to sleep, already set up the servers, and set everything, be ready to publish when I would wake him up at like four a.m. And he would be up editing it to make some presentable, shareable edit out of it so we could get it up on the web by like nine a.m. or eight a.m. in the morning.

The Duke of URL (writing in *Suck*): *Suck* is more than a media prank. Much more. At *Suck*, we abide by the principle which dictates that somebody will always position himself or herself to systematically harvest anything of value in this world for the sake of money, power, and/or ego-fulfillment. We aim to be that somebody.

Joey Anuff: So that's why we were there overnight.

Carl Steadman: There was the little downside where I'd get up in the morning and everyone else was already there, and I'd be charging around in my pajamas. But *Wired* had the best showers in the world, because you would turn on the hot water and it would run forever, it would never run out of hot water.

Kevin Kelly: One person, two people having something to say every day: In that sense *Suck* was the beginning of the blog. It proved that a single interesting person or two could have an audience, and you don't need a whole team of people. And that would change the economics of publishing, because it wouldn't take that much advertising or whatever to support one or two people.

Gary Wolf: Carl had a very distinctive, extremely dark and ironic voice. Everything that he said was exactly what he thought, so heavily inflected with irony that it was nullified.

Webster (writing in *Suck*): The Web is the first of the new "interactive media" to give us so little for so much: it demands our attention and forces us to interact, but only as so many mindless automatons basking in the phosphorescent glow of our "terminals," developing RSI and serving as cancer hosts.

Gary Wolf: Joey was an extrovert and extremely funny and lightning fast with a gigantic range, especially in popular culture and scatological terms of abuse. Both of which came in handy.

The Duke of URL (writing in *Suck*): There's something exciting about the breaking of news on the Web that can make an otherwise bullshit-quality story smell sweeter than Glad Potpourri-in-a-Spray.

Kevin Kelly: The supersnarky—that's the style. A thousand sites do that now, but at the time, it was sort of naked and unabashed. No apologies: There it is. It was sort of like seeing somebody in their birthday suit. It was novel and it was effective in that sense. But their style of humor was not my style. It was very satirical and insider-y and immature. So none of that was working for me at all. I had really zero interest in the content of what it was doing.

Jack Boulware: The early digital publications were all pretty earnest. They came out of this *Whole Earth* futurist-Utopian kind of vibe. The Well certainly reflected that...

Steven Johnson: I had founded *Feed* in New York with some friends in May of '95. We were the first digital-only magazine. We were coming out of this academic- and pop-culturally aware, somewhat left-y zine culture. And then *Suck* appeared.

Jack Boulware: Doing snark.

Steven Johnson: I just remember thinking, *Oh*, that's *how to do it!*

John Battelle: *Suck* was the best writing: The voice was unique, it was honest, and it was dead-on about the culture, and it was critical and snarky and I was just like, *Fuck, we need this in the magazine!* But it didn't really translate to a magazine. It really was native to the web.

Steven Johnson: They were using the hyperlink as a form of shorthand humor, so the joke would only become fully apparent to you if you actually clicked the link. So they would say "unlike those East Coast idiots" and "idiots" would be a link to something I had written. The use of hypertext as humor I don't think anyone had done before—it was just a whole new way of doing commentary. Then the other thing that they were doing was making fun of too-serious technology people. *Wired* hadn't done that. *Wired* took that world very seriously. And *HotWired* was a reasonable extension of the print brand, but it just wasn't as interesting as *Suck*. We got this voice that we hadn't heard before, *Oh, it was now possible to make fun of the sacred cows of Silicon Valley!* It was very liberating.

Gary Wolf: So what they decided to do was just take aim at the readership that was close at hand: That's the people who were working on the web. So they take aim at Netscape and Marc Andreessen.

Webster (writing in *Suck*): If it's all too seldom that people know when their fifteen minutes are up, it's rarer still when these same expired starlings have a clue as to what meager gimmickry (quote unquote talent) thrust them into the limelight. And if ever there were a one-trick pony on the Web, it's Marc Andreessen…

Joey Anuff: After Netscape popped and it became possible to become a millionaire off of the internet, or a "Mozillionaire" as we dubbed it, that became sort of a big target.

The Duke of URL (writing in *Suck*): The key to understanding Netscape, it seems, is to read through their PR sophistry about "application platforms" and label their product as what it really wants to be: an Operating System. Y'know, like Windows 95, except cross-platform. It's only fitting that Netscape would strive to emulate the most conspicuously shady tactics of the Microsoft juggernaut.

Jamie Zawinski: I loved *Suck*. Those guys were awesome. They were hilarious. And I sent them an e-mail saying, "Oh my God, this is amazing." And then they sent me a T-shirt. That's pretty funny.

Gary Wolf: And they could see if the shit that they said was being read based on looking at the server logs. They can see where these incoming pageviews are linking from—the traffic sources. Now you have this crazy feedback loop that is incredibly addictive. In which the effects of what they were doing were visible to them within twenty-four hours. This is like making crank phone calls on a mass scale.

Carl Steadman: I just really fell in love with this genre. *Suck* would both participate in the culture of celebration of these internet millionaires, but in the process ridicule them.

Gary Wolf: In short order everybody was reading *Suck* every day. *Suck* did more traffic than the entire rest of the *HotWired* site combined. And nobody knew that it was them!

Louis Rossetto: So I didn't even know who Carl and Joey were! They got sucked up in that wave of needing bodies to do the back-end stuff that needed to get done. On the *Wired* side of the office there were seventy or eighty people. On the *HotWired* side there were 110 people in the same amount of space. And they were literally crammed shoulder to shoulder. Stuff is all over the place and wires are dripping off the ceiling. But it just seemed like, *Okay, that's what's going on here. It's just crazy.* And then people start saying, "Have you seen *Suck*?"

The Sucksters (writing in *Suck*): As you rub your eyes and pinch yourself to verify the irreality you see, the thought dawns on you: with all that luscious lucre slithering across palms and winding its way into Swiss bank accounts, what if there were opportunities for *someone like you* to play in the same economic sandbox as the new media Snuffleupagi? The trick is to put down your crusty bong for ten minutes and draw up some sort of plan. The good news is that once your outline's been hashed out, you may actually be able to sell *that*—no further work required.

Carl Steadman: The *Suck* theme was always "sold out," which is a wonderful pun.

Gary Wolf: If you read the portrait of the web industry, the early web industry, through *Suck*, what you see are extremely immature, semidesperate, profoundly overconfident morons on the make setting themselves up as industrial titans. Right?

The Sucksters (writing in *Suck*): For now, consider the basics: how long do you want to commit for? What's your exit strategy? If your motives are purely financial, how many figures? Who are the ultimate victims of your swindle, besides yourself? Microsoft? Pathfinder? *Suck*?

Gary Wolf: And what is that portrait? Where is that portrait coming from? There's a lot of easy targets all around them. And it's also them. And this is where the beauty really of *Suck* is.

Louis Rossetto: They basically discovered what the web was about.

The Sucksters (writing in *Suck*): The magic formula for cheap content? It's called point-of-view, baby.

Louis Rossetto: *Suck* was the first blog. It was like a phase change.

The Sucksters (writing in *Suck*): Opinions are like assholes, sure: just make sure yours smells sweeter.

Howard Rheingold: The beginnings of internet meme culture came from *Suck*.

Louis Rossetto: And we were amazed that this was happening under our own noses—and we didn't even know it!

Culture Hacking

The cyberunderground goes mainstream

*T*he decade between the introduction of the Macintosh in 1984 to the twin IPOs of Netscape and Pixar in 1995 was a lean one when measured by Silicon Valley's now-favorite metric—business success. Yet when seen through a cultural lens, the decade is profoundly rich and kaleidoscopically colorful. The Well kicked it off by emulsifying what had been an uneven and partial mix of Silicon Valley's technical culture and the San Francisco counterculture. Then the underground magazine Mondo 2000 gave the new scene a voice and a name: the cyberculture. Wild "cyber-" parties, salons, and happenings held the scene together until the midnineties, when the web and Wired took it mainstream. What was once a subculture specific to Silicon Valley and San Francisco became the basis for a new global internet culture.

Fred Davis: *Mondo*'s circulation list was relatively small, but if you looked at it, there were a lot of America's great thinkers, writers, painters, and designers.

R. U. Sirius: Timothy Leary was an enthusiast for what we were doing right from the beginning and we were enthusiasts for him.

Fred Davis: Alison Kennedy was basically using up her trust fund to subsidize the magazine as a vanity publication, I guess you'd say.

R. U. Sirius: And at the time Jaron Lanier was courting someone who'd become his wife, and she lived with Alison. So we learned about his virtual reality project pretty early, before a lot of other people had heard about it.

Michael Naimark: You can't have something called virtual reality without it being a circus. At the time it seemed totally obvious that Tim was going to join the circus—and he slipped in like a hand to a glove.

Louis Rossetto: He was in his cybervisionary stage.

Dan Kottke: Tim was promoting anything and everything related to new technology. Remember S.M.I².L.E? Space Migration, Intelligence Increase, Life Extension?

Jaron Lanier: If you talk to Tim about it, and he's no longer with us, he said that actually this plan was set in motion years earlier by William Burroughs, who told him that the computer people would eventually remake the world. And it was really important to connect with them as soon as it started to happen. And so Tim felt that he had sort of marching orders from Burroughs from way back.

Fred Davis: William Burroughs's dad was the Burroughs of Burroughs Computer. That was a big company back then.

R. U. Sirius: Leary was very much a futurist—him and William Burroughs, who was always a sort of futurist. He was the only futurist of the Beats, really. And also, Stewart Brand.

Fred Davis: Leary was into mind expansion. There was a chemical way of mind expansion. And now, there was an electronic–digital way of mind expansion.

Mitch Altman: Virtual reality is the ultimate psychedelic toy, right?

R. U. Sirius: Tim told us that he thought Jaron was like the smartest person in the world at that time. And Jaron thought Timothy Leary was the smartest person in the world at that time, which he probably doesn't remember.

Jaron Lanier: Tim's nickname for me was "the control group," since I was the only person he knew in the scene who hadn't taken LSD. I used to get a lot of flak for being the straight man in the scene, because I didn't take drugs and everybody took drugs.

R. U. Sirius: In those days it was very much a psychedelic drug scene but it was a very intellectual, psychedelic, drug scene.

Jaron Lanier: We forget that in those days the drug culture was huge, dominant, and the technical culture was this little tiny, tiny, tiny bubble.

Jamie Zawinski: I would characterize it as more the art culture than the drug culture, but obviously there's like a 90 percent overlap between those two.

Fred Davis: *Mondo* was throwing fantastically wild parties at Alison's house that were completely full of hallucinogens: people doing mushrooms or they were doing acid or they were doing ecstasy.

Michael Mikel: Once in a while somebody would come in and bring in some drugs that had just been made—not even illegal yet!

Fred Davis: And the lightweights like me were just smoking pot.

Michael Mikel: So it was a pretty wild scene over there.

R. U. Sirius: We had this mansion in Berkeley and there'd be an awful lot of people, even early on, coming to these parties. I remember one that looked like a small rock festival. Every space on the back lawn was taken up and almost everybody there smoked DMT. That was kind of the thing in the mideighties. DMT is just the wildest carnival ride of the psyche that one can possibly imagine. It's like having a million universes mainlined into your spinal column for three minutes. It's amazingly incomprehensible but at the same time sort of—refreshing.

Mark Pauline: They would just have these huge, unbelievable parties. Hundreds of people at these parties and just all these people doing work in 3-D graphics, virtual reality, smart drugs. All the scientists and weirdos and artists all together. And all these gazillionaire tech people at the time in the eighties would be showing up at these things.

Michael Mikel: The things that were discussed and talked about were fascinating. I was connected to a guy there who was deeply involved in the idea of time travel. We had some deep discussions. We even did some experiments—mental experiments—about the subject. There was this whole scene in this great old house in Berkeley. Almost every weekend there was some kind of gathering there. That was the *Mondo* house.

Fred Davis: It was an exciting, heavy party time when the people who were inventing the future wanted to get together and talk. The tech nerds and the cyberpunks or whatever wanted to get together and talk about this shit. We were immersed in it. It was changing our lives right before our eyes. There was kind of a party circuit back then. *Mondo* was throwing parties. I was throwing parties. There was the Anon Salon. Survival Research Laboratories did their thing at a bunch of different parties.

R. U. Sirius: SRL was this insane performance art thing.

Jamie Zawinski: SRL was a very underground art troupe who built robots that they put on performances with where the robots tore themselves and other things apart.

R. U. Sirius: They were building these sort of Frankensteinian monsters and setting them loose to brush up against each other.

Fred Davis: These weren't little wimpy robots that you wound up and walked around on your desk or something. These were fucking mean fighting machines: Fighting robots! Warring robots! Flame-throwing robots! That was, again, part of the sci-fi culture–hack, technohack, counterculture rage-against-the-machine thing.

Jamie Zawinski: There was a dystopian sci-fi feel to a lot of this stuff. Obviously we were reading the same books.

R. U. Sirius: Now, you have the commercialized version of this: *Robot Wars* and all that sort of stuff. SRL was real.

Max Kelly: The SRL guys had a studio down in Bayview, which at the time was not the place to be. We'd regularly find cars on fire, and people would run up and try to carjack you all the time. The thing was you could do whatever you wanted down there and no one would give a shit. And these crazy white guys are sticking a rocket engine—a V2 rocket engine—out the side of their garage and turning it on. Test firing! A V2 rocket engine! Okay, it might have been a V1, but it was definitely an old Nazi rocket engine.

Dan Kottke: It was a jet engine, running on kerosene. They had it on a sled, and they would fire it up. And it was very, very loud and exciting. I used to go to those SRL events. They were always fun, they were always word-of-mouth, they were always illegal.

Jamie Zawinski: It was incredibly illegal. They were not safe environments. Hearing damage was a real issue.

Gary Wolf: It was superhallucinatory. It was frightening. There are almost no words for it.

R. U. Sirius: Oh my God! They would take it out onto the street...

Mark Pauline: You mean *Illusions of Shameless Abundance*? That was one of my favorite shows!

Gary Wolf: That show scared the shit out of me: I'm breaking out in a sweat right now thinking about it! It was so far beyond...

Mark Pauline: In 1989 I had a bigger crew than usual: forty or fifty people, plus I had gotten a grant from the Fleishhacker Foundation, so I had their name to throw around, so I was able to talk the people who ran this parking lot South of Market into renting me the space.

Jane Metcalfe: South Park was full of winos and drug addicts and homeless people. They had oil drums filled with garbage that they burned. It was a really sketchy neighborhood.

Louis Rossetto: All of South of Market was like that.

Mark Pauline: There's this beautiful part under the freeway between Third and Fourth Street and between Bryant and Harrison. There the freeways are very high—fifty feet—and it's almost like a cathedral under there. It's just this huge open space that's basically a block long and a half a block wide. It's very flat and it's open so it's just a fantastic location for a show.

Jamie Zawinski: You'd go down to an abandoned parking lot where these guys were going to build some giant flame-throwing robots that would tear each other apart. They're not selling you a ticket. This is not Disneyland.

R. U. Sirius: It was really dangerous stuff.

Jamie Zawinski: There was going to be a giant robot spitting flame and if you're in the way you're going to catch on fire. And then everyone's going to run, you know?

Mark Pauline: I picked all sorts of machines that I thought would be good for the theme. One of the things I did was locate a tree chipper that was in some other city's maintenance yard and remove it back to San Francisco and set it up so it was remote-controlled. I put a bigger fan in it, too, so it had more blowing going on. But you know the V8 engine had the ability to chop up four-by-fours, big boards, trees, whatever you want to put in there. So I made the hopper bigger for it and we collected, oh, probably four or five tons of rotten food from the produce market and from other places. We made a huge cornucopia filled with rotten food and had all sorts of other kind of like passive dumping systems built around the freeway. So if you bumped against them they would load the hopper full of rotten food. And we set it up with a long chute so it could project the stuff it chopped further than a regular one. It really smelled pretty bad. It was like all kinds of horrible stuff in there. It was shooting chopped-up rotten food all over the audience. That was one of the things going on.

Gary Wolf: Then they lit that tower of pianos on fire!

Mark Pauline: There were pianos, about twenty-five pianos stacked against a part of the freeway, and at one point the pianos caught on fire and the Highway Patrol shut the freeway down because the flames got high enough that they were coming over the top of the freeway, fifty feet above us. So it was pretty out of control. We also had a huge flamethrower that was making forty- or fifty-foot flames down underneath there also. All this stuff was happening at once, and huge huge fires, smoke and just a lot of chaos, a lot of explosions and just a lot of crazy stuff. There was too much chaos to really pay close attention. I just know that the fire was melting things in the control booth from a hundred feet away! It happened because there was a ceiling. There is a phenomenon that happens with infrared radiation called "flame focusing" where if there's something on top and on the bottom the infrared can't radiate up or down, so the flame felt much more concentrated than it would have if it was outdoors. That also damaged the freeway pretty badly, but fortunately the earthquake came a couple months later and so Caltrans just said, "Don't worry about it. We don't even want to admit that you did this. So we're just going to let it slide." So they never charged us to fix it because of that earthquake, which was lucky. But the interesting thing about it was nobody ever told us to stop. The police never said a word afterward. The fire department never said a word. The highway patrol never came down and said, "Hey, we had to stop the freeway because of your fucking fire. What the fuck?"

Fred Davis: That punk attitude was a San Francisco attitude: The punks were raging against the machine in the best way they knew how.

Mark Pauline: I always have considered myself a comedian. So it was all about trying to find the entertainment value and the humor in all these things that just really were faceless and humorless like industrialization, commercialization, high technology, military weapons systems. I always have tried to look at those things and find the other side of it. *What about this is entertaining?* I would try to mine those things.

Fred Davis: SRL was another type of rage against the machine, which is what the phone phreakers were doing.

Mark Pauline: I had an Oki 900 that was cracked. We used to call everywhere in the world all the time. We used to call friends in Europe. We would rack up hundreds and hundreds of thousands of dollars' worth of free calls.

Fred Davis: And desktop publishing was a rage against the publishing machine and whatever.

Mark Pauline: Jane and Louis were the only rich people, wealthy people, who ever helped SRL out ever.

R. U. Sirius: We all felt connected. Mark Pauline was doing his mechanical thing but he would come into all the virtual reality parties. There were all kinds of virtual reality parties. There were all kinds.

Kevin Kelly: Cyberthon was in '90. I was inspired by Stewart Brand to see what you could do with a happening. And once I intersected with Jaron I became aware there was a bunch of VR stuff happening: There was Fakespace, there was the NASA folks, the folks at Autodesk.

Jaron Lanier: Oh. There were a zillion things like that. And Kevin and Stewart and Barlow and a variety of other sort of vaguely psychedelic-era intellectuals were always putting together weird events and happenings.

Kevin Kelly: And so my idea was to bring together all the current demos and prototypes for VR and let anybody who wanted to try them try them. All at once. That would be cool.

Jaron Lanier: There'd be like this incredible pressure: "You have to open it up and give everybody demos!" Well, how many people? "Oh, there's only going to be two thousand." All right, we'll give ten people demos. And they are like, "No. We have to have at least..." It was just constant pressure and arguments. I mostly remember being beleaguered by that scene.

Howard Rheingold: Cyberthon happened because the *Whole Earth Review* wanted to have an event around virtual reality. There was no money, and we wanted to have all this stuff happening that would take two or three days. And Kevin, who is a genius at this stuff, said, "Well, why don't you just have it around the clock for twenty-four hours?"

Kevin Kelly: We ran it from noon Saturday to noon Sunday. We sold tickets on The Well; we announced it in the magazine. We tried to have as many interesting people come as well. So Bruce Sterling was there, William Gibson, Tim Leary, Robin Williams, Jaron Lanier was there, John Perry Barlow. And of course all the VR people. And I thought the interesting thing would be in the middle of the night. People were tired and it was kind of crazy and dreamy. That was part of the happening aspect of it.

R. U. Sirius: There were all these rooms with all this tech stuff being presented. Everybody had their VR stuff there. It was very lively. It went on all night.

Kevin Kelly: There was no music. There weren't any crazy antics. People didn't trash anything. The Hells Angels did not show up. There were no drugs. It was trippy enough.

R. U. Sirius: I smoked DMT with Wavy Gravy out by the side of a van. I gave William Gibson some vasopressin—it's a smart drug similar to cocaine. You just squirt liquid vasopressin up your nose. Cocaine releases vasopressin in the brain. It doesn't have quite the same powerful effect for some reason. Gibson liked it, but now he swears that he remembers not getting off.

Fred Davis: I think William Gibson and Neal Stephenson had a lot to do with the whole virtual reality thing. A lot of these cyberculture people were sci-fi enthusiasts, and books like *Neuromancer* and *Snow Crash* were the cultural icons. You had to read *Snow Crash* because if you didn't, you weren't up to speed on what the real thinkers were thinking about the future of tech. The future is people logging into these VR worlds and getting caught in them, that's what *Snow Crash* was about. We still talk about that today. *Snow Crash* became the iconic cyberculture book. It described virtual reality in a more elegant way than what Jaron was doing.

Chris Caen: The parties were thrown out of a collective frustration that we couldn't make *Snow Crash* happen, and I joke that no book has cost the VCs more money than *Snow Crash*. Because at the time we were trying to jam these virtual worlds down a 9600-baud modem to a Pentium 60 processor, Pentium 90 if you're lucky, and we just didn't have the horsepower to do it, but damn, we were trying.

David Levitt: Plus there was a whole rave culture growing up partly out of this, which left us scratching our heads.

R. U. Sirius: It was a very rapid, almost unbelievable adoption of the rave culture by the San Francisco counterculture. That kind of surprised me a little bit. You wouldn't think people who were either growing up with the Dead Kennedys or growing up with the Grateful Dead would immediately rush out by the thousands and fill large clubs to dance to this music with no voice and no center of attention.

Brian Behlendorf: Raves were pretty obscure in March of '92, which was when I started the SF Raves list. It seemed to be a gathering point, or a concept that a lot of young, creative, future-minded people seemed to be attracted to.

R. U. Sirius: In the early nineties a guy named Mark Healy who had been writing for *i-D* magazine in London came over and started doing articles on the cyberculture in America. He interviewed me and he interviewed Terence McKenna, the psilocybin guy, who was very much into virtual reality. And he decided to move to the United States and try to bring the cyberculture together with the rave culture. He started this thing called Toon Town.

Jane Metcalfe: At the time the Toon Town people were living in a house in South Park and doing the best drugs, having the best parties, and experimenting with visuals. So they brought a lot of that energy into our office.

Brian Behlendorf: *Wired* was in that building at Second and South Park and then moved to 510 Third Street. Every time we expanded we had a rave, basically, starting in the new space we were taking over, before we moved desks in or painted the walls or whatever. These would be all-night, not-permitted kind of parties.

Justin Hall: When we moved out of 2 South Park, I threw a Halloween rave to celebrate, taking advantage of the fact that even though we'd moved to Third Street, we could still use the old space. They gave me a $1,500 budget to acquire food and drinks and stuff. I ended up I think spending five hundred or more dollars of it on LSD microdots that I handed out to people at the door. It was the launch party for *HotWired*.

Brian Behlendorf: I think electronic music was a filter then as well. People who were into that tended to be programmers, tended to be very eager to embrace new technology. In the nineties we were still optimistic about technology and the future, and spending more time online.

R. U. Sirius: It's like this whole cultural thing came together. It was a psychedelic celebration of technologically oriented music. You'd have these spaces where people would come and be in a virtual reality and drink smart drinks and play with other tech toys. There wouldn't even be any virtual reality there sometimes, you know? Just people partying about this idea of another place for minds, I guess.

Po Bronson: It was a rave scene somehow under the guise of a tech thing, because all the tech people wanted to rave.

Justin Hall: There was a strong overlap between the people who organized and deejayed these parties and the people who had some internet connection. Obviously, that was a small group at the time, but it also felt very open because it was sort of like, "Oh my God, this internet thing!"

Captain Crunch: I got involved with the SF Raves scene. I started giving out rave reports. Man, once I got hooked into dancing and candy flipping, I just couldn't stop the rave scene. Candy flipping is taking Ecstasy and acid at the same time.

Brian Behlendorf: I had set up a simple website for the SF Raves mailing list at this time, and it was just the back archives of the mailing list and maybe a few deejay sets and some scanned-in flyers and things like that, because it was very early days for the web. SF Raves was starting to generate traffic, so I went and bought a piece of server hardware and I parked that box at *Wired*, because I could hang it off the bandwidth *Wired* had, and no one would notice. I gave it a different name—Hyperreal.

Justin Hall: Hyperreal.org, which was Brian's web domain, was where so much raving culture happened. There were flyers posted for parties. The fact that he hosted the SF Raves mailing list meant that he was the hub for the rave community in the San Francisco Bay Area. So if there was a party happening he knew about it, and there was a good chance he would be driving out with them. Today I think there's a lot of people in San Francisco who sort of live that life where there is a mailing list that tells you about a party. But it was very fun to be doing that when I was nineteen and when that was the portal to a world that was just being created.

Brian Behlendorf: Hyperreal became this kind of independent incubator for interesting little projects that I started to get into. It became home for erowid.org; it became home for burningman.com.

R. U. Sirius: And then, again, there was this huge adoption of Burning Man by the tech culture. Really, suddenly, it became this massive thing that everybody had to do.

Joey Anuff: The SRL, rave kind of vibe coalesced into the Burning Man thing.

Tiffany Shlain: Burning Man was a big through line! We all went there for years.

Amy Critchett: So Michael Mikel, who was one of the founders of Burning Man, was a handyman for *Wired*. So it definitely was a presence in the early days of *Wired*.

Michael Mikel: I was involved with Burning Man from pretty close to the beginning. I brought in a lot of tech people. The main thing I was doing was just inviting everyone I knew, and putting them on the mailing list.

Amy Critchett: My crew, we started going in 1994.

Justin Hall: Burning Man in 1994 was just a mass of people who were all parked in the middle of the desert in any which way. I don't remember where people pissed and shit. There certainly were no Porta Potties. There were no organized roads. There were motorcycles and there were dogs and there were fireworks and there were probably guns. I remember dropping acid that had Beavis's and Butt-Head's faces printed on it. Then this man burned and I jumped over it in my toga and Birkenstocks or whatever. It was the core essence of what Burning Man is today just with two thousand people or something.

R. U. Sirius: You couldn't plug in. You didn't have cell phones there then or anything like that. It wasn't really a tech event.

Michael Mikel: It isn't really a tech event, but by its very nature it appeals to tech-minded cyberpunk people. It's a complete alternate reality, a different world.

Justin Hall: My favorite thing that year is someone brought out like a truck-load of like sixteen old upright pianos and they were all piled on top of each other, and people at all hours of day and night were just hammering on these pianos. Like pulling the pianos apart and then beating on the pianos with piano parts. It was the most amazing cacophony.

R. U. Sirius: I think it shows some of the flexibility, creativity, imagination, and strangeness of tech culture in the nineties, in that virtually everybody adopted it.

Joey Anuff: I fucking hate Burning Man! It's my least favorite thing that has ever existed. And I see it as like the most goddamn poser-tastic thing that has ever happened to California since like Wavy Gravy.

R. U. Sirius: I think that the core energy of the cyberculture counterculture thing seemed to dissipate in about '94.

Joey Anuff: One of the things that happened in the first year of *Suck* was the Netscape IPO. And that was also game changing because just in terms of supplying a wealth-creation narrative, which turns everybody's head and becomes the dominant narrative in every instance.

Jamie Zawinski: And then, when the money came in, let's just call it July '94, the industry exploded. Suddenly, there's another quarter-million people in this industry who are nineteen years old, twenty years old, and haven't met these old hacker guys, and their experience with computing is completely different. These new people who got into the computer industry? If it was 1982, they would've been traders. They would've gone into finance because "Oh, that's challenging work. And man, you can make a lot of money doing that if you get hired by the right firm." It's that attitude. Programming is the safe job. It's the safe job for smart people.

Po Bronson: Kevin Kelly and Stewart Brand had lent the early technology this incredible gravitas and nobility and high-mindedness. The ultimate thing was your mind—mind expansion. Prior to the dot-com, it was this philosophical aspiration where ultimately the money was irrelevant. There was never a conversation about money.

Joey Anuff: It turned into a money story.

Jamie Zawinski: So the story is that we won the lottery, we made a bunch of money, and because of that, we sucked all of the creativity out of the computer industry and turned it into the finance industry, and made everything terrible.

R. U. Sirius: That's when the scene down in South Park becomes a whole different kind of thing. The upscale tech geeks were moving in.

Jamie Zawinski: There was a big boom and suddenly there were a lot of people moving to San Francisco who weren't actually interested in what San Francisco was. They came here and they wanted to make it be more safe and suburban. It was the South Bay–ification of the city

Jack Boulware: Maybe people were tired about reading about smart drugs and this and that in *Mondo 2000*? You know it was all really cool for a long time to get a stud put in your dick or whatever it was. But computer geeks were moving in because there were jobs and opportunity, and they're walking around the city going, "Why are you guys doing this? I'm not interested in this at all." A younger group of people is always going to be suspicious of what came immediately before. At some point it will be a cliché that they will just laugh at. Generations come and go.

R. U. Sirius: In a way, it's sort of a movement from the hippies and the punks to the hipsters, if you will. I used to call them the "golden nose rings." A lot of them were like sweet hippie kids but like with no edge, no struggle in them. The golden nose rings had no politics. They just kind of go along with whatever's happening. "Go with the flow," I guess—and make lots of money. That seemed pretty weak to us. *Mondo* was very punk influenced, very sort of left-anarchist influenced. Whereas, you know, then *Wired* came along and had more of the right-libertarian vibe to it. In some ways, that's representative of the change that went on there.

Erik Davis: We don't think much about magazines anymore, but they really were markers of identity. And the transformation of magazines reflected something about what was happening on a deeper level in culture or even in people's identities. If you look at the *Whole Earth Catalog*, it has a collage logic. That reflects the way it's laid out but it's also a kind of constructivist idea, manifested in design—you take bits and pieces and you combine them together to build culture. Then comes this desktop machine that allows you to control typography and this good laser printer that can make something that doesn't look janky. There's a new cultural logic. This new cultural moment is not collage-y; it's not bits and pieces just sort of jammed up against each other. With the new media tools we can actually sort of invent a new environment. It's smoother, it's more global, more virtual in a way, because you can make a worldview appear more present than it necessarily is, and then you have people identifying with being cyberpunk and living out their cyberpunk experiences. So it actually kind of engenders itself. That's what *Mondo* did and *Wired* brilliantly recognized that and absorbed elements of it, and then it took off in a more mainstream way. So both magazines actually invented, in many ways, the culture that they end up writing about. And that recursive loop, that self-reference, is essential to technoculture: That's what it is. This is the moment when technology—not economics, not history, not new social norms—starts driving the culture.

BOOK THREE

NETWORK EFFECTS

We shape our tools, and then our tools shape us.

—MARSHALL McLUHAN

The Check Is in the Mail

eBay's trillion-dollar garage sale

*T*he web, suddenly accessible by ordinary people thanks to Netscape, sparked a gold rush. Almost everyone had the same big idea: to sell things on the internet. Up in Washington State, Amazon sold books. But down in Silicon Valley, something stranger and more surprising was being sold. What we now know as eBay started out as a philosophical experiment, utopianism in the form of a website. Back in 1995 General Magic employee Pierre Omidyar was a young, ponytailed idealist: He believed in the power of markets to improve people's lives. So over Labor Day weekend he hacked together an online marketplace. It was powered by free (and flaky) open-source software—and the honor system. Strangers mailed each other money and stuff with no guarantee that they would get what they were expecting. The founding conceit was that people were basically good, and if you gave them the opportunity to be good, they would prove to be so. Remarkably, people on the internet were good. The checks were in the mail. Even more astonishing is how thoroughly the often barely functional site trounced the auction-site competition. Something about eBay made people loyal, too.

Jeff Skoll: In 1995 when I graduated from Stanford Business School, I was living in a house with five guys that were sort of internet focused, even though '95 was still a little bit early on the internet trail. I knew Pierre from just hanging around, and when Pierre talked about the idea of eBay—or AuctionWeb, as Pierre called it back then—I wasn't quite so sure. Pierre said, "We're going to build this thing, and it's going to get people selling things to each other." And there were other players coming along. Amazon was around, there was a CD site called Music Boulevard...

Brad Handler: There were fifty other sites that were in the actual space that eBay was in.

Mary Lou Song: One site comes to mind which actually took physical possession of the goods: So basically the seller has an item, they send it to the company, the company puts the item up for bid, and then turns around and delivers it to the buyer. That was OnSale.

Jerry Kaplan: OnSale was very much what Overstock.com is today, but with an auction component. That was the concept. We ran a test and somehow got some publicity and basically said, "Get on your computer, dial into America Online, connect to this thing, and you can put in bids." We didn't know if it was going to work, and people loved it. They went nuts over it—of course it was a total nerd-ball crowd. That was early 1995. But when I first took this idea around to the venture capital community, nobody thought that it was a good idea at all! You have to understand that the computer community at that time was kind of countercultural, that's where most of those people came from. And so for them the opening up of the internet for commercial purposes would be a little bit like the selling of advertising on PBS.

Jeff Skoll: There were things in the early nineties that were considered taboo. You don't charge people for news, right? News is free. But we had the philosophy that free is bad, because nobody knows how to associate value to what you are getting for free. So you actually have to charge, even if it is a small amount—a penny. We thought that was an important line to lay down from the start.

Mary Lou Song: Pierre has always been interested in economics.

Brad Handler: Pierre is very libertarian.

Mary Lou Song: Although much wiser and more spiritual than that.

Jaron Lanier: The eBay people were a later generation of Silicon Valley people, but there was still a continuity there. They were still hacker-y and kind of nerdy.

Jerry Kaplan: Pierre Omidyar was working at General Magic, and we wanted to hire a good engineer and he came in for an interview, and we obviously told him all about what we were doing at OnSale, and he thought that it was really cool. We offered him a job, and he said, "No, I think I'm going to stay where I am and work on this thing on the side," so he began to take the concept that we had, and apply it to person-to-person trading.

Pierre Omidyar: I wanted to create an efficient market where individuals could benefit from participating in an efficient market, kind of level the playing field. And I thought, *Gee, the internet, the web, it's perfect for this.* This is more of an intellectual pursuit, you know, than anything else. It was just an idea that I had, and it started as an experiment, as a side hobby, basically, while I had my day job.

Jeff Skoll: There was no prototype. It was in Pierre's head. And I ended up taking a job out of school with Knight Ridder and Pierre kept his previous job at General Magic. I said, "Pierre, I just don't see it. I'll work with you on weekends and see what happens, but I'm not terribly optimistic about what one guy with an idea in this giant space with all these destroyer boats around can do."

Steve Westly: Pierre just took it three levels deeper than anybody else. Pierre's a true genius for figuring it out.

Jeff Skoll: Part of proving that this would work was building a fairly open system and trusting that people would do the right thing.

Pierre Omidyar: The whole idea there was just to help people do business with one another on the internet. And people thought that it was impossible, because how could people on the internet—remember, this is 1995—how could they trust each other? How could they get to know each other? And I thought that was silly. You know, it was a silly concern because people are basically good, honest people. So that was very motivating. It was, *Gee, I'll just do it. I'll just show them. Let's see what happens.*

Jeff Skoll: The site turned on—as an experiment—on Labor Day of 1995.

Mary Lou Song: Pierre had a broken laser pointer which was the first thing he sold.

Mark Fraser: Somebody pointed me to a brand-new website which turned out to be eBay, and I was amazed to discover a broken laser pointer that was listed, and thought, "Hey, I can make that work!"

Mary Lou Song: To Pierre that was the sign: "If I can sell a broken laser pointer, this is good for everything."

Jeff Skoll: Almost right away, people started transacting. They started sending money around.

Jim Griffith: I remember the day I started using eBay, it was May 10, 1996. I was living in Vermont at the time, rebuilding computers and selling them locally to people who wanted to get online, for a couple hundred bucks.

And I needed parts. So I bid on this chip and I get it a few days later, that's my first purchase, but I'm fascinated by this site—fascinated by both the concept of people getting together and buying directly from each other with no real middleman, and by this guy Pierre, who shows up every night on his little chat board to ask questions: "How do you like what's going on? Do you have any suggestions? I just had a long day at General Magic." He was working at General Magic at the time.

Pete Helme: By 1996 things at General Magic weren't going well.

Steve Westly: And he's got the thing set up in the back room of his apartment—it's a two- or three-bedroom place, and the back room is eBay. It's twenty-five cents to list and a teeny little fee, and you send your check to this P.O. box. And after six, eight, ten, twelve weeks, there's a check every day. Then it gets exciting—because it's no longer two or three checks, it's like fifty.

Jim Griffith: Pierre created a rudimentary billing system which would track all of your successful listings and sales and then would invoice you for the amount of money you owed eBay—the commission and for listing. In the early days if you got way behind you might get shut off, but it was basically the honor system.

Steve Westly: And one of the first people we hired was a guy named Chris, whose job it was to go down to Wells Fargo and sign all the checks and deposit them, and that is how eBay worked.

Mary Lou Song: Poor Chris, so there wasn't PayPal at the time, right? So every day the mailman would deliver a garbage bag full of envelopes, and Chris would have to open them by hand, every single day.

Chris Agarpao: Well, first the bags would start out real small, like stacks of maybe thirty, and then sixty, gradually as we grew, and then, you know, they were really heavy bags—tons and tons of mail.

Pierre Omidyar: And it just kind of grew. Within six months it was earning revenue that was paying my costs. Within nine months the revenue was more than I was making on my day job.

Chris Agarpao: It wasn't all checks. Sometimes it was nickels and dimes, that they'd tape to index cards. So that was kind of funny. But it all added up to be something big.

Steve Westly: The company was profitable every quarter of its existence—every month of its existence—from day one. Now I'm not sure if any other company could say that.

Pierre Omidyar: That's kind of when the lightbulb went off. *Knock, knock, knock. You've got a business here, do something about it.* So that's when that really started.

Jeff Skoll: I said to Pierre, "The odds are long against this working, and it's certainly going to be up against a lot of challenges in the future, but yes, maybe we have a chance!"

Jim Clark: The internet, prior to the point when we first brought out Netscape, was complete mayhem. People worried about doing any kind of private transactions. In those days, there was no security whatsoever. The hardest sell I had was people saying, "You mean you are going to send your credit card numbers around on the internet?" I said, "Well, it is going to be done securely." I hired this PhD from Stanford who was a security expert, Taher Elgamal, and gave the problem to a small team. These guys cooked up what became known as SSL: Secure Sockets Layer. It is a way of establishing a secure link using the browser, and it became the most widely used technology on the web, and the reason we did it was so that you could have a secure connection so that you have a commercial transaction.

Jeff Skoll: We left our jobs mid-'96 and started working together full-time. We worked out of Pierre's living room, and you know it's me and Pierre trying to figure out what to do—and Pierre's fifteen-year-old cousin, who was playing video games all the time.

Michael Malone: Now keep in mind that Jeff at this point was in his mid-twenties. He had gone to Stanford B-school. He was still living on a couch with some other guys in a house.

Chris Agarpao: Pierre knew how to start the business, but he needed someone who could run it. Jeff was the business guru, and I wouldn't say uptight, but he kind of knew his stuff. He was all business. Pierre was laid-back.

Jeff Skoll: When Pierre and I started working together, he showed me the link where you could send an e-mail to customer service. I said, "Okay, but who reads that?" And he says, "It just goes away!" I'm like *hmmmm*. One of our fundamental debates was about if the system could take care of itself, and I felt it actually couldn't. I'm an electrical engineer with an MBA, I wanted to put some meat on the bones here. My thinking was *Wow, this could be a very big company, or it could also be roadkill very quickly*, and I wanted to make sure that we were ready for when this thing took off.

Brad Handler: Pierre was very focused on building out a community and trust and felt that the community could self-police, that it can take care of itself if we give the community the basic tools to do so. Jeff was more for rules, and that there needed to be enforcement of those rules for the commerce to flow. Pierre talked about how he was going to build a community, and I think either one on his own would have failed. So they were very fortunate to have found each other.

Mary Lou Song: In 1996 I had just graduated from Stanford and I met Jeff and Pierre. It was just the two of them in this teeny-tiny office on Bascom Avenue in San Jose. And I said, "What are you doing? What is this?" And Pierre said, "Well, I'm building a community where people buy and sell with each other." And I asked all the right questions about what's the business model, how are the goods transferred, what kinds of goods…

Jeff Skoll: Mary Lou was employee number six.

Pete Helme: I got a call from Pierre in August or September saying, "Pete, I need help! I need a bright engineer. I can't do this on my own. Can you come?" I thought, *Why not?* I think Mary Lou had just started.

Steve Westly: It was a pretty young group. People would walk around in shorts. People would be sitting around in lawn chairs at the senior staff meeting.

Brad Handler: Mary Lou was the glue that sort of held everything together. She was the leader of the band.

Mary Lou Song: But what I loved about the conversation with Pierre and Jeff was everything about that conversation came back to the people, and to the community and contributing to a greater good, and I loved that because it was so refreshing.

Brad Handler: Every other would-be competitor made you sign in with a credit card in order to get access to the inventory. We were the only one that would let you view the inventory without first registering with a credit card. eBay treated people like people in a community. The others treated you like a credit card.

Jim Griffith: There was this little chat area that Pierre had put up, and it was just a very, very simple, actually kind of crude scrolling chat. It would hold one hundred posts, and every time you added a post, the oldest post would scroll off. There was no archive, but it was an ongoing conversation, with some very interesting people.

Mary Lou Song: There was The Well at the time, but we were enabling community around commerce. Our founding CTO, Mike Wilson, had worked at The Well.

Steve Westly: Pierre, Jeff, Mary Lou, and Mike, they should be historic people, because they are the ones that built the foundation for social media, by figuring out how to create online reputations and intense community. It's hard to explain now, because it feels a little bit obvious, but back then, everybody said the internet is cool *because* it's anonymous. They did gambling anonymously and adult sites anonymously, and everything was anonymous, and people kind of liked that. But Pierre and Jeff were saying, "No. No. No. No. No. What you want to do is make everything public in this increasingly anonymous world, and we're going to allow people to create reputations so you can trade in a fair and safe way, with people you don't 'know' in the traditional way." The great thing about eBay is there is nothing anonymous on eBay.

Pierre Omidyar: I created the feedback forum, which I'm very proud of because it has been copied gazillions of times, and it was my idea. It allows people to kind of rate each other and give feedback on how their transactions went. And if they don't like something that somebody is doing, and enough people don't like it, that person is automatically kicked off of the system. That worked very well in the early days.

Jeff Skoll: I wish we had patented that, because that reputation system became quite pervasive and useful all along the way. And in the early days, that was the key.

Mary Lou Song: Pierre had made it so that if you got to ten positive comments, you would get a gold star. So I just created a star system. After you get a gold star, you get a red star, then you get a green star at fifty, you get...I can't even remember the rainbow of colors, I just made up the system. I remember posting it to the announcements board and that I had created this new star system for them and instead of getting a "Hey, thank you, that's great!" I got "Your colors suck, and your numbers suck and who put you in charge of this?" And the harder they pushed back on my stars, the more I dug in my heels, until somebody shared my personal e-mail address with the community on the forum, and I would come in to the office at you know eight o'clock, open up my e-mail box, and it was full of stars, stars, stars, stars, conversations about stars. Pierre and Jeff had a policy that if a community member e-mailed you, you personally had to respond. You couldn't just ghost them: You couldn't not respond. So I would spend the better half of my morning responding to all of these e-mails about the color of the stars, and so after about three days of this, I went to Pierre: "I have to talk to you about the stars." He says, "Yes, you do." And I said, "Well, I guess it would be good if I went back and apologized?" He said, "Yes, that's a good start. And then what are you going to do?" And so I went back and I posted an apology and said, "I'm going to take suggestions for stars' colors and numbers for the next week. We're going to take a vote on it and then we'll implement it." So the star system wasn't our star system. It was a star system designed and created and implemented for and by our community. And after that pretty much everything that we built was touched by the community.

Chris Agarpao: Jeff hired Griff so he could talk to the community, because Pierre had all these jobs and he needed somebody to lighten the load.

Jim Griffith: They hired me to do e-mails. But within a few weeks, the explosive growth started. So about two weeks into this, it went from thirty, to fifty, to a hundred, to about 150 e-mails a day. It was word of mouth: There was no marketing.

Chris Agarpao: Mary Lou was in charge of marketing.

Michael Malone: I got a phone call from this gal named Mary Lou and she said, "Mike, you remember me," blah blah blah. I said, "Yeah," and she goes, "I'm in a new start-up out here, and I'd really like you to come out and meet these guys. I think they've got something really interesting." And that was this company called eBay. It was just Pierre and Jeff. And Mary Lou, but I didn't really see Mary Lou much after that. I was mainly just seeing these two guys. And it turned out they had read my stuff over the years in the *San Jose Mercury News* and elsewhere and they just kind of wanted my advice.

Jeff Skoll: I was looking for someone who could be a guide. Mike was quite the futurist.

Michael Malone: So we had a bunch of meetings and we'd discuss the long-term business strategy of the company. And then we would pool our money, and we'd usually throw in five bucks each, and we'd go to Fung Lum's and we'd get a big bowl of wonton soup to share. Then we'd go back and meet a little bit more. We talked about a lot of stuff. And I gave them the metaphor of the medieval crossroads.

Pierre Omidyar: If you think about it, commerce and trade is at the base of all human activity, and it's a bit of an exaggeration, but I like to talk about how, you know, in the old days people would bring their stuff to market and they'd do business and then they'd go back to their hillside homes or wherever. And eventually they were doing this enough that you had to build a wall around them to protect them, and that was the birth of cities and so forth.

Michael Malone: And that kind of became the metaphor of the company.

Pierre Omidyar: And again, gross generalizations and simplifications, but fundamentally everything we do in human activities is related to trade and there's something, I think, that's wired in human beings that drives us to commerce.

Michael Malone: Later on eBay tried to promulgate the Pez story as the founding story of the company because it sounded so good.

Pierre Omidyar: The Pez thing, right. Yeah. My wife—who was my fiancée at the time—whenever she hears about it she rolls her eyes. "Tell them I'm a management consultant. Tell them I have a master's degree in molecular biology. I am not just this little Pez candy collector."

Mary Lou Song: Okay, so this is the real Pez story that I was told. That at the time that Pierre was thinking about all of this, he and Pam were in France driving through the French countryside. They stopped at a French country store, where she found French Pez, and she scooped it up. They pay for it. They get in the car. He starts driving. She's going through all her stuff that she's just bought and realizes that she forgot one. She missed one, so she insisted that he turn the car around...

Michael Malone: And Pierre said, "Let's figure out a way to find these things," blah blah blah blah, "and buy and sell them." It wasn't quite a myth—I mean I'm sure there was a Pez dispenser or two in there.

Pierre Omidyar: That was part of the inspiration, but frankly it was a small part of it for me...The birth of the idea is definitely a media-enhanced story.

Michael Malone: Basically they were just looking at "How do we build a place where you can do auctions?" And once again they weren't the first. There were other auction sites out there.

Steve Westly: OnSale was the big dog; it had been funded by Kleiner Perkins. It was downright scary.

Pete Helme: And eBay was a tool for people who liked to do flea markets!

Pierre Omidyar: And garage sales and that kind of thing. So that flea market base has really evolved into a market for pretty much everything.

Michael Malone: So why did eBay win?

Mary Lou Song: The economic genius of what Pierre came up with was that we never interfered with the velocity of the transactions. We facilitated, we helped people find each other and connect, and our job really was to make sure that buyers could find sellers and sellers could find buyers and make that transaction as seamless as possible. If we had taken possession of their items, we would have interfered with that velocity.

Jerry Kaplan: We had a different business model, so Pierre was applying it in a different way. I certainly wasn't delighted about it, but I don't think I sat there and thought, *Oh my God, this guy is trying to knock us off or steal from us.* Remember the kind of countercultural "ideas are free" kind of attitude? Plus we didn't think eBay had the same commercial potential as OnSale. It's like *How big are flea markets, versus how big are stores?*

Steve Westly: When I told people that I had seen the future and I was going to work for an online flea market, they wouldn't say anything, but their eyes would be saying, "I think he has lost his marbles."

Pete Helme: I don't know how we hired people in the early days. They were like, "You sell Beanie Babies and laser pointers? I don't get it." Nobody was excited about us.

Mary Lou Song: And then McDonald's ran a Teenie Beanie promo. They were Beanie Babies that fit in those little packages inside of Happy Meals. So that was perfect. All of a sudden you put out a couple of press releases: "Beanie Babies here! This is where they're at!" Then you get a lot of attention and a lot of people coming to buy and sell Beanie Babies.

Chris Agarpao: That was 1996. I collected them myself just because my coworkers were collecting them. They were like, "Dude, this is going to be big! You just got to keep the tag on." Everybody just bought them and kind of put them on display all over their cubicles and so on. Pierre had a whole collection, Furbies, too. Gosh, you name it, he had it all.

Pete Helme: I don't understand Beanie Babies, and I still don't today, but it was one of the groundswell movements that got eBay noticed. Beanie Babies was one of those huge things—and porn, but we don't really talk about that.

Steve Westly: And playing cards, trading cards, baseball cards, hockey cards, football cards, and you know a lot of people were very dismissive, because Beanie Babies don't sit up there with commodities like oil and automobiles and PC or semiconductor sales. It was seen as pretty kludgy.

Pete Helme: People didn't think of eBay as a big technology thing, and at the time it was like, "Yep, I'm not doing anything cutting-edge here," until we had our issues later with scaling and crashing, and then we learned how much you really had to know to do this kind of stuff.

Steve Westly: But I was pretty good with numbers, and I could see the sell-through rates and the growth rates. I saw what others outside the company didn't, which is that there are a zillion people who are fanatical and passionate about tens of thousands of niche areas—we only put categories which the community asked for—and I was looking at these categories being requested, thinking, *You've got to be kidding, I don't even know what Depression glassware is. No one is going to buy this stuff!*

Joey Anuff: When eBay came around I couldn't think of anything better to do than buy every single toy my parents ever told me I couldn't get. Like Bionic Bigfoot from *The Six Million Dollar Man*. Still in the box! It was a fucking miracle.

Steve Westly: It's insane. There are millions of people who love this stuff. It was wild, and they were all in communities, community-ing with each other, and they were becoming pals and all along they're becoming deeply committed to eBay. We had built ourselves into the middle of a very powerful storm.

Pierre Omidyar: For the first, I think, three years—at least the first two full years of our history—we grew at 20 to 30 percent every single month. I don't think any other business has seen…I mean, every business in the start-up phase sees that kind of growth for a short period of time, but for such an extended period of time? And so as we were doing projections in terms of, "Okay, so this is what we've seen in the past, what can we project for growth next year, next quarter or whatever, so we can do budgeting and all that," we would only say, "Well, this can't last, you know, this can't last. There's no way you can grow this fast." So clearly there's no way I anticipated it, and even as we were growing—even with smart people—and I finally hired businesspeople to actually look at this thing.

Michael Malone: They brought in Meg Whitman.

Chris Agarpao: Pierre hired her to take it to the next level, and kind of stepped down from his position.

Brad Handler: Pierre gets credit for being able to say, "I'm not the guy to do the next step," and that's very hard for a founder to do. Jeff also had to give up some operational control to the new CEO.

Jeff Skoll: Neither Pierre nor I ever dreamed about being a captain of industry in Silicon Valley. We both believed in the company, but we both recognized we didn't have the experience at that time to maximize what eBay would be for the world.

Michael Malone: They got out of the way fast, almost too fast. There are stories of Meg saying, "Pierre, don't you care what I'm doing here?" And Pierre saying, "No, I trust you," and walking away.

Chris Agarpao: That's when it all changed. The next level was a more corporate, bigger phase.

Brad Handler: I was a lawyer. My job was to be the first real grown-up, I was the first person who ever said, "No, you can't do that," to anything. And so the very first week I was there I created a system to work with people who owned the intellectual property to goods that were being sold illegally, as well as to work with law enforcement for goods that had been stolen.

Steve Westly: eBay has this stunning idea: that they will enable you to sell anything to anyone. But there's these wild downsides, like, one Saturday I get a call from our engineering department saying the FBI is calling and they want to know why we have an operational rocket launcher for sale.

Brad Handler: The kidney, the virginity: That kind of stuff was all in the press. They were all just publicity-seeking pains in the butt, as far as I could tell.

Steve Westly: And I'd say, "Well, I'm terribly sorry, sir or madam, but we get ten thousand items posted every minute and we have strict standards and any time there's something inappropriate, we take it down."

Brad Handler: eBay would have been crushed if they didn't take the regulations seriously. The businesspeople and the technology people at eBay believed that the center of the universe was Silicon Valley: I'm not sure they could find Washington, DC, on a map! But we eventually got everybody on board with the fact that "being on the internet" is not the answer to why you don't comply with the laws of any particular jurisdiction.

Jeff Skoll: In the original peer-driven laissez-faire libertarian version of the site, we relied a lot on users to identify things for us that were problems. So if it's illegal, and you let us know about it, we'll take it down. If it's some sort of fake item or whatever, we'll take it down. But the law at the time—before 1996—was that if you meddle and go in and take down an item of your own accord, you are going to get sued if you miss a similar item. So we stayed away from that big-time.

Brad Handler: The backbone of eBay's defense against liability, as well as the backbone of every online community, is the notice-and-take-down regime established by the Digital Millennium Copyright Act. The DMCA established the rules of the road on how copyright was to be enforced online, and what we were able to get in there was this notion that as long as you take it down, you won't be held liable. That's a big deal, because without that, it would have been hard for eBay to prosper.

Pete Helme: And at some point in '98, around the time Meg Whitman started, we surpassed Visa and MasterCard for the number of transactions per second, which was a turning point for us. That was pretty cool, but on the other hand it's like, "Oh, we have to deal with this . . ."

Maynard Webb: They had so much success that for like a quarter or two they didn't let any new users sign on, because they couldn't handle the volume while they retooled.

Brad Handler: At the same time, we're trying to go public . . .

Jim Griffith: It was 1998 and the internet bubble was still really big, but it kind of went into temporary suspension while the bottom fell out of the Asian currency market. All the scheduled IPOs for that spring and summer were canceled.

Steve Westly: And people said, "It's a bad market, you can't go out, you're in the online auction business and no one knows what the hell that is."

Jim Griffith: Goldman Sachs had scheduled the IPO for October of '98.

Steve Westly: We're there with literally the most serious people from Goldman Sachs, and they're saying like, "How big do you think the Beanie Baby playing-card sports-card trading-card industry is? One hundred million? Two hundred?" And that's when I jumped up and said, "Well, you know we now have a thousand categories and we have an average eleven hundred items per category and we have a sell-through rate of X, and we're adding categories at this rate, and by the way, we're going into more geographic areas, and so we think this is a multi-billion-dollar company," and we were able to convince the people from Goldman Sachs that we might be on to something.

Jim Griffith: We were going to be the first tech IPO after the Asian currency market fiasco and there was a lot of speculation: Is this going to be good or bad?

Brad Handler: There was a long wait before the first trade. It was more than an hour after the market opened, and when the first trade came across the tape I think it was somewhere in the fifties. We all realized that our lives had all irrevocably changed. In that instant everyone in that room was probably worth more than their parents had ever made in their lifetimes.

Chris Agarpao: We had a big party. There was a conga line through the building! We could fit the whole company on one floor back then.

Jim Griffith: It was a great IPO, and that was an interesting summer, it really was.

Maynard Webb: I get my first call from eBay in 1999, in the June-ish time frame. And this isn't well known, but they had had a history of problems on their site.

Jeff Skoll: This goes back to some of Pierre's early choices to build on shareware: shareware software, shareware compilers, shareware databases. He did this because in part for costs, and in part because it was easy to find people that could help build stuff into shareware. But it didn't scale. When you have so many simultaneous users, bad things would happen. Either the whole system would go down, or you would have conflicts between the databases. When we didn't have any competition it was a nightmare, but it didn't kill the company. But when people started to know about the company it was a nightmare and it could have killed the company.

Maynard Webb: It was fits and starts all the way, and then they had this big catastrophic outage on June 10 of 1999.

Steve Westly: I was used to this, because I'd have to be handling marketing, and the press would start calling and asking when the site would come back up, and I'd talk to engineering and we'd say, "Fifteen minutes," or "Forty-five minutes," but that day there was an outage to the extent where engineering for the first time ever said, "We don't know." I said, "What do you mean, you don't know?" This was a bad one.

Pete Helme: It made the whole site unusable, and nothing worked, and people had to scramble, all the engineering teams got together, and Meg was there, rallying the troops: "How do we figure this out?"

Jeff Skoll: Part of the problem was that the only person that knew how the system worked was Mike Wilson. Pierre had worked with him before at The Well. And Mike worked his tail off: He was in the office seven days a week. And Mike for the first time in three years took a vacation. He went off to the Caribbean somewhere and he was sort of out of touch for a week, and that was the week the system went down. Pierre had long since lost track of what was going on, Meg didn't have a clue—that's not her background—and neither did the other engineers in Mike's group. The holy grail was held by Mike, so nobody really understood what was wrong.

Maynard Webb: The stock tanked, Meg spent the night there, and it was really bad. People were wondering whether they could keep this thing alive.

Jeff Skoll: We called in folks from SUN and Oracle and IBM and other consultants, and they worked nonstop for days.

Mary Lou Song: Meg brought in cots and toothbrushes and people didn't go home and were dying for sleep and we couldn't figure out what was going on. We were all scared, it was like, "What the heck?" There were TV crews in the parking lot. It was all hands on deck, and it felt like we were just trying to save ourselves from who knows what.

Pierre Omidyar: It was front-page news. We had CNN satellite trucks in the parking lot. It was big, big. It was, "The world is watching, this company is gone. It's going away."

Mary Lou Song: We started a telephone tree. Everybody who wasn't an engineer was calling customers and making sure they were okay.

Steve Westly: I just told people the truth: The site had grown too quickly, we did not know when it was going to be back up, and we were doing everything humanly possible to deal with it.

Jeff Skoll: There was a moment, about four or five days into it, when the technical team said, "We don't think that we can bring this up." And everyone just sat in silence for about twenty minutes. If you have a mindset where you are always looking for a black swan no matter how good the news has been, then you are always afraid of that black swan. Well, that was our black swan. It was, "Yes, the company is dead," and we're all in shock, but then a new breath of life came in. Right around then a junior Oracle programmer said, "I've just found something, I wonder if this could be it?" It was a line of code that was just off. And sure enough, that was the problem, and after we made the patch, the system started to come back up, gradually.

Pete Helme: It took us a day or two to figure out how to get everything rebooted and get the backups in place and all that.

Jeff Skoll: Now Mike was still away, and he had heard that there were problems and he was on his way back, but by the time he made his way back, Meg had called Maynard, and then that was the end of Mike's tenure.

Maynard Webb: Data corruption is a horrible thing; we don't need to go into all these fields of what went wrong, but what became clear is there needed to be a different path, and I was one of the people she reached out to.

Mary Lou Song: And then I think it happened again another month later.

Maynard Webb: And I was like, "I can fix this." Meg just reeled me in.

Pierre Omidyar: Meg really woke up to the fact that infrastructure and technology was critical and just really built that organization out.

Jeff Skoll: We struggled with the old system a little while more, while Maynard built a new system to take its place, and that was that.

Chris Agarpao: So Maynard was just awesome. He was like the scientist of all the technology back then, and he knew how to run it.

Maynard Webb: It was a crazy time, a time of very much worry at eBay because everybody thought the whole thing was just going to die and this wonderful dream that had been created was not going to be realized.

Pete Helme: I don't think anybody had an idea what to do in these early days of the internet and we were one of the fastest-growing companies, but Maynard knew what to do, or had a better idea than we did, for sure.

Maynard Webb: And Meg had promised, based on the June outage, that we would do a free listing day, which meant we took the fees away, and everybody would go crazy for twenty-four hours, and our site traffic would double or triple in volume—which is not necessarily a recipe for greatness when you've had scale issues.

Pete Helme: We ended up spending a lot more money on more machines and more hardware and more networking and stuff like that, yes. We had a room dedicated to monitoring all the systems. It looked like an air traffic control room, with all the monitors on the wall and stuff like that.

Maynard Webb: And so, nine days after I officially started was our free listing day, we were like Scotty from *Star Trek*—you know: "Captain Kirk!

Captain! She's going to blow!" But it worked, and the marketing folks were so relieved that they started coming in banging drums, thanking us.

Jeff Skoll: If there is an unsung hero, it's Maynard. He saved the company.

Michael Malone: And they just kept growing and growing.

Steve Westly: The stock split twenty-four to one from the initial IPO. It turned out to be a thousand-to-one return on the venture investment. Still one of the biggest in history.

Michael Malone: The next time I saw a company grow that fast it was Google.

The Shape of the Internet

A problem of great googolplexity

As the nascent web was taking off two graduate students at Stanford— Larry Page and Sergey Brin—watched from the sidelines. They weren't interested in using the internet to buy and sell stuff, or to read and publish stories, or even to score Grateful Dead tickets. They had a better idea: They would use it to get their doctorates. For Page and Brin aspired to be doctors of computer science, and the world wide web was the field's unstudied frontier. Yet studying the web was easier said than done. They would have to first capture it—download the web in its entirety—before the PhD-minting process could begin. In 1996 the web was still small enough that downloading its entirety to a stack of hard drives in a dorm room was possible, but just barely. They had to limit themselves to collecting the links and tossing the actual content of the web pages themselves. Yet that limitation didn't matter much, because Page and Brin weren't interested in the web's content, but rather its shape. And to two aspiring computer scientists the shape of the web looked like an equation—with approximately four hundred million variables and three billion terms. The equation proved solvable, but again, just barely. It spat out a number corresponding to each web page, sorting them like so many cards. When Page and Brin looked at the order that their equation had forced onto the web, they were surprised to find that they had inadvertently solved the hardest problem in computer science. They had made the internet logical, useful—searchable.

Heather Cairns: Larry and Sergey were PhD students at Stanford. That's where I met them. They were both young to be in a PhD program. I think they graduated high school a few years early.

Scott Hassan: Normally what happens to a first-year is that you get there and then a month later you take these things called qualifier exams, or "the quals," right? At the time there were like ten exams that you had to take, and most people fail three or four of them. Then for the ones you fail, you have to take those courses in order to bone up on that knowledge. And so that takes a couple of years. Then you start working on the PhD in year two or three. Well, supposedly Sergey was the first kid to pass all of them. He passed every single one of them, perfectly! And so the computer science department didn't really know what to do with him, and so he just played around.

David Cheriton: Back in 1994 or '95 I remember Sergey was Rollerblading in the Computer Science building—Rollerblading around the fourth floor with some of my graduate students.

Scott Hassan: So Sergey and I are good friends and we would go around picking locks and things like that. We could open any door in the whole entire place!

Heather Cairns: Sergey would come into my office with bad paintings because he knew I had a history in art and ask me what I thought. They were abstracts. He'd just kind of have a black blob on a brown background. He was probably trying to imitate Rothko or something, I'm not sure. I told him to keep his day job, but you've got to admire the spirit. Sergey was sort of a showboat, definitely an extroverted kid.

Scott Hassan: Then the next year, Larry comes as a first-year PhD student, and Larry is very different.

Heather Cairns: Larry's an introvert.

Scott Hassan: I don't think he passed all his qual exams like Sergey did, but Larry's very driven, right?

Larry Page: I was interested in working on automating cars when I was a PhD student in 1995. I had about ten different ideas of things I wanted to do.

Heather Cairns: Building a space tether to propel people into space was also an ambition.

Terry Winograd: So, the basic idea of the space tether was you put a rock out in space, in orbit, swinging around the Earth with a string all the way down to the ground, and then it's an elevator. You can just climb up the string like Jack and the Beanstalk, right?

Heather Cairns: Yes, the space tether. In those days they were still talking about it. I never thought it was serious but apparently it was.

Terry Winograd: They had this desire to be quirky. They were Burning Man before Burning Man. They just enjoyed speculating. "Oh, could we build a space tether? What would it take to build a space tether?"

Sergey Brin: I was chatting with Larry a lot. We got along pretty well.

Terry Winograd: Sergey and Larry were looking for projects they wanted to do. And so they got connected into the Digital Libraries Project because they were interested in that.

Scott Hassan: I was full-time at Stanford as a "research assistant," but really I was just a programmer. I was in a little, tiny little cubbyhole and working on the first parts of the Digital Libraries Project.

Terry Winograd: This was the era when online information was things like Lexis/Nexis, science journals, and medical journals: big institutions that had important information that was siloed. If you wanted to look up something in one of those information sources, you paid them, you got their application, and you went to it. As the internet became more widespread, it became obvious you wanted something that was able to deal with information from a number of these sources, so you weren't stuck with only one.

Sergey Brin: I was interested in data mining, which means analyzing large amounts of data, discovering patterns and trends. At the same time, Larry started downloading the web, which turns out to be the most interesting data you can possibly mine.

Larry Page: I had one of those dreams when I was twenty-three. When I suddenly woke up, I was thinking, *What if we could download the whole web, and just keep the links and . . .*

Scott Hassan: . . . *surf the web backward!* Mainly because it seemed interesting. You can say, "Oh, I'm on this page, what pages point to me?" Right? So Larry wanted a way to then go backward to see who was linking to whom. He wants to surf the whole net backward.

Sergey Brin: It was basically just collecting this data and seeing what we could do with it.

Larry Page: Soon after, I told my advisor, Terry Winograd, that it would take a couple of weeks to download the web—he nodded knowingly, fully aware it would take much longer but wise enough to not tell me.

Scott Hassan: So what Larry did was he started writing a web crawler. So what a web crawler does is that you give it a starting page and then it downloads that page, looks through the page, finds all the hyperlinks, then it downloads them and then it just continues to do that. And so that's how a web crawler works.

Terry Winograd: Downloading the entire web was their goal. Of course in those days "the entire web" meant something very different. And there was a technical question of how much space it would take? How many machines? How much bandwidth? I mean, one of the things we were fighting is the amount of just plain network bandwidth, out of Stanford, that it took to do all that crawling.

Scott Hassan: So if you were to write a crawler in a very simple form, it will be super-super-slow, because you'll be fetching a page and you'll be waiting for that thing to come back, right? At the time there were millions of web pages already, and so if it took you a minute to download every page, there would be more pages being created per minute than you would download per minute, so you would never finish.

Terry Winograd: Getting hundreds of thousands of pages and downloading them was a big deal.

Scott Hassan: So in the fall of '95 for some reason I started hanging out with Larry in his office. It was a really big office down near Winograd's, and he was in there working on it. And at that time I got really careful helping people with their projects because I found that once I started helping them, I started becoming their programmer. So I didn't want to help Larry specifically with his program—but he was having problems. He had written his crawler in a language called Java. Now, Java had just come out of SUN and was extremely buggy, especially running on top of Linux. So I carefully just helped Larry fix the Java runtime environment and the Linux kernel—rather than helping Larry with his actual Java code.

Terry Winograd: There's a pecking order in graduate school. So, PhD students are expected to do "original research" and master's students are not. They're expected to take courses and learn what to do and be helpers on programs. Scott was a master's student.

Scott Hassan: At the time, Larry was trying to download a hundred pages simultaneously. And I was fixing some of the bugs that he was having with Java itself, and this went on for weeks, if not months. And I remember thinking, *Wow, this is insane!* because I was spending a lot of time fixing this underlying tool. And so one weekend, I just took all his code, I took his whole entire thing, and threw it all out, and rewrote the thing that he's been working on for months very quickly—over a weekend—because I was just sick and tired of it. I knew I could get the thing working if I used a language I knew very well called Python. I wrote it in such a way that it could download thirty-two thousand pages simultaneously. So Larry went from barely downloading a hundred, to doing thirty-two thousand simultaneously on a single machine.

Terry Winograd: Scott was a programmer. I was not in the meeting so I don't know, but the basic model is Larry said, "Okay, we need a piece of code that does X, Y, Z," and Scott went off and built it.

Scott Hassan: I was really happy to show it to him on Monday, but Larry took one look at it and goes, "Great, it looks like you have this problem here, you have this problem here…" He pointed out like three different problems immediately. So very quickly it turned into him telling me what's wrong with it, and then me fixing it: the thing that I was trying to avoid in the first place.

By March 1996, Hassan's new and improved crawler had churned through fifteen million web pages: not the entire internet, but a sizable chunk. It was now possible to surf the web backward. The question then becomes: Why? What could back links tell you? What was this information good for?

Larry Page: Amazingly, I had not thought of building a search engine. The idea wasn't even on the radar.

Scott Hassan: So at the time he started doing this, there were already a few search engines out there.

John Markoff: The Yahoo model was actually to try to structure the information on the internet by sorting it—by hand—into categories. Then people created search interfaces that would allow you to type an arbitrary string and get back a list of results.

Brad Templeton: Then Alta Vista launched in December of 1995 and it immediately became the popular favorite. Everybody switched very quickly, because of its speed.

Scott Hassan: And so nobody wanted to do a search engine, because that wasn't "research." At Stanford you had to do something fundamentally new, so you couldn't do something that's already done, because that's not cool, right?

John Markoff: There were so many search engines at the time. They were all over the place. Building the crawler and downloading the web was not Google's breakthrough. The breakthrough was PageRank.

Terry Winograd: I can remember Larry talking about a random walk on the web. "Random surf," he called it. So, you're on the web at some page and it's got a bunch of links. So you pick one of them at random and you go there. And then you do this again and again with a zillion bots. So, if everybody's doing this, where would you end up most of the time? The point is if lots of people point to me, you're going to end up with me more often. I'm very important—so I'm getting a lot of traffic. Then if I point to you, you're going to get a lot of them even if there's only one link from me to you: You're going to get a lot because I'm getting a lot. So think of this traffic moving through this network, just statistically. Who would get the most traffic?

Scott Hassan: Larry came up with the idea of doing random walk, but Larry didn't know how to compute it. Sergey looked at it and said, "Oh, that looks like computing the eigenvector of a matrix!"

Sergey Brin: Basically we convert the entire web into a big equation, with several hundred million variables, which are the page ranks of all the web pages and billions of terms, which are the links. And we're able to solve that equation.

Terry Winograd: You can get fancy about it in formal terms. But in informal terms, PageRank was the implementation of that intuition.

Scott Hassan: So Sergey just saw that and was like, "Okay great, I'm going to need a computer with four gigabytes of main memory to compute this." So at the time to have a computer with four gigabytes of main memory was crazy, but it turns out that there was one computer in the computer science department that did have that, and that was in the graphics lab. They had a really, really big machine that had a whole bunch of memory and they were using it for graphics. So Larry got access to that big computer and we basically ran the algorithm on it for a couple of hours, and once it computed it, it was done.

Sergey Brin: Every web page has a number.

Larry Page: Then we were like, "Wow, this is really good. It ranks things in the order you expect them!"

John Markoff: It's a very simple idea: You saw the most popular things first. PageRank was an algorithm that looked at what other humans thought was significant—as demonstrated by other people linking to them—and used that as a mechanism for ordering search results.

Sergey Brin: And we produced a search engine called BackRub. It was fairly primitive, it only actually looked at the titles of the web pages, but it was already working better than the available search engines at the time in terms of producing relevant results. If you're searching for Stanford you get the Stanford home page back, for example.

Scott Hassan: So then I sat everyone down and said, "Hey, let's build a *full* search engine!" Right? And both Larry and Sergey thought it was going to be a lot of work. I was like, "No, no, no, actually it's not that much work. I know exactly how to do it."

Butler Lampson: A search engine has two pieces. One piece is the piece that crawls the web and collects all the pages and the other is the piece that indexes them.

Scott Hassan: If you look at the back of a book, sometimes there's an index, and essentially that's what you do when you build a search engine, is you do that, but you do that for every word.

Butler Lampson: And then of course nowadays there is a third piece which does the relevance part. It has to figure out what answers to show in response to the query.

Scott Hassan: Very quickly, in six to eight weeks, we were able to build the whole structure of Google. It was mostly just Sergey and I from two a.m. to six a.m. in the morning. We just worked on it in the middle of the night, mainly because if I worked on it during the day I would get yelled at by my boss, because building a search engine was not considered research.

Sergey Brin: By 1997 we are already developing a bunch of other technologies, too. For example, hypertext analysis, which is where we started asking questions like: Where are all the terms on the page? What are their font sizes? Which things are headers? And even: What is the text of pages which are closely linked to this page? What do they say? PageRank was the importance component, but we combined that with this relevance measure, based on this hypertext analysis, and the two together really functioned well.

Scott Hassan: We got the search engine to a certain point and then Larry built this little interface. You go to this web page, and then on top of the web page there was a single box, very similar to Google's search box today, right? It's just a single box, and beside the box was another drop-down box, and "Which search engine do you want to use?"

Brad Templeton: There was a bunch: Excite, Lycos, AltaVista, Infoseek, and Inktomi—that one was done at UC Berkeley.

Scott Hassan: So, you could select one of those other search engines, and then you'd type in your little query and you'd hit Search, and what it would do, it would split the screen in half. On the left-hand side it would just pose that query to the search engine you chose, and then on the right-hand side it would then pose it to our search engine, so that you could compare the results side by side. So Larry set up all these meetings with all the search engine companies to try to license PageRank to them.

David Cheriton: I think it must have been the '97 time frame that Larry and Sergey came to me about the idea of licensing the search technology, and I told them that I didn't think that this was likely to work, because the reality is it's often difficult to find anybody in a company who is going to say, "Gee, this is really important and I'm willing to say we're too dumb to have developed this ourselves, so we'll license it from you!"

Scott Hassan: Larry wanted to finish his PhD, so he just wanted to license it to whoever wanted it.

David Cheriton: So anyway, they went off on that particular venture of trying to license it, and I think some of the interesting early stories of Google date from then—who could have bought the whole thing for $2 million or something of that flavor—if they had decided to pounce on it.

Scott Hassan: I remember going to this one meeting at Excite, with George Bell, the CEO. He selects Excite and he types "internet," and then it pops up a page on the Excite side, and pretty much all of the results are in Chinese, and then on the Google side it basically had stuff all about NSCA Mosaic and a bunch of other pretty reasonable things. George Bell, he's really upset about this, and it was funny, because he got very defensive. He was like, "We don't want your search engine. We don't want to make it easy for people to find stuff, because we *want* people to stay on our site." It's crazy, of course, but back then that was definitely the idea: Keep people on your site, don't let them leave. And I remember driving away afterward, and Larry and I were talking: "Users come to your website? To search? And you don't want to be the best damn search engine there is? That's insane! That's a dead company, right?"

Sergey Brin: Ironically, toward the end of the 1990s most of the portals started as search engines. Yahoo was the exception, but Excite, Infoseek, HotBot, and Lycos began as search engines.

David Cheriton: Search kind of got steered off the path once the business-people got involved. They decided that the model that they understood was a newspaper. And so in a newspaper, search is a minor thing, it's not the business. It's the index that says where the sports section is and so on.

Sergey Brin: Searching was viewed as just another service, one of a hundred different services. With a hundred services, they assumed they would be a hundred times as successful.

David Cheriton: So I think it was approximately a year later that they came back to me and said they weren't getting anyplace with the licensing, and I didn't say, "I told you so," but I think I felt a little bit smug inside.

Terry Winograd: By that time, I think it was clear that they had traction in the real world and they wanted to put their energy into making that grow, not into doing an experiment that would satisfy some thesis committee.

Sergey Brin: It was the summer of 1998. At that point we were just scroung-ing around to find resources, we had stolen these computers from all over the department, sort of. We'd assembled them all together, but they were haphazard, like a SUN, an IBM A/X computer, a couple PCs.

Heather Cairns: They were taking servers off the loading dock. They were crashing them with traffic just through word of mouth.

Larry Page: We caused the whole Stanford network to go down. For some sig-nificant amount of time nobody could log in to any computers at Stanford.

Heather Cairns: And they were, essentially, nicely asked to leave because of that.

Larry Page: Stanford said, "You guys can come back and finish your PhDs if you don't succeed."

David Cheriton: They thought they had a big challenge of raising money, I thought that money wasn't the big problem, and so I kind of put myself in the spot of having to prove this, so I contacted Andy Bechtolsheim.

Sergey Brin: One of the founders of SUN Microsystems, and a Stanford alum.

David Cheriton: Andy said he was interested, and we arranged to meet at my house, and Andy pulled up in his Porsche and they put on this demonstration of the Google search engine, and there was a certain amount of discussion.

Larry Page: It seemed like they really wanted us to start a company.

Andy Bechtolsheim: The question, of course, is, "How do you make money?" And the idea is, "Well, we'll have these sponsored links and when you click on a link we'll collect five cents." And so I made this quick calculation in the back of my head: *Okay, they are going to get a million clicks a day at five cents, that's fifty thousand dollars a day—well, at least they won't go broke.*

David Cheriton: Andy just got up and walked back to his Porsche, if not ran, from the porch to the Porsche, that is, got the checkbook, came back, and wrote them this check.

Sergey Brin: He gave us a check for a hundred thousand dollars, which was pretty dramatic. The check was made out to "Google Inc.," which didn't exist at the time, which was a big problem.

Larry Page: We didn't have a checking account, we didn't have a company, we didn't have anything.

David Cheriton: Andy just handed them the money: "Let's work out the details later; let's get going."

Sergey Brin: We hadn't really discussed valuations and stuff like that. He figured it would pay off, and he was right. We finalized all the details on the round after that. I guess we figured if we didn't agree later, that it would be a loan. He liked us and he just wanted to sort of push us forward.

Larry Page: It was pretty unreal.

Sergey Brin: It was like, *That was pretty easy!*

Brad Templeton: Then they went to Burning Man.

Ray Sidney: Sergey put up a Burning Man logo on the website. It was the first Google Doodle.

Marissa Mayer: It was more of an out-of-office notification than anything else—it said, "We're all at Burning Man."

Brad Templeton: There was a Google contingent that camped at Burning Man. I remember saying something rude to Marissa that I shouldn't have said about wanting to see her naked. She won't remember that, I hope.

Marissa Mayer: Remember, we were all young, and we were all, in addition to being coworkers, friends.

Scott Hassan: I was responsible for the shelter and Sergey was responsible for the food. So he went to the Army Navy supply store and just bought all the rations—MREs. They were pretty fascinating. You would pour water in this little bag, and it had some sort of chemical in there, and it got really, really hot—and cooked it. So you didn't even have to have a stove, you didn't have to have anything! And we drove his car to Burning Man and just went around.

Jamis MacNiven: In '98 it was only about fifteen thousand people and really relaxed and cool and there was a lot of space—very spread out.

Scott Hassan: We didn't really have a camp, we had little tents and saw all the sites and stuff like that.

Brad Templeton: In the early years Sergey would just actually sleep in whatever camp he found himself in at the end of the night.

Google is incorporated immediately after the return from Burning Man.

Heather Cairns: They handed me a folder full of checks for like $100,000, $200,000 from Andy Bechtolsheim, Jeff Bezos, David Cheriton. They sat in the back of my car for weeks because I couldn't get out of work in time to even get a bank account opened.

Ray Sidney: I had never worked at an early stage start-up before, and you know what? It was intense. My first week at Google I did two all-nighters. We saw this big opportunity and at the same time there was so much in doubt, so we wanted to do whatever we could to make it work, and so we worked hard. We had visions of greatness.

Kevin Kelly: I had a chance to meet Larry Page because his brother, Carl, worked in the same building as *Wired*, and so I was at a party with Larry, and this is at a time when Google was pre-IPO and they actually didn't have a business model, they weren't selling ads...

Heather Cairns: We didn't have a business plan, and they would tell me that their actual mission statement was "to rule the Earth." I'm thinking, *Well, whatever you want, just make sure to sign my check and I'll go on my way when it crashes and burns in a couple of years.*

Kevin Kelly: I said, "Larry, I don't get it. What's the future of search for free? I don't see where you're going with this..." And Larry said, "We are not really interested in search. We are making an AI." So from the very beginning the mission for Google was not to use AI to make their search better, but to use search to make an AI.

Heather Cairns: *To rule the Earth!?* Here we are. There are seven people inside of someone's house, working out of bedrooms, and that's what they were saying then.

Ray Sidney: The first Google digs was half of Susan Wojcicki's house, including a garage.

Heather Cairns: We were allowed to use Susan's washer and dryer that was in the garage. But we were working out of bedrooms; we weren't in the garage. That's the folklore, because every start-up is supposed to be in a garage.

David Cheriton: I'm tempted to say it was like a frat house, although I'm not sure I've ever been to a frat house. It was a long way from a professional company—frankly, once I invested I really didn't have the sense for some time whether these guys were really taking the company routine seriously or not, so I used to say, "Well, I spent $200,000 on a T-shirt."

Heather Cairns: The parties would be rocking—like, they'd be rocking by anyone's standards, let alone an office party standard. We'd have a hundred people come and we have props from movie theater companies. And we had a hot tub, too, so you can take it from there.

David Cheriton: The office they had on University Avenue was a step in the right direction.

Brad Templeton: It was this office space in downtown Palo Alto that had all the giant balloon chairs and stuff which became their motif.

Marissa Mayer: The lava lamps were a thing because they came in every color of the Google logo. The bouncy balls were a posture thing, but also a fun thing.

Charlie Ayers: The first time I met Larry and Sergey, Google was still head-quartered over the bicycle shop on University Avenue. It didn't look like a business whatsoever—it looked like a bunch of kids in their midtwenties, you know, just screwing around.

Marissa Mayer: Charlie started when we moved over to the Shoreline campus.

Charlie Ayers: I remember going in for an interview and Larry bounced on by on one of these big balls that have handles on them, like you buy at Toys"R"Us when you're a kid. It was just a very unprofessional, uncorporation attitude. I have a pretty good understanding of doing things differently from the Grateful Dead—I've worked on and off with them over the years—but from my perspective, looking from the outside it was an odd interview. I'd never had one like that. I left them thinking that these guys are crazy. They don't need a chef!

Heather Cairns: I was very surprised that they hired this ex–Grateful Dead chef, since clearly everything that goes with that is coming with Charlie. Talk about a counterculture person!

Charlie Ayers: Larry's dad was a big Deadhead; he used to run the Grateful Dead hour talk show on the radio every Sunday night. Larry grew up in the Grateful Dead environment.

Larry Page: We do go out of our way to recruit people who are a little bit different.

Charlie Ayers: There was no under-my-thumb bullshit going on where you all had to dress and look and smell and act alike. Their unwritten tagline is like: *You show up in a suit? You're not getting hired!* I remember people that they wanted showing up in suits and them saying, "Go home and change and be yourself and come back tomorrow." A lot of companies wouldn't have done that, but they were bent on being the new local color of the Silicon Valley. Before Google, Apple were the local-color guys, because they had the skull and crossbones flag flying on the roof. But then Google wanted to do that fuck-you-to-the-Man thing. Again, they were doing things no one else had done before in Silicon Valley. They were doing what they knew was right. I always thought of them like that. They're really cool like that.

Larry Page: We find that we have a really fun culture. We have a dog policy which we got from Netscape. It's like a two-page document explaining what you can and can't do with your dog at work.

Heather Cairns: We said it was okay to bring pets to work one day a week. And what that did was encourage people to get lizards, cats, dogs—oh my God, everything was coming through the door! I was mortified because I know this much: If you have your puppy at work, you're not working that much.

Larry Page: We take the whole company on a ski trip to Tahoe every year.

Douglas Edwards: We would go up to Squaw, and attendance was pretty much mandatory. That became the company thing.

Ray Sidney: The very first ski trip was in the first part of 1999. That was definitely a popular event over the years.

Charlie Ayers: I come from the world of the Grateful Dead, so I know how to party, and so on the ski trips in Squaw Valley I would have these unsanctioned parties and finally the company was like, "All right, we'll give Charlie what he wants." And I created Charlie's Den. I had live bands, deejays, and we bought truckloads of alcohol and a bunch of pot and made ganja goo balls. I remember people coming up to me and saying, "I'm hallucinating. What the fuck is in those?" It was like something out of *The Hangover*—hot girls drooling all over themselves and people passed out on the chairs.

Larry Page: All sorts of fun things.

Charlie Ayers: Larry and Sergey had like this gaggle of girls who were hot, and all become like their little harem of admins, I call them the L&S Harem, yes. All those girls are now different heads of departments in that company, years later.

Heather Cairns: You kind of trusted Larry with his personal life. We always kind of worried that Sergey was going to date somebody in the company...

Charlie Ayers: Sergey's the Google playboy. He was known for getting his fingers caught in the cookie jar with employees that worked for the company in the masseuse room. He got around.

Heather Cairns: And we didn't have locks, so you can't help it if you walk in on people if there's no lock. Remember, we're a bunch of twenty-somethings except for me—ancient at thirty-five, so there's some hormones and they're raging.

Charlie Ayers: HR told me that Sergey's response to it was, "Why not? They're my employees." But you don't have employees for fucking! That's not what the job is.

Heather Cairns: *Oh my God: This is a sexual harassment claim waiting to happen!* That was my concern.

Charlie Ayers: When Cheryl Sandberg joined the company is when I saw a vast shift in everything in the company. People who came in wearing suits were actually being hired.

Heather Cairns: When Eric Schmidt joined, I thought, *Well, now, we have a chance. This guy is serious. This guy is real. This guy is high-profile.* And of course he had to be an engineer, too. Otherwise, Larry and Sergey wouldn't have it.

Sergey Brin: Larry and I searched for over a year, and managed to alienate fifty of the top executives in Silicon Valley. Eric was experienced and the only one who went to Burning Man.

Eric Schmidt: We had all been at Burning Man together.

Sergey Brin: Which we thought was an important criterion. He's a great cultural fit. We hang out together. We discuss and decide on stuff together. More companies should look at cultural fit.

Charlie Ayers: A lot of people internally in the company were really happy to see him coming because he was now the official old guy. Before Schmidt, you'd look around the building for an adult, and you're not seeing too many of them.

Heather Cairns: One of his first days at work he did this sort of public address with the company and he said, "I want you to know who your real competition is." He said, "It's Microsoft." And everyone went, *What?*

Ray Sidney: In one of his gigs previous to Google, Eric was the CTO of SUN Microsystems, which was a high flyer until, essentially, Microsoft ate all their lunch. Because over the years regular PCs became so much more capable that it became very different for people to justify some crazy expensive workstation. It's the same thing that happened to Silicon Graphics, right? SGI had these even higher-end systems, and then basically PCs caught up. So Eric had the experience to know.

Heather Cairns: And Schmidt said, "Microsoft is a monopoly, and as long as we can stay under the radar, we have a chance. So we don't even want them to look at us. We don't want them to know about us." We're like, "But they're not a search engine!" He's like, "It doesn't matter. As soon as we get on their radar, they're going to try to crush us."

Terry Winograd: I can remember some higher-level meetings I was in, which were about what Google could do that would stay under Microsoft's radar. In fact, "Canada" was the code word for Microsoft, because it was big and up north, right? There was basically a sense that if Microsoft decided Google was a threat they could squash it, and they wanted to make sure they didn't trigger that reaction.

Ev Williams: There was quite a bit of angst and existential concern that the next version of Windows was going to have search built into the OS. And how were we going to compete with that?

Heather Cairns: So, I remember thinking, *Huh, wow. He thinks we're a threat to Microsoft. Are you kidding me?* So I think then that speech made me realize that maybe we had more gravity than I understood.

Marissa Mayer: It was a bigger vision than we had really tangibly talked about before. That was a big moment for us.

Free as in Beer

Two teenagers crash the music industry

*I*n *the late nineties dot-com hype and hysteria overtook Silicon Valley and the nation. If not for Monica Lewinsky, the web would have been all any journalist ever wrote about—and thus the old media drove legions to the new. Millions bought a modem, installed Netscape Navigator, and signed a contract with an internet service provider. Then they finally logged on for a firsthand look at cyberspace—and often logged off feeling a bit disappointed. One could buy something off eBay or from one of the other e-commerce sites that were popping up like mushrooms. One could Google various topics in an infinite library of ridiculously uneven quality. One could snicker over webzines like* Suck. *And that was about it. There was just not all that much fun to be had online—until Napster came along in 1999. Napster was a search engine—like Google—but one that was specifically optimized for finding music files. It opened up a Pandora's box: all the world's music, ripe for the picking. But was Napster a tool for stealing? Or sharing? The battle over the meaning of Stewart Brand's maxim "Information wants to be free" heated up until it nearly became a shooting war. Napster pitted young versus old, the future versus the past, Silicon Valley versus the entire rest of the world.*

Shawn Fanning: It's hard to explain where things were at back then.

Jordan Ritter: This was a long time ago. MySpace was cool. Google hadn't taken off. There was no Facebook. None of that stuff had happened yet.

Shawn Fanning: I was eighteen. I hadn't seen much of the world…

Ali Aydar: I'd known Shawn Fanning since he was fifteen. For him I was kind of the older guy that he looked up to and asked advice…even though at the time, you know, I was twenty-three.

Mark Pincus: Sean Parker came and worked for Freeloader in the summer of '96 when he was fourteen. We were the only internet company in DC, and Freeloader was, for a minute, the first viral phenomenon on the internet. Then we got acquired just as we were moving to San Francisco. We felt like San Francisco was the Motor City of the internet, and if you were going to be on the internet you had to be there.

Eileen Richardson: I was a partner of a venture firm based in Chicago, and I believed the internet was going to change the world. The other partners told me that this was just another wave of technology and it soon would be gone. And I said, "I disagree. Bye-bye. I'm going out to Silicon Valley." That was '98.

Ali Aydar: It was the dot-com boom, and I'm in Chicago languishing away at a bank, and I have a degree in computer science from Carnegie Mellon. *Why am I at a bank in Chicago? I need to be in Silicon Valley, I need to be in the dot-com boom, and be taking advantage of it in some way.* So I came out to Silicon Valley in August of 1999.

Jordan Ritter: I was a paid hacker. I lived in downtown Boston, which was the center of hacking in the United States at the time: It wasn't New York, it wasn't LA, it wasn't even Silicon Valley. Boston was the seat of it, and I was in the middle of that.

Sean Parker: Shawn Fanning and I met in the computer underground. We're living in suburbia, interested in computer science, and hanging out with a lot of people online who shared our interest.

Jordan Ritter: Fanning's online handle was "Napster," and Napster is in this hacker community called w00w00, and he was just like, "Hey, there's a .exe, go check it out," and a couple of us download it and use it, and there's nobody on it, but it works for the ten of us that were on it.

Shawn Fanning's program immediately got noticed: The first people to jump on the Napster bandwagon were his friends Sean Parker and Jordan Ritter. Parker started working on a business plan, while Ritter rewrote much of the code in order to make sure it could scale. Sure enough, the users started coming—and immediately after that, the money. Silicon Valley investors were desperate to find the next Netscape, the next eBay, the next big score.

Jordan Ritter: So I'm working on this stuff. As we started to work on it, we all got bit by the bug, and maybe it was May when it was starting to take off:

Maybe there were a hundred users. I'm feeling good, right? I was young, and I'm half stoned. I did some of my best work stoned, it was great. Actually, most of Napster was written stoned. I'm just going to be honest about that. But I was writing code before there was a company, before there was equity, before there were salaries. So I am a technical cofounder. We didn't have that language back then, because nobody gave a fuck about engineers back then. Parker and I barely ever interacted. Parker plainly was a groupie, he wanted to be part of this elite, underworld group of hackers. But he could never make the cut, which was hard for him, because we were the outcasts. We were the social outcasts. How the fuck do you get rejected by the outcasts?

Mark Pincus: Sean Parker was a Unix programmer—or he said he was. What he had, and still has, was an amazing nose for what's going to be the next viral thing.

Shawn Fanning: Sean was trying to make it a business and legal while we kept it up and running and were scaling and treading water. It was one day at a time.

Jordan Ritter: Then there was a day, I think it was June, where I got a message from Shawn that said something to the effect of, "Holy shit! John incorporated the company—and he dicked me out of the ownership!" And the blood drained out of my face. Shawn didn't have a dad growing up; John was his uncle. He was absolutely the car salesman type. He could bully you into doing anything. And John, without telling Shawn, incorporated the company and gave himself 70 percent of it. And if there was one mistake, one fucking mistake that sank us, it was that one. Because when it came time to raise money, no one wanted to deal with John and his big-swinging-dick attitude, so we ended up settling for second best, or third best, or fourth best, fifth best, time and time again. That really set us up for failure from the very beginning.

Mark Pincus: Sean e-mailed me the business plan for Napster and he said he was working with this other Shawn, Shawn Fanning. He said he had this music download service that was exploding, and every time they added more servers it would grow and they needed more money so they could buy more servers. And that's usually a good sign that you want to invest: Everyone is eating the dog food, they just need more of it. So I gave them a hundred thousand dollars, and it was one of the first checks that they got.

Shawn Fanning: We eventually took money from John Fanning's friend Yosi Amram.

Eileen Richardson: Yosi was a guy that I had known for years in Boston and who had moved to Silicon Valley not long before me. And that's when this man, Yosi Amram, said, "Hey, check this out. I think that you'd be really interested." And I went home and downloaded Napster. And I remember just going, "Oh my God, this is it! This is the missing key." So Yosi put in a couple hundred thousand dollars and I put in a couple hundred thousand dollars so that we had some money.

The money came with three conditions. One: that John Fanning get out of the way. Two: that the company move to Silicon Valley. And three: that the company get some professional management help—so-called adult supervision.

Shawn Fanning: Sean Parker and I moved out to Northern California. We hired some people and it became a company.

Ali Aydar: So I had been in Silicon Valley for maybe three weeks when I get a phone call from John Fanning, who said, "You're not going to believe me, but Shawn has gotten some angel funding for his project and they made a move out to San Francisco and you should really, you should really meet them." So we met in a restaurant in downtown Palo Alto. And Shawn had told me the concept to develop Napster about ten months earlier, when he was still in school, and at the time I had told him, "It's not going to work, and the reason why it's not going to work is, you're going to have a hard time getting people to open up their hard drives. Because people are going to be concerned about viruses." The thought that people are just going to open up their hard drives to some program that they downloaded off the internet was incredulous to me, and for Napster to really work, everyone would need to do that. So they're describing it to me at this restaurant in downtown Palo Alto, and now I take some time to understand, because I fundamentally still had the same problem with it. So Shawn explained more about it technically. How he got some initial adoption by just putting a server up through his uncle's cable modem and because it got so much traffic the first time he put it up how it crashed the cable modem. And both Sean and Shawn were really animated about why this is such a big deal: It's not just us young people who are into music but our parents and our grandparents. Music touches everybody, and for everybody, young or old, there is some music that they like, and so that means the addressable market for this thing is every human being on the planet. And I slowly begin to see why this is going to be huge.

Sean Parker: And, look, if you grew up online, you knew that in a matter of years everyone would be sharing MP3s.

Jordan Ritter: We had our MP3 libraries that were totally legal, and we shared them with everybody. We shared with each other, and we were like, "Yes! This is cool!" and we all got excited about it.

Sean Parker: There was no moment of doubt about the inevitability of what was happening. And also there wasn't any arrogance; it was like, "Oh, we're just building a better front-end interface to something that's already been taking place."

Eileen Richardson: Napster sort of took what existed and just made it 1,000 percent better.

Mark Pincus: It was just like *The Matrix*—everyone's computers are just connected directly to each other and it's one gigantic hard drive. You typed in "Madonna, Like a Virgin," and it's like: "4.5 million computers are connected right now, and there's 12,125 versions of that file available, pick one and download it."

Hank Barry: It was miraculous—on the order of inventing a warp drive in your own basement.

Ali Aydar: So here's the deal, here's how Shawn fundamentally came up with the concept. So he's in his dorm room at Northeastern University in 1998. The web was a few years old. Web pages were a bit arcane relative to today. Everything was HTML based, there's no Flash, there's no interactivity.

Shawn Fanning: At the time most of the music downloading services on the web were pretty much designed in the same way that traditional search engines were.

Ali Aydar: And if you go to those search engines, you could type in artists' names and song titles and look for music, and oftentimes you'd get search results that looked like they led you to MP3 files.

Eileen Richardson: MP3 is just a format—which had to happen before Napster could even happen. We had to have music in that format.

Hank Barry: MP3 is a compression standard that could take one track on a CD, which is about forty megabytes, and make it into a three- to five-meg file.

Ali Aydar: If we were all still on modems, this wouldn't have worked. It would have been too slow.

Hank Barry: Of course it was in the dorm rooms of America that you had the first beginnings of broadband adoption.

Shawn Fanning: And at the time I was in a sort of corner room, so two adjoined rooms with a common area, so we had a total of five students in there. One of them was a huge music fan, and he would pretty much skip class on a regular basis to download music. When he would find a site that was up, he would just skip class.

Ali Aydar: So Shawn observed his roommate looking for music, and Shawn got into the act, too: They'd type in an artist-and-song-title combination, and might get a hundred results, but ninety-nine out of a hundred of them are dead links!

Shawn Fanning: The reason was because where traditional search engines are usually indexing content on dedicated servers, in this case it was a lot of people's personal music collections on personal computers. So what would happen is someone would turn their computer off after it was indexed.

Ali Aydar: So you'd click and get "File not found." You can't download anything because there's nothing there. And Shawn observed his roommate getting really frustrated, and he himself was frustrated. So you've got all these college students—college students are really into music and listening to music—and now you've got this internet thing in dorms: It's a perfect mix.

Jordan Ritter: Ingenuity requires three basic things: adversity, scarcity, and necessity. And when those three things happen, you invent new things.

Eileen Richardson: What people didn't understand then was that Napster was not a technology, it was an invention.

Ali Aydar: So Shawn thought, *Why can't there be a central server that keeps track of whether that guy is on or off?* And if such a server existed, and you searched against the content that that person has available, if that person is on, then you will see that piece of content in a search result. If it's off, then you won't see it in a search result.

Sean Parker: I remember having conversations with Shawn saying, "Look, we just need to build a central index. So every time I log on, it tells the index all the files I have and makes them all searchable, and in order for me to download, I have to share—so there is a little bit of quid pro quo. And it has to be so easy to use that my grandmother could do it."

Hank Barry: He went down and bought a book called *Introduction to Visual BASIC*. Napster was the first software program he ever wrote.

Shawn Fanning: It wasn't the first piece of software I'd developed; I'd written a lot of software before. It was my first Windows application.

Hank Barry: And he just sat there, would not give up, he was dogged, and he pulled this thing together: He got it technically together.

Ali Aydar: Okay, so now, so that's all fine and dandy, but here's where the stroke of brilliance was—Shawn found an ingenious technical solution to that case, such that if I was downloading a file from you and the download gets cut off, for whatever reason, the Napster server would find another person who had that exact same file and seamlessly start downloading from that other person. As the downloader, I wouldn't even know the difference. I wouldn't even know this was going on! It would just happen automatically, and I would get a complete file every time. So the reason why Napster really worked is because you got complete files, every single time. Really, to have come up with the concept of Napster? Sitting in a dorm room? I mean it's almost like a fable. It's almost unbelievable. But then when you really understand what was going on? When you peel the technical onion a couple layers deep? *Wow!* It really was incredible. It was a eureka moment.

Jordan Ritter: Fast-forward about two months: The company actually got formed, equity got served out, and executives were hired. All this was happening on the other side of the country and I didn't know what was going on. So I fly out and land, and for a boy who has never been to the Valley before: *Holy shit, this is magical!* The Valley at night is this amazing mind-bending, -blowing experience. There's mountains right there, and there are lights peppered across these mountains, and there's water on the other side, and *Wow!* So I get picked up by Fanning, Parker, and Ali and go straight to the Napster office.

Ali Aydar: It's just a little four-story building in downtown San Mateo. We were on the top floor.

Jordan Ritter: I walk into this room and *Wow!* What a clusterfuck of a feeling that was. I could feel the hair on the back of my neck standing up, like, *This doesn't smell right—this doesn't feel right.* Fanning has already lost control. But God bless him: I'm not expecting a fucking eighteen-year-old to have control of the situation. He's eighteen! There's Parker, who looks like he's high and jittery, probably on cocaine. There is Eileen, who had a really nice smile and a really ebullient energy, and all these other executives I had never heard of or met before. I really don't trust these people, so I'm really tentative, but Ali I like: He is peeps, he is kin, he is community. And Ali and I spent basically the rest of the entire evening and morning talking about circumstances and people and history and how we might move forward.

Ali Aydar: We all spent a year in that building. That's where we spent the time building the bulk of Napster.

Moby: A friend of mine showed me Napster early in 1999. I felt very high-tech with my Netscape browser and my dial-up modem, and it was interesting, but I didn't have the prescience to understand what it augured for the future.

Eileen Richardson: I remember being so excited once we hit one million users. we had a little celebration in the office. We knew that Napster was off to the races.

Sean Parker: Although in an aesthetic sense it was terrible: This ugly, poorly built Windows application that looked terrible in the same way that MySpace looked terrible.

Hilary Rosen: The first time I saw Napster was in the summer of '99. My head of antipiracy came into my office, pulls up Napster, and says, "Pick a song," and I think I picked something current that week and there it is! We were astounded. It was so beautiful and simple. And we thought, *This is incredibly cool—but it's obviously illegal.*

Hank Barry: It's very fashionable for people afterward to say, "Oh yes, it was a really cool thing and we really saw the potential and gosh, we were just so troubled that it was just sort of apparently illegal." That's a fraud. That's a falsehood. That's spin.

Hilary Rosen: By the time it came to our attention, it had already been sort of turned into a minibusiness. And so our first communications were with Eileen Richardson.

Eileen Richardson: I said, "Well, tell us where we're infringing. Let us know. And then let's talk again and see what we can do." And she said, quite snobbily I thought, "Just get *Billboard* magazine and open it up and there is the Top 200. You're infringing on all of them," or something like that. I can't really do anything with that, I really can't. That's sort of where that conversation went. She said, basically, "Shut this shit down!" And I was like, "Shut what down?" We didn't own anything; we don't serve it up!

Shawn Fanning: I really didn't think that there would be so much controversy around the legality because we were just linking to the content and there were so many other search engines at the time that were providing links to that content.

Ali Aydar: Napster was perceived by the investors, the founders, myself, to be legal because Napster's central server did not hold any copyrighted content. There was no content on our servers. Moreover, no content ever traveled through our servers. The content traveled from one peer—a "peer" is a computer, sitting out there somewhere—to another peer.

Hank Barry: Napster was the first application based on that interchange of data out on the edge of the network.

Sean Parker: Prior to the release of Napster, prior to that pivotal moment, the web was one-way. It was a client-server model, you accessed information that was stored on a server. It was very much this broadcast model where individuals passively consumed what had been published. It wasn't a two-way street. But the moment Napster launched, you were fully utilizing the capabilities of the internet. Everybody was sharing content. Everybody was downloading content. Everything is interactive. That was the original potential of the internet. Napster was ahead of its time.

Hank Barry: Nowadays, it's the norm, it's Facebook, it's what an internet application is, but it really did not exist before Shawn did it. Everything up until that point had been like media, like, *Oh, we're going to play a program and you're going to watch it.*

Ali Aydar: So the only thing Napster's central server had was information about the names of files people had, that's it. I couldn't even tell you what was actually on those files. They were just file names: fundamentally the same thing Google has—with the exception that Google also indexes words on a page. At the time the thinking was that this is just a search engine like every other search engine. So to us, that's all we were creating at Napster. The copies were being made by individuals, and individuals were choosing to copy; to us, the choice was being made by the users to make the copy.

Eileen Richardson: We had a lawyer weigh in and a memo that said, "You could use this stuff for personal use. You can share it like you could share it at a VCR or on a cassette tape. You make a mix tape for your friend." That was the legal analysis.

Hilary Rosen: Once they got some copyright legal advice, they realized that the contributory infringement claim was the one that they were liable for, not direct infringement, since they weren't the ones uploading the music. They had created a software program that contributed to infringement.

Hank Barry: One of the big cases in the contributory infringement area is a case called "Betamax," which was a case that the motion picture studios brought against Sony back in the seventies and early eighties, around the video recorder. The thought was that if people could make recordings of television programs in their home, then "Oh my gosh, that's it, we're done!" And so the studios sued Sony, and it took seven years actually, but it went up to the Supreme Court. The decision there was, "Yes, this particular technology can be used to infringe copyright, there's no question about it. It also has a lot of beneficial uses, and when we're facing these dual-use technologies, we're going to err on the side of protecting the new technology." In the Betamax case they found that Sony was not liable for contributory infringement by manufacturing VCRs that had Record buttons on them. That was really the fundamental theory of why Napster was *not* liable for contributory infringement.

Jordan Ritter: It's a great pitch, isn't it? It's not wrong. It's not incorrect, but it is not the whole truth. If you give to one hand, you are taking from another.

Ali Aydar: So at that point I actually asked the moral question. Which was: *Wait a second, are we not stealing? I mean I'm downloading this for free, like, something that I would normally have to buy?*

Jordan Ritter: You know, we didn't show up with some business model to disrupt the music industry, we did some cool shit that took off and here we are, right? We're just a bunch of punk-ass kids!

Ali Aydar: And the argument that they convinced me with was, "Look, if we don't do this, somebody else is going to do it, and you know us, we're not trying to steal anything, we're trying to build something, and we believe that this is an opportunity to do it the right way, and we believe that the record labels ultimately will want to sanction this system, because ultimately, if we don't do it, somebody else is going to do it or others are going to do it, and it's going to be harder for them to control it. This system is something that the record labels can participate in and ultimately have control over. Why *wouldn't* the labels love this?" It was that naïve.

Sean Parker: Shawn Fanning fundamentally believed that the company would be successful, and that we would ink deals with the record labels that would advance the music industry and prevent its certain destruction.

Moby: But the labels were doing all that they could to pretend that the future wasn't really happening.

Hilary Rosen: The day before the Grammys, the heads of all the record companies would come to the board meeting, and I literally had a screen put up and a PC brought into the room and we played Name That Tune with all of the heads of the record companies trying to stump Napster. And virtually every title that someone came up with was immediately available. Napster was that jukebox in the sky that we had been talking about for years. Shawn Fanning achieved what the record companies had been unable to achieve in partnership with the technology companies that they had been working with for the two years prior, which was a fast, easy way to distribute music online.

Sean Parker: At the time, record labels were mapping out the most complex digital rights management systems you could ever imagine, and were in a never-ending conversation with their technology people about how to create a perfect system where no piracy could ever take place. Meanwhile the barbarians are at the gate: Rome is burning, and it's over.

Hilary Rosen: Everyone at the Grammys that night started really freaking out. For some of them it was the first time they really saw the disruption in the distribution system that they were facing.

Shawn Fanning: We made every effort to try and work things out, but it became clear that their goal was kill Napster.

Hank Barry: What those people did from day one was engage in a systematic effort to stomp this thing out, a systematic effort.

Ali Aydar: The labels actually filed the lawsuit pretty early: December of '99.

Hank Barry: There's no good faith around the edges here, there's no "Gee, aren't those kids cute, boy, this is Silicon Valley innovation and yes, we saw how great it was and we really love it but we, you know, gosh, we just have our obligations to do these things to serve the artists, and it wasn't really us doing them…"

Hilary Rosen: The Recording Industry Association of America had been a sort of behind-the-scenes trade association for the music industry, then all of a sudden we became public enemy number one. People took their free music very, very seriously.

Ali Aydar: Napster was never about "music is free," it was never about "screw the man," never, not even close! The grand master plan was to bring the industry in. But after the labels filed a lawsuit, they wanted to paint us in this light, when we weren't that way at all.

Hank Barry: The RIAA went to the editorial boards of all the major newspapers saying, "Look, Napster is your enemy. Napster is not a good story for you, Mr. Newspaper, don't look favorably upon Napster, because the next thing that's going to happen is we're going to have Napster for newspapers, so you need to come out against these guys." They had a whole tour, they went around the country to talk to other people who were beneficiaries of copyright laws and anticopying laws, with a view toward coming after us, and indeed, with a view toward coming after me personally, going to prosecutors and saying, "You really ought to arrest Hank Barry."

Moby: In 2000 or 2001 the RIAA was basically calling people who downloaded music genocidal murderers.

Eileen Richardson: One day, I had a friend who I trusted and knew and thought was smart, call me and say, "I just need to tell you, the CIA's on its way over to take all your laptops." I'm like, "What the hell! The CIA?" But I believed it. And the same morning Howard Stern called, "I want you guys on the radio show." So, in the same day, it's Howard Stern and "The CIA's coming to get your stuff"—it was insanity. A media firestorm.

Hilary Rosen: I had regular death threats. I had the FBI guarding my house, I mean it was crazy. It wasn't just a court case; it was a cultural revolution.

Sean Parker: There is some truth to the idea that Napster was much less of a company and more of a revolution. Despite all the potential liability, nobody really had anything to lose. Half the company were a bunch of radicals who were not there for the job. They understood that it was a social revolution, a cultural revolution.

Jordan Ritter: We were literally the very first one to go meteoric, to go interstellar.

Sean Parker: Most artists were kind of cool about it. Some were so angry with their record labels that they didn't complain. Some, like Billy Corgan and Courtney Love, were adamantly pro-Napster.

Moby: I saw the major labels as being big, hulking, stifling behemoths. They practiced institutional theft. When you made a record for a major label, they would give you a couple hundred thousand dollars to make the record. And once the label recoups the money, the label then owned your record—your master recordings. Imagine you take out a loan from a bank to build a house, you pay back that loan—and the bank owns your house! The labels never treated musicians terribly well unless they were making hundreds of millions of dollars for the label. And so I saw Napster as being this cool, progressive, almost vaguely punk-rock new paradigm. I was very much on the side of Napster.

Sean Parker: And then other artists just had a visceral negative reaction to it in part because they didn't understand it. They thought we had some giant server farm sitting in Palo Alto or San Mateo, California, that was just giving their music to the world, which also wasn't the case.

Eileen Richardson: Also we had the fans on our side, so we could say, "Would you like to talk now?" That was the only way the labels were going to listen.

Moby: The labels had treated musicians and consumers so badly, for so long, who really was going to shed a tear if they suffered?

Eileen Richardson: They were scared. Deathly scared.

Sean Parker: It went from ten thousand people to millions of people in a year.

Eileen Richardson: We hit twenty million users. It was a crazy time.

Sean Parker: You're sort of living your normal life, and everything is kind of proceeding in a relatively normal way, and then all of a sudden out of nowhere, you're picked up by the tornado and spun around and then you're not in Kansas anymore. You're in this totally strange, altered reality. You go through this door and you turn around and look back, and the door is not there anymore. There's no turning back.

Hilary Rosen: It was crazy for Shawn Fanning too, I mean he was the cover of *Time* magazine, and you know there were *60 Minutes* profiles and there were ten cameras every time we went to court, there were congressional hearings—I think four or five of them!

Jordan Ritter: There was a decision made in the company very early on about what was marketable to the outside world. Because here's me, Jordan, here's Fanning, and here's Parker. Right, does the world like this notion of two nineteen-year-olds blazing a path through Silicon Valley,

with a somewhat wiser, older twenty-one-year-old? Or is it just simpler to just talk about the two kids and frame it that way? Eileen and her team made a choice, that what we really want is one or two faces, and it can't just be Fanning, because Fanning isn't a good public speaker, therefore it's two, and it's Parker, and that's how he got the microphone. He's a good speaker. But Parker himself? He never wrote a line of fucking code! And in fact the engineers were discouraged from venturing outside—physically discouraged—so we were aware of things going on, but didn't really see it until it arrived at our fucking doorstep in the form of Lars Ulrich.

Lars Ulrich: We're suing Napster for one reason and one reason only, because they exist to pirate music. Nothing more, nothing less.

Eileen Richardson: Metallica was a band that became famous through the sharing of mix tapes, so when they said that they were going to sue us, I thought it was a joke initially, but it was not a joke.

Lars Ulrich: The ideal situation is clear and simple: to put Napster out of business.

Ali Aydar: There was a new album that they were working on, and one of the songs on that album somehow got on Napster. And that was when Lars, their drummer, who is sort of their spokesman, got upset.

Howard King: The song was not finished, and artistically that was disappointing to say the least, and then the thought was, *Wait a second, we sell our music—how can people get it for free?*

Ali Aydar: And that caused them to go and load Napster and search for other songs, other songs that they hadn't completed yet, and songs that they had released, and they found everything on Napster. And this made them very upset, and so they wanted that content taken down.

Lars Ulrich: It's not just about money at the end of the day. It's about trying to put your foot down before this whole internet thing runs amok.

Howard King: Napster had a policy that they thought, in their youthful ignorance, complied with the Digital Millennium Copyright Act. It was the "as long as you give us notice of who is infringing your copyright, we agree to take it down" policy. So we're bulletproof, you can't do anything to us.

Lars Ulrich: They dared us to come and prove to them that people were trading Metallica around.

Ali Aydar: One of the things that the copyright act allows for, is that if a copyright owner sees their content being distributed without their

permission, they're able to give notice and the search engine or platform or ISP then must take it down. And that was the legal umbrella we were operating under.

Howard King: We had a service monitor Napster over the weekend to see how many Metallica songs were being illegally distributed, and we came up with I think it was 322,000 incidents of people illegally posting Metallica songs. And rather than send over the names we let the press know that Lars Ulrich, the drummer of Metallica, would be delivering these names to Napster's headquarters!

Lars Ulrich: They refused to take Metallica off their list, so they asked for it.

Howard King: We were going to bring them in an armored car because it was almost being sarcastic: So somebody doesn't rob us of the 322,000 names. But we learned that you can't rent an armored car. It's a security thing, apparently. So we take the Suburban.

Sean Parker: The San Mateo County police department had their motorcycle squad there, there were helicopters swirling and news crews. It was covered on MTV and a lot of other media outlets live.

Jordan Ritter: We're like this little dinky company at the top floor of this shit building, and here's a massive media presence, everybody within the company is like, "Don't go outside, don't talk to reporters, don't talk to people because they might be reporters pretending to be people," and so I do it anyways. I totally went outside and started bumping elbows.

Sean Parker: Shawn and I snuck outside and were watching from across the street, just watching the spectacle.

Aaron Sittig: It was just a big sort of circus of people waiting for Lars to show up, wondering what was going to happen. Mostly supporters of Metallica, some detractors.

Jordan Ritter: There was a group of guys, really cool guys who were doing another start-up who had put together this massive poster that said, "RIAA = Master of Puppets."

Howard King: We drove down whatever street that is to Napster's headquarters, and we turned the corner, and there must have been five hundred newspeople there—I thought there would be ten reporters there. So again, we pull around the corner and see all the satellite poles coming off the trucks and all the press. It was a mob scene. It was incredible. Until then we had no idea what a big issue this was.

Lars Ulrich: We were not quite prepared for the shit storm that we became engulfed in.

Eileen Richardson: They pulled up and started bringing out all these boxes of paper.

Jordan Ritter: I think it was four boxes of usernames printed out in landscape mode, so you've got an even shorter part of the paper, one name per line on paper, and it was hysterical because like, "Hey, if you actually wanted to resolve this, you could have e-mailed it to us and we probably would have taken action and fixed it."

Howard King: Sending over a disc with 322,000 infringements is what a lawyer would do. Why go quietly into the night? Why not let the world know what a disaster Napster was for artists? That was the whole point.

Ali Aydar: If we're going to sit there and enter every IP address in and take it down, or every filename in it and take it down, that would have taken months. But if they give us a CD, we'll take it right down. Which is what they did. So the printouts were just for show, and they did a whole press conference in front of the front door of the building, and it was a whole scene.

Aaron Sittig: Lars started to grandstand a little bit and just yelled about how this is a terrible thing, and they couldn't believe the betrayal of all these people who downloaded their music.

Jordan Ritter: Right, and they got this little dog standing by them, and then somebody put a Napster sticker on the dog! It was so fucking crazy, it was cool, it was unknowable, un-understandable, but exhilarating and wonderful and you really didn't know what to expect. What was going to happen next? Is someone going to lose their shit and start a fistfight?

Lars Ulrich: It started off as a street fight, and then about a month later it was like, "Whoa! Look at this!" And then we were a little bit like a deer caught in headlights.

Sean Parker: It was the ultimate weird spectacle. All of a sudden we were in the middle of the media maelstrom, but from our perspective it doesn't make sense.

Howard King: So then Lars, Lars's manager Peter, and I went in and delivered the boxes.

Aaron Sittig: So of course we ran back inside the building and back up to see what was going to happen next.

Ali Aydar: I didn't really care much for Lars, because he's a fucking drummer, but whatever, and then here he is: the poster child for the boot coming down on our neck.

Sean Parker: Lars walks into the office with his black sunglasses on, and he basically just walked past us and delivered the boxes.

Ali Aydar: And now like we're all kind of like looking at them, and it was a weird, interesting dynamic.

Aaron Sittig: Lars's demeanor had totally changed. He went from being an angry guy downstairs, to being the sweetest guy. He's really nice. He shook everyone's hand.

Howard King: And we met Sean Parker and Shawn Fanning and we shook hands, and they were in awe of Lars. They were very respectful.

Sean Parker: At that time if you were a rock god, you were from another planet.

Aaron Sittig: We were all starstruck and happy to meet him, because we're all music fans. And so we had a good conversation and it was really nice, and he dropped off the paperwork and they left.

Howard King: I don't remember it being a long discussion, it was something like Lars saying, "Man, you know, you can't do this to artists. This is what they live on!" I think his phrase was something like, "We got enough money to heat the pool and put gas in the Suburban, so this is for the other artists who need to sell records in the future."

Jordan Ritter: Lars made a comment that I really appreciated. It was like, "You're not the evil Man trying to take my shit from me, or trying to destroy the world, you're just, you know, some really bright, cool kids." I'm paraphrasing but this is what he said.

Howard King: You know it looked like these kids had a hobby and it took off and they were in kind of awe of what they created: "Oh, look at it, it's just sharing, it's like you lent an album to your next-door neighbor, they'll never get us!" Then Napster brought some adults on board.

Eileen Richardson: The day Metallica came was like the day we switched over from me being CEO, to Hank Barry.

Sean Parker: Hank Barry was this copyright attorney from Silicon Valley.

Hank Barry: You could disagree with Metallica, or agree with them in terms of their position, but for me the most disappointing thing about that whole episode was that it ended up being about money. It was sort of cast as a battle on behalf of artists everywhere. But "artists everywhere" didn't get any money out of it, although Metallica did, because eventually we settled the case.

Sean Parker: Lars and I have become really good friends. I was joking with him the other day, that he's the only artist that actually made money off Napster.

Ali Aydar: So there are two lawsuits, one from Metallica and Dr. Dre, and one from the five major record labels at the time, both of them accusing us of vicarious copyright infringement. Which essentially means facilitating the massive infringement of their rights on their copyrights. The labels filed a motion for something called summary judgment, and summary judgment means "Judge, you decide right now, that Napster is guilty, without hearing all the evidence, without going through a proper trial, because it's so obvious." And so in those situations, the judge has various options. She could refuse that and go straight to trial. She could hear some arguments and then make a decision, or she could say, without argument, "Yes, Napster is guilty." She decided to listen to arguments.

Hank Barry: And so everything at Napster, all the two years of litigation, was around this question of preliminary injunction. There was never any kind of a trial. There was never any evidence presented in the case, none. Although there was this gigantic poster of Sean Parker's e-mail.

Hilary Rosen: It was Sean Parker's e-mails that really ended up having them lose those early court battles.

Sean Parker (in an e-mail): Users will understand that they are improving their experience by providing information about their tastes without linking that information to a name or address or other sensitive data that might endanger them—especially since they are exchanging pirated music.

Sean Parker: Their point was that, "Well, we know all along that users were going to be pirating music." That was beside the point. The point was that it was totally fine for us to operate a service where users shared files! It was no different than Google linking to a file.

David Boies: The use of the term *piracy*—I think gave the judge a way of focusing on something other than how the technology works.

Jordan Ritter: Parker's e-mail, presented in discovery... The fucking dick!

Hank Barry: You know it looks bad, feels bad, it smells bad...

David Boies: And, as I tried to urge the judge, the question of whether the activity is lawful or unlawful is a matter for the courts to decide. It's not supposed to be decided by the exuberant outbursts of teenagers, no matter how talented they may be at designing software. But I obviously didn't get that point across adequately.

Ali Aydar: The labels painted us as bad actors, as people who don't have good intentions, and instead of responding with, "We're operating under this area of the copyright act and here's why it's legal," David Boies decided to make the "Well, we don't really know what's being transferred" argument, right? And the labels are saying, "Come on, Judge, these guys totally know, it's totally our content for free, these guys are totally facilitating this massive copyright infringement."

Hank Barry: Boies is an amazing lawyer, but the bad news is he's not a technologist, and he was in the position of having to make a technological argument, and at the end of the day David never really understood it.

Ali Aydar: And the judge from the bench, without even going back and thinking about it, said, "Napster is guilty."

Hilary Rosen: The judge essentially said that it was *not* incumbent on us to tell Napster what was infringing; it was incumbent upon Napster to get permission for the works on their site. And that was a double problem for them. First they had to obviously get permission, but they also had to have technology that could read the digital files, so that they were not distributing infringing files.

Shawn Fanning: It was really surprising to me that we were being asked to function differently than all of the web-based search engines.

Hank Barry: And then we were trying to deal with the injunction and the outside reports were saying, "You know, despite your best efforts you're really not doing a great job on complying with the injunction."

David Boies: The judge had said she did not want to shut the system down. And yet, the scope of the injunction inevitably required the system to be shut down.

Ali Aydar: And of course the press was on top of this, and so the shut-down date was published everywhere. So everybody knew we were shutting down on a certain date, so before that date, our traffic spiked.

Hilary Rosen: We won the first round in court, and they came to us and said, "Now we want to be licensed." They wanted these sort of blanket license fees, but at the time there was no all-you-can-eat model and when the Napster team tried to create one, the pricing that they had come to the table with was really unrealistically small, and we had refused to allow them to continue to operate while they were trying to accomplish this, and ultimately they just ran out of time and money.

Sean Parker: Having now launched Spotify, I can attest to what it means when you first think that you're going to get a deal done with the record labels, and you're three weeks away from getting those licenses and launching. You're really two years away.

Hank Barry: And so we decided to stop the sharing.

Jordan Ritter: So Ali and I are in the break room, on the first floor, and I'm sitting there depressed, he's sitting there depressed, I'm like, "Ali, what the fuck are we going to do?"

Ali Aydar: Apparently I turned it off. I have no recollection of actually doing that, I have no recollection of that day, maybe it's one of those things I just shoved deep into my unconscious, I don't really know. I mean, I would be the person to do it.

Jordan Ritter: Napster had to end, because that's what happens at the end of every revolution. They gather up the leaders, they line them up against the wall, and they fucking shoot them.

Sean Parker: I didn't care about Napster as a company, I just wanted the idea to survive. I wanted unlimited access to all the world's music; that was the public good.

Jordan Ritter: And the idea lives on, but the progenitors of that idea have to die.

Hank Barry: Afterward we found out that there were all these OpenNap servers still out there, and so the network was actually still working, even though we had shut everything off at our place. It's kind of a wild outcome.

Ali Aydar: You had all of the offspring of Napster: LimeWire, Morpheus, Grokster—everybody's getting their content from all these other sources, and not Napster anymore. So Napster doesn't exist, but millions of people are still using all these other services.

Howard King: After Napster closed down we had another six, seven, eight years of fifty other bootleg sites where you were able to get the same music. We did not end bootlegging. We lost the war, even though Lars was in it for only this one battle.

Sean Parker: So Napster shatters into a million jagged little pieces, all of the users scatter all over the world to these decentralized services that take its place—services that they are incapable of litigating against because they have no corporate structure and no nation-state behind them.

Jordan Ritter: We blazed the path, and they would build cities on the soil stained with our blood. We had everything against us, and yet we still had the one thing, that we can see now with hindsight, we had that social empathetic connection: Facebook taps into that, YouTube taps into that.

Sean Parker: The story of how you take a completely un-indexed hypertext world and turn it into an information resource that anybody can tap into, that's the Google story. And then at the same time you have the social story, this is what people are using to share music in their dorm rooms and this is what people are using in order to connect with one another, and Napster is way early to the party, but in some strange way doing both of those things.

Ali Aydar: Prior to Napster, it was all about "Who is the adult in the room?" If there is a young founder, there is an adult CEO. Napster was the first case where you've got young founders, and the VCs came in with the "adult supervision," and it was an utter disaster.

Jordan Ritter: I know why we failed, and it had nothing to do with the tech, and it really didn't have anything to do, in my opinion, with the industry. It was a political game; we didn't play it in a political way, it was a game, and we didn't know how to play that game, and Hank Barry sure as shit didn't know how to play that game.

Ali Aydar: Now, it's all about the young founder remaining the CEO, and why is that so important? It's important because nobody can understand their company, their product or their service, at a visceral level, better than the founder. Nobody is going to care more than the founder, and now VCs understand that, and now the thinking is, "Hey, we'll just surround the founder with people that are adults who have skills in areas that they don't have skills. Coach them, help them, give them whatever resources they need, but they're going to stay the CEO." And that would have been a totally novel approach in 1999. It would have been completely ridiculous for any of us in that room to be the CEO. But it probably would have been the right thing.

Sean Parker: I think at that point I was nineteen.

Jordan Ritter: You know the iPod would have eventually come out and we probably would have done deals with Apple, and instead of Madonna getting prime position on iTunes it would have been like local band X, because we know what music you like, because we know the ones that you download!

Shawn Fanning: From our standpoint it was hard to wait. Especially coming from a generation where you can get basically anything you want instantly. It's frustrating when you want to make a change and you see a fix but you can't make it happen. So sometimes you push too hard.

Sean Parker: You have to learn how to take it step-by-step. Even if you see the ultimate outcome, you've got to be willing to understand other people and how to sequence all the steps to get to that outcome. My end of the bargain at Napster, having gone from a programmer to helping run the company and handling label negotiations and the legal stuff, that part of it was a failure. My whole contribution to Napster in some sense was a failure. And so I didn't feel like I had finished the job that I set out to do until I got a chance to help negotiate the licenses for Spotify. That's when things kind of came full circle. We now have a service that looks basically identical to what we were trying to build at Napster.

Hilary Rosen: Napster was very prescient, but everyone was greedy.

Sean Parker: There was so little greed on the part of the people who worked at Napster. We would have given the company to the record labels! Like, you can just have it! We just wanted the idea to survive. We knew that we were the best chance that the record business had of a seamless and orderly transition to a future where artists and labels and publishers all would have been paid—because we had every user all in one place. I made the argument to the heads of every record label that was willing to listen. They didn't listen to us.

Doug Engelbart: Silicon Valley's original visionary and the inventor of the modern "interactive" style of computing—not to mention the mouse

1

In 1950 I got engaged. Getting married and living happily ever after just kind of shook me. I realized that I didn't have any more goals. I was twenty-five. It was December 10 or 11. I went home that night, and started thinking: *My God, this is ridiculous.* I had a steady job. I was an electrical engineer working at what is now NASA. But other than having a steady job, and an interesting one, I didn't have any goals! Which shows what a backward country kid I was. *Well, why don't I try maximizing how much good I can do for mankind?* I have no idea where that came from. Pretty big thoughts.

Alan Kay: the theorist and intellectual behind the "graphical user interface"—the innovation that made it possible for mere mortals to use computers

2

Businesspeople should be shot. They always say, "We are in business to make money." And I say, "Well, not really, you just want to make a few million or billion." But the return from PARC is about thirty-five trillion! Count those extra zeros and tell me what they are really doing. They are just trying to be comfortable.

Nolan Bushnell: the maverick man-child who founded Atari, mentored Steve Jobs, invented the video game business, and pushed Silicon Valley into the popular culture

3

I've seen how technology has moved from *Pong* to what we're playing today, and I expect the same kind of pathway to virtual reality and I think that twenty years from now we will be shocked at how good VR is. I like to say we are at the "*Pong* phase" of virtual reality. Twenty years from now, VR is going to be old hat. Everybody will be used to it by then. Maybe living there permanently.

Steve Jobs: a native son who made a pilgrimage to India seeking enlightenment as a young hippie, only to find that his true calling was the computer business

4

This is the only place in America where rock and roll really happened, right? Most of the bands in this country outside Bob Dylan in the sixties came out of here: from Joan Baez to Jefferson Airplane to the Grateful Dead. Everything came out of here: Janis Joplin, Jimi Hendrix, everybody. Why is that? That's a little strange when you think about it. You also had Stanford and Berkeley, two awesome universities drawing smart people from all over the world and depositing them in this clean, sunny, nice place where there's a whole bunch of other smart people and pretty good food. And at times a lot of drugs and a lot of fun things to do, so they stayed.

Steve Wozniak: the hacker-genius who, in his spare time, designed and built his own computer in order to play video games at home—inadvertently spawning an industry

5

Look, I came up with the product that made Apple! If Steve Jobs had started without me, where would he have gone? Keep in mind, Steve tried to make four computers in his life—with millions of dollars—and they all failed: the Apple III, for marketing reasons; the Lisa, because Steve didn't understand costs; the Macintosh, which wasn't really a computer, just a program that looked like a computer and led to big problems later on; and the NeXT.

Alvy Ray Smith: the cofounder of Pixar who grew computer animation from a few color pixels on a screen into its first full-length film, *Toy Story*

Life with Steve was awful. A lot of people jump to the conclusion I got fired. But Steve didn't have the right to fire me. Steve was the chairman of the board, so Ed would've had to fire me, and Ed wasn't about to fire me. And so even though I was there for another year I knew I had to get Steve Jobs out of my life, because he was a foul bullyboy down underneath it all. So what happened in there is that Disney finally came. They knocked on the door just at the right crank of Moore's law and they said, "Let's make the movie that you guys always wanted to make." This is the big dream, right? So we go off and make the movie and, by the way, I left right in there. What finally freed me up to leave was the movie happened.

Howard Warshaw: the creative force behind Silicon Valley's biggest bomb, *E.T. the Extra-Terrestrial*, a game that will forever live in infamy

7

Nobody had ever done a game in less than six months on the VCS, and I had to do a game in five weeks. I was used to working under pressure, but this was just crazy. The CEO of Atari was betting a lot of his career on making this thing happen. Later, *E.T.* kind of emerged as this thing that destroyed the industry. How much worse can you get for a game than to eliminate the medium that it was released on? And I'm thinking, *Wow! What an amazing thing: I can destroy a billion-dollar industry with 8K of code!* Right? That makes me feel huge. I mean that's the ultimate! That's like the anti-matter release. But it's also ridiculous, absurd.

Andy Hertzfeld: the programmer-hero behind the first Macintosh who went on to prototype the equivalent of a working iPhone—back in the 1990s

8

In Silicon Valley there are two really common sets of values. There are what I call financial values, where the main thing is to make a bunch of money. That's not a really good spiritual reason to be working on a project, although it's completely valid. Then there are technical values that dominate lots of places where people care about using the best technique—doing things right. Sometimes that translates to ability or to performance, but it's really a technical way of looking at things. But then there is a third set of values that are much less common: and they are the values essentially of the art world or the artist. And artistic values are when you want to create something new under the sun. If you want to contribute to art, your technique isn't what matters. What matters is originality. It's an emotional value.

Bruce Sterling, Brenda Laurel, and Steven Levy: The Hackers Conference, which was first held in 1984, is where Silicon Valley technical types started to recognize themselves as a culture

Left to right: Bruce Sterling, Brenda Laurel, Steven Levy.

This was the moment where the consciousness of that time really became something that you could pick up on. Just like in the gay movement where there was a time when people just felt incredibly alone, that there was no one else like them. There was this consciousness that came out there: *Oh, this is who I am!* I think it's fantastic that now in high schools people can be themselves whether they are gay or a hacker, right? The other thing was the laughter. Every session was joyous. People were really funny. There was so much laughter throughout. I don't think I've ever been to a conference where people laughed more. It was a shared humor out of a shared experience there. And even though some people had never met each other, it was like you were part of a crew that had worked together for years and years. It was like inside jokes with people you'd never met before. They were also just genuinely funny.

—*Stephen Levy*

Stewart Brand: the Zelig Silicon Valley avant-gardist who first put the phrase "personal computer" in print, popularized the word *hacker*, and created The Well, the first wildly successful experiment in social media

I had seen online teleconferencing. And I had seen things that had made it wonderful and things that had made it terrible, and there were also bulletin board systems around at that time which had also established a certain amount of behaviors that were wonderful and not so wonderful. So, based on those experiences, we designed what became called The Well to reflect what had been learned about online discussions at that point. I priced it to be very inviting. We made it easy to make conferences, and people would invent conferences. Anybody could start a conference. I wanted hackers inside the system, so we invited them in and writers, journalists, all got free accounts and that was our marketing. Initially it was just sort of friends of the *Whole Earth Catalog.* People learned how to deal with trolls. If you respond to them, they will make the flame even brighter. So, The Well kind of grew and got a life of its own and established a certain amount of online practice.

Jaron Lanier: the coiner of the phrase "virtual reality" and the first to sell a complete VR rig—googles and gloves paired with the hardware and software to make it all work

11

My whole field has created shit. And it's like we've thrust all of humanity into this endless life of tedium, and it's not how it was supposed to be. The way we've designed the tools requires that people comply totally with an infinite number of arbitrary actions. We really have turned humanity into lab rats that are trained to run mazes. I really think on just the most fundamental level we are approaching digital technology in the wrong way.

Survival Research Laboratories: an outlaw art troupe that catalyzed the Valley's "cyberculture" with terrifying spectacles of robotic carnage and mechanical mayhem

12

There were pianos, about twenty-five pianos stacked against a part of the freeway, and at one point the pianos caught on fire and the Highway Patrol shut the freeway down because the flames got high enough that they were coming over the top of the freeway, fifty feet above us. So it was pretty out of control. We also had a huge flamethrower that was making forty- or fifty-foot flames down underneath there also. All this stuff was happening at once, and huge huge fires, smoke and just a lot of chaos, a lot of explosions and just a lot of crazy stuff. There was too much chaos to really pay close attention. I just know that the fire was melting things in the control booth from a hundred feet away! It happened because there was a ceiling. There is a phenomenon that happens with infrared radiation called "flame focusing" where if there's something on top and on the bottom the infrared can't radiate up or down, so the flame felt much more concentrated than it would have if it was outdoors. That also damaged the freeway pretty badly, but fortunately the earthquake came a couple months later and so Caltrans just said "Don't worry about it. We don't even want to admit that you did this. So we're just going to let it slide." So they never charged us to fix it because of that earthquake, which was lucky. But the interesting thing about it was nobody ever told us to stop. The police never said a word afterword. The fire department never said a word. The Highway Patrol never came down and said, "Hey, we had to stop the freeway because of your fucking fire. What the fuck?" *—Mark Pauline*

Jane Metcalf and Louis Rossetto: the cofounders of *Wired* magazine who captured and popularized Silicon Valley's distinctive voice and point of view

13

I was at Columbia in '67, '68, '69, and '71, and of those four years in college, two of them were wiped out by eruptions in the spring. It felt like the world was coming unglued. And if you had eyes you could tell that there were issues in society that weren't right. But the analysis from the left just seemed wrong. It was like refried Marxism and this colonial grievance bullshit and that was so inauthentic and their prescriptions— what they wanted to have happen— so wrong! So then I started looking for other stuff. And I think I'd read Ayn Rand and then from there you realize there's more than Objectivism. There's this libertarian strand and then you realize that libertarianism is really deep. It goes back to Pierre-Joseph Proudhon and beyond and has all sorts of other manifestations in American history. And I just got farther and farther into that and realized this is a kind of good way of thinking about things.

—Louis Rossetto

The world was changing. And so everybody's seeing it from their own worldview. Some of them thought it was changing because we were all doing the same drugs. Other people thought it was changing because we were using the same digital tools. And so there were people who could surf in between those worlds and so there were some common threads there in terms of contributors.

—Jane Metcalfe

Sean Parker: one of a long line of bad-boy Silicon Valley entrepreneurs—but the first to destroy a traditional industry

14

There was so little greed on the part of the people who worked at Napster. We would have given the company to the record labels! Like, you can just have it! We just wanted the idea to survive. We knew that we were the best chance that the record business had of a seamless and orderly transition to a future where artists and labels and publishers all would have been paid—because we had every user all in one place. I made the argument to the heads of every record label that was willing to listen. They didn't listen to us.

Mark Zuckerberg: the cofounder of Facebook who came to California as a teenager with little more than an idea and quickly rose to become Silicon Valley's most powerful man

15

Mark sat down with me and described to me what he saw Facebook being. He said, "It's about connecting people and building a system where everyone who makes a connection to your life that has any value is preserved for as long as you want it to be preserved. And it doesn't matter where you are, or who you're with, or how your life changes: because you're always in connection with the people that matter the most to you, and you're always able to share with them." I heard that, and I thought, *I want to be a part of this. I want to make this happen.* Back in the nineties all of us were utopian about the internet. This was almost a harkening back to the beautiful internet where everyone would be connected and everyone could share and there was no friction to doing that. Facebook sounded to me like the same thing. Mark was too young to know that time, but I think he intrinsically understood what the internet was supposed to be in the eighties and in the nineties. And here I was hearing the same story again and conceivably having the ability to help pull it off. That was very attractive. *—Max Kelly*

Biz Stone and Ev Williams: the dynamic duo that discovered what the written word meant for the web by building the presses that power its blog posts and its tweets

16

Stone is wearing glasses; Williams, the bow tie.

Tons of other businesses are driven by people who like to transact, to "do business" or "make money"—I mean you can be on Wall Street and just transact. What's different about Silicon Valley are those people who are driven by the creation. —*Ev Williams*

Only in Silicon Valley can you be like, "Yeah, we would like $10 million and we'll sell you a percentage of our theoretical company that may one day have lots of profits and if we lose all the money we don't have to give it back to you. And maybe we'll start something else." In what crazy world does something like that exist? Like wait, you can just blow the money and then you don't owe it back, just wash your hands clean, done? "Sorry about that. Sorry, I spent all your money. Oh, well." I mean it's just crazy. And not only that, here's another scenario. Here's what some people do. They say, "We need $25 million, but you know, just so we can stay focused my cofounder and I each need $3 million of that money in our bank accounts. Then we don't have to worry about bills so we can really focus. Okay?" And then they blow the money and they say, "Oh, well, that didn't work out, but we're still keeping the $3 million each, so now we're rich." What the hell? That is crazy. So, it's a crazy world. This is like some kind of nutty place where you can do that kind of stuff. —*Biz Stone*

The Dot Bomb

*Only the cockroaches survive... and you're one of the
cockroaches*

Wired's *first issue predicted digital revolution and it showed up right on
schedule, engulfing the Bay Area at the start of the new century. Wave
after human wave washed up in Silicon Valley and flowed to its most exciting outpost—San Francisco, the unofficial headquarters of the dawning web.
It was kids and carpetbaggers; MBAs and coders. And they all wanted to meet
each other, preferably with a drink in each hand. These were the "dot-commers"
who worked for "internet companies"—the web was so new at the time that the
net itself was seen as a type of economic activity, like mining. Indeed, it was a
new gold rush, and by the turn of the century the entire region erupted in a
speculative frenzy—with plenty of money left over for parties. It didn't make
much sense to the generation of hardware engineers that had actually put the
silicon in Silicon Valley, but the newcomers were having a lot of fun and billions
were pouring in. So why question it? And then, one day in March 2001, the
music stopped.*

Sean Parker: When you think about the history of Silicon Valley, if the first
era is personal computing, and what it took to establish personal computing
as a ubiquitous platform, then the second movement is the rise of the web.

Jamis MacNiven: In 1991 I opened a restaurant in Woodside, Buck's, that
has become pretty well known. In 1992 there was one mention of Buck's in
InfoWorld: Bob Metcalfe said he was having "a power breakfast at Buck's."
In '93, there was a mention in the *Economist* that John Doerr, whoever that
was, was having breakfast at Buck's. Netscape had a lot of their early meetings at Buck's. That was the most important thing that ever happened
here, because it allowed all of us to go online seamlessly.

Peter Thiel: So Netscape comes along in '93 and things start to take off.

Jamis MacNiven: In '94, three TV crews came in. In '95, 150 TV crews came in.

Peter Thiel: It was Netscape's IPO in August of 1995—over halfway through the decade!—that really made the larger public aware of the internet. It was an unusual IPO because Netscape wasn't profitable at the time. They priced it at $14 a share. Then they doubled it. On the first day of trading the share price doubled again. Within five months, Netscape stock was trading at $160 a share—completely unprecedented growth for a nonprofitable company.

Jamis MacNiven: All of a sudden, everything broke open like in *The Wizard of Oz*. We'd been in black-and-white and then it all turned into color, and that happened in '95.

Jamie Zawinski: Netscape was the poster child. We were the first big IPO. The time when the industry changed from one thing to another? That was us.

John Couch: The money was coming too fast and furious. I started interviewing people and they started to say, "Oh, I'm going here for two years because they're doing an IPO." The Bill Hewletts, and the Fairchilds, and the Groves of the world had been replaced by twenty-year-olds who were not focusing on building a long legacy.

Po Bronson: There's an old Silicon Valley lifestyle that comes off of the old NASA base in Santa Clara that was confused by the dot-com thing, right? They built satellites and stuff and what was happening up in San Francisco just didn't make sense to them. They were like, "What's their technology?" They were like, "HTML? It's not like you can build a company on that!" And so they were like, "They're not that hard-core."

John Markoff: The San Francisco scene really emerged with the web.

Tiffany Shlain: At the time there were only sixteen million people online—and a lot of them were in San Francisco making and creating this nascent medium. It was so exciting.

Fred Davis: So, why did the internet get started in San Francisco? Why did that become the center of all the internet things? Because we had the talent already sitting right here, from the multimedia industry! The multimedia industry drew a lot of artists and creative types and interactive people to San Francisco. So we were *already* building connected multimedia applications—but we had to build them all on a CD-ROM. They all lived in a tiny, six-hundred-megabyte universe. And now, if you could take the principles of HyperCard and interconnect everything, it was like, "Wow, this is amazing. This is going to change everything."

Tiffany Shlain: It was so exciting.

Jim Clark: From our going public in '95 to the year 2000—five years in there—there was an eruption of historic proportions.

Fred Davis: Wall Street called it "the New Economy," and indeed it was. It was that big.

Jeff Rothschild: I remember people used to talk about the New Economy, and the New Rules. "Those are the old rules, where you value the company based on margins and return. That stuff doesn't apply anymore." I think the people saying those things believed them, I really do.

John Markoff: Silicon Valley began to move north. It began to be about software instead of hardware. It began to be about commercializing and disrupting normal business.

Joey Anuff: It was wave after wave of fortune-chasing opportunism. And it took different shapes. There was a distinct 1995–1996 era which was much more literal. There was the Spot, which was going to be *The Real World* for the web. There was Amazon. They were very basic ideas, and some of them were completely inconsequential like the Spot, and some of them more world-defining like Amazon.

John Markoff: Silicon Valley had been an insular world dominated by engineers, and now it was touching the mainstream.

Tiffany Shlain: The first Webby awards were in March of 1996, and I started working on them the year before. I wanted to create something that was just as edgy and crazy as the web was at that time.

John Markoff: The Webbys was doing something that was an imitation of Hollywood in Silicon Valley.

Tiffany Shlain: That first year I brought in Cintra Wilson as the MC, and she was the one who said, "Let's limit acceptance speeches to five words," and I was like, "That's a friggin' brilliant idea!" She came out in a dominatrix outfit, and if anyone went over the five words, she would whip them.

John Markoff: It was just so odd and off-putting to me—even though I am a judge.

Joey Anuff: From 1997 to 1998, it got to a level where there were so many companies with so many IPOs, the idiotic rhetorical question of "If you know so much why aren't you rich?" became overpowering at that time. "How can we make a buck off of this when every jackass is getting rich?"

Jamis MacNiven: We had people coming in to Buck's looking for work just so they could meet venture capitalists. We had a guy come in once, Nan Omo was his name, and he was an Eskimo who weighed about three hundred pounds and was not very tall, so he had a spherical suit on, and was walking around with a copy of the *Industry Standard* magazine.

John Battelle: The *Industry Standard* was an idea I had at *Wired*. I was the editor at *Wired* that brought the business stories in. And in '97 I thought it was getting really interesting, but it just seemed like *Wired* couldn't contain all the stories. So I conceived of a magazine with a weekly cadence, that was part *Economist*, part *Variety*.

Jamis MacNiven: Nan Omo was looking at pictures of venture capitalists, spying the room, seeing someone, going up and slapping down something the size of an old Manhattan phone book, saying, "This is the new Amazon, you've got to look at this!"

Chris Caen: At the time everyone wanted to be the Amazon of fill-in-the-blank. "We want to be the Amazon of pets!" Pets.com!

Jamis MacNiven: He chased Steve Jurvetson out across the parking lot. So I had to grab him and threaten to drop him in the creek if he didn't stop. He goes, "Oh, but it's my dream!" People came here with this big dream, and it was just off the hook.

Chris Caen: So imagine there is more money than God, and it has no place to go, but there are a lot of cocktail napkins with things scribbled on them, and those are what are called start-ups. I'm being a little facetious but not by much. The joke used to be at the time, this being '97 through 2000—the golden era—was basically all you had to do was stand at the corner with a cocktail napkin, and VCs would throw money at you from a passing car. I loved it.

The dot-com scene was clustered around South Park, a seedy neighborhood park in San Francisco's industrial South of Market district. Wired maga-zine had its headquarters on the park, as did many of the most notable dot-com companies.

Steve Perlman: South Park is this small little oval-shaped park that's largely grassy except for a few swings and a couple of picnic tables.

Yves Béhar: I had either lived or worked on South Park since 1993. Artists and makers lived around South Park. Technical people at that time lived in Silicon Valley.

Po Bronson: South Park felt authentic for a while—this rustic industrial corridor full of garages, corrugated-tin roll-up doors, no heat.

Yves Béhar: Then during the dot-com boom it went from being known as Heroin Park to having Japanese tour buses driving around the loop and taking pictures.

Steve Perlman: You couldn't find a square foot of grass to sit on and cross your legs and eat your lunch—it was that packed.

Yves Béhar: It was the epicenter of the dot-com boom: We had eGroups, 21st Century Internet, which was a VC firm, BigWords. We had Pets.com a block away.

Scott Hassan: I was mostly doing eGroups at the time, and Yahoo purchased the company for $450 million, which is not bad for two or three years of work.

Yves Béhar: All these companies were right there. People have always been looking for where the internet is—and for a little while the internet was "at" South Park.

Po Bronson: Masses of young people who didn't know even why they were coming were coming. There was this sense of a whole generation just showing up here, unannounced. And I wanted to tell the little people's story, so I was doing things like staying at the airport, and in the middle of the night, at two in the morning, the flights from different countries would arrive and I would stand outside customs and I would look for people who looked like they didn't know what they were doing, and say, "You've heard of *Wired*, right? I'm a *Wired* writer!" I met some people like that and I would follow these people on and off—just follow the string for a while.

Patty Beron: Back then pretty much everyone that you met was like, "Oh, I'm in the tech industry," or "I'm trying to get into it." It was like, the thing. At first we would go to these boring networking events, but everybody still wanted to go because it was like, "Oh look, there's Craig Newmark from Craigslist!" you know? Like, *how exciting.*

Po Bronson: I thought I was telling this story about the little people. They're just getting chewed up alive. They're living on each other's couches and they're not having any fun, and relationships were being broken up and the parents aren't talking to them and they can't even get a job in this thing. And I'm telling this story and then there's this guy Ben Chiu. He was just another naïve kid that I just found, another naïve kid from Taiwan who had no real chance. He and his six-person start-up had no chance. Then all of a sudden Ben Chiu has been bought out for $35 million. And I couldn't fucking believe it! I was like, *This really is insane!*

Ev Williams: There was a lot of talk about who was getting funded and who was doing biz-dev deals with who and then, of course, the parties and the events. What's interesting is how that information came out, because there wasn't any social media at all. There was a nascent blogging world being built: just a handful of hand-rolled web-logs—this was before they were called blogs.

Biz Stone: I was reading EvHead.com, and really impressed by the guy. Totally in sync philosophically with his thinking. I was completely sold on this idea that blogging was the true democratization of information.

Peter Thiel: Launch parties became so important that someone put together an exclusive e-mail list that published rankings of the various parties going on that day.

Patty Beron: So I had this website, SFgirl.com, and pretty much everybody in the tech community knew about it. We listed all the dot-coms, we had community forums where people involved in start-ups would come and discuss what was happening. I don't know who decided to start having these launch parties, but that became a thing—going to dot-com parties. The new dot-com would get all this publicity and the more extravagant their event was, the more publicity they'd get.

Ev Williams: Every night there was a party to go to, where you could drink for free and eat for free and there would be lots of young people—half of whom were brand-new to town.

Patty Beron: The more extravagant the event was, the more publicity they'd get: the more write-ups, the more people talking about them, like, "I heard that blah blah blah had pig races at their party," and "This or that party had ten-foot ice sculptures and vodka luges," and things like that. It just became a thing to one-up the last party and get attention.

Ev Williams: I remember leaving some party with a bottle of champagne. It was like, "You're leaving? Have a bottle of champagne!" In those days, it wasn't a ridiculous gift...

Jamis MacNiven: There was a guy—the name of the company was Pixelon—he raised fifteen, eighteen million dollars, had no product, went to Las Vegas, hired the Who and, I think, Kiss. He basically just threw away all the money. And put all his friends up at a hotel until the regulators came in and arrested him. It's, like, that wasn't unusual.

Patty Baron: There was a party where the Barenaked Ladies played, and then we went to another party and the B-52s played.

Max Kelly: There was a company called Icebox Studios which was doing online animation series. They did this one show called *Mr. Wong*, which was amazingly, psychotically offensive. This company got funded, had no revenue, had high production costs—and they still threw this huge bash. Run-DMC played! There were all these celebrities that they invited and they had a walk where photographers were out there just like a movie premiere. The whole shebang went on for, like, twelve hours.

Chris Caen: Oh God, the parties were fabulous; I mean parties every night!

Jamis MacNiven: Draper Fisher Jurvetson once rented one of the biggest buildings in the world, Hangar One at Moffett Field—it was where the USS *Macon* zeppelin was—and filled it with amusement park rides. There must have been eleven thousand pounds of shrimp and it was raging. Barbara Ellison, Larry's ex-wife, used to throw parties with zebras and elephants and giraffes. There was a party once that had army tanks driving around with naked women. One guy had a huge party and hired these four party girls, to just walk around completely naked, in Woodside, on a Sunday afternoon!

Po Bronson: There was always this high-low tension that was always trying to make you feel like you hadn't totally sold out—like you add some funk and you get rid of the guilt. Like everybody would wear a tuxedo with sneakers, or serve Kobe beef on a cheap ol' napkin. It was some form of apology or some sort of a desperate cry to not be just about the money.

Jamis MacNiven: It was all very interesting, and we were younger then, and we thought it would go on forever.

John Battelle: There was this sort of reality distortion field around the entire internet, that was like, "Everything is changing, and I want a ticket to the change. Where do I buy a ticket to the change?" And so if you wanted to buy a ticket to the change in, say, retail, you would buy a share of Webvan. If you wanted a ticket to the change in travel, you would buy a ticket to Expedia or Priceline.

Joey Anuff: Day trading on the internet with E-Trade is just a matter of clicking something to buy or sell, so it brought this twitch element to retail investing. And if you're me, a fucking idiot who doesn't know anything about the history of the market, for all you know you might as well be playing *Dig Dug*.

Carl Steadman: So you have people investing in ideas, some of them good, some of them bad, and because what you get in a frenzied economy that is in search of IPOs are ideas like Pets.com. They were willing to ship you a forty-pound bag of dog food, for no shipping fee, at a tremendous discount. It made no sense, there was no underlying business plan. It was this tulip mania. It was the exuberance, irrational exuberance.

Ev Williams: This company would IPO and that company would IPO. We were all swept up in it.

Tiffany Shlain: There was so much talk about IPOs that we had an Initial Pumpkin Offering for Halloween. Everyone came out to South Park. I still have all the schwag from that; it was very funny. It was a ridiculous time.

Carl Steadman: My marker for a long time was Paris, 1968. It was this time when students and workers and especially the farmers and, to some extent, city goers joined in this general strike, and if you read the accounts from the people at that time, they felt that anything was possible. And in the mid- to late nineties, there was also that sense there's just this miracle happening and anyone with a sense of history understood that we were in a moment which was not sustainable.

Ev Williams: It really felt like a moment.

Joey Anuff: There was a point that you could buy a stock and the next day it would be up four times in value.

Jim Clark: Silicon Graphics was never anything more than ten times earnings. In 2000 there were companies trading at one hundred times earnings!

Jeff Rothschild: There were people who invested everything in these growth stocks, and they were buying companies at a hundred times earnings, waiting for it to go to two hundred times earnings, or in many cases, they were buying things at three hundred times earnings hoping it would eventually be five hundred...

Max Kelly: These kids were going crazy, and trying to figure out how to leverage their shares and all this other stuff, and of course they all got fucked.

Carl Steadman: At *Suck* we celebrated the craziness because we were living in a world that was not going to exist again. That was my take at the time, because it was not as though people were going to listen to reason.

Max Kelly: All of them almost universally got fucked, as I recall.

Patty Beron: We kept hearing, "Oh, the dot-com bust is coming" but we were like, "What does that mean?" We didn't really know. Was everyone going to lose their job and be begging on the streets? We didn't really have any idea.

Scott Hassan: It was a superhot time in the Valley, things were really hopping, and there was so much money flowing in that I think the VCs were just very inundated by money, too, and they needed to actually get that invested, because they can't take a carrying fee if it's not invested, and so they have to invest it.

Jamis MacNiven: Just before the crash, John Mumford at Crosspoint Ventures, who had raised a billion-dollar fund, said, "You know, I can't responsibly invest, so I am not going to do my fund, I am going to give my investors their money back." And all these other VCs are going, "No! No! You can't give the money back! That wrecks the whole game!" That's when the money dried up and the ideas stopped and everything ground to a halt.

Chris Caen: Also, people started realizing that it was kind of a circular economy. That it was the VCs funding companies that invested in other companies that were bought by other companies...

Jeff Rothschild: I was skeptical about the valuation of these companies, because, you know, I thought about it and said, "Well, the only way that they could continue to be this valuable is if other people were acquiring them, and that can't go on forever, you eventually don't have any more bigger fish." Companies were getting such big valuations that nobody could acquire them.

Chris Caen: It was a circle jerk. It was all the companies in Silicon Valley giving each other the same dollar and having it go around the room. And so once the tap turned off for one company, and you multiply that across the whole ecosystem, the disaster was almost preordained.

Jeff Skoll: In January 2000 was the merger between AOL and Time Warner. So I remember Pierre and I sitting in the little conference room when we hear that news, and Pierre—who is superwise—said, "It's probably going to be the crash of the internet." And I was like, "Why?" And he said, "Well, this is the first time you are having Yankee dollars buying Confederate assets, right? And up to now those Confederate assets have been priced at whatever they've been priced at because of a dream and a prayer, but now you have real hard assets being placed against those numbers and it's going to crash the NASDAQ and all that." And he was right.

Ev Williams: The stock market crashed in March. But it wasn't immediately obvious that everything was over. There was a suspension of disbelief. It was like Wile E. Coyote running off a cliff—he's still running for a while. In fact we raised the first money into Blogger in April of 2000.

Biz Stone: I was like, Damn it! Why didn't I do that! I was so blown away by Blogger. I was like, *Blogging is going to be amazing.* I just got the shivers thinking about all the people who weren't coders who would now be able to have a voice on the internet.

Ev Williams: We raised $500,000 and hired seven people. We assumed we could raise more later in the year, and then as that money started running out and it got to be the end of the year, we realized, *Oh, there's no more money!*

Brad Handler: Then in March of 2001, everything just started to implode, companies were failing.

Brad Handler: Webvan came and went. Pets.com—whoever thought that was a good idea?—went. The Globe had the highest run-up and then a year later was bankrupt.

John Markoff: Pets.com and Webvan—nobody believed in those. They had the air of excess right from the start.

Andy Grignon: We were doing stupid stuff: Pets.com and Webvan and Excite@Home. It was the early development of the service economy.

Jack Boulware: It was crazy. All this money that was pouring into the Bay Area—ridiculous schemes like Beenz.com, WebVan, Kozmo—and everybody would profit from it. Limo drivers were accumulating fleets of like ten or twenty cars. The strippers were all buying houses. Then—

boom—all of a sudden it was gone. People were left looking at one other: "Did we just get played? What happened?"

Po Bronson: I was having fun roasting this place, then I met Ben Chiu—I can't even remember what he was doing—but I saw all that money and I'm like, *Maybe it's somewhat for real?* And I lost my critical mind in there.

Brad Stone: A conventional wisdom formed right after the crash that these companies were bad or ridiculous or excessive, and I think that's unfair. Because we've seen all these businesses play out, by and large. The dream of Pets.com seemed ridiculous when the company went bankrupt, but it certainly doesn't now. Chewy.com was just sold to Petsmart. It was the biggest e-commerce sale in history!

John Markoff: So was Webvan ahead of its time? Well, there's this experiment called Instacart which seems to be working.

Brad Stone: They weren't bad ideas. The ideas were sound. Ultimately the market liked those ideas. It was the wrong time. It was too early.

Andy Grignon: Those ideas became successful later on—Amazon ultimately birthed out of that thinking—but we did a lot of stupid things. We spent a lot of money on making things that weren't ever going to be profitable.

Fred Davis: There were a lot of bad companies that needed to go. My joke company is eggsbymail.com—it just isn't a good idea to start with. So, shaking them out would be good, if it was a correction. But it wasn't. It was a complete turning off of the monetary faucet.

Marc Andreessen: In the dot-com crash, all the dot-coms go out of business. Then all of a sudden all the big companies were like, "Oh, this internet thing? We don't have to worry about it after all. If the dot-coms are going away, we were right. This internet thing wasn't going to be a big deal, and so we can put all these projects on hold or slow them way down." Our market just cratered.

Chris Caen: The speed with which it happened was the shocking thing, and it happened when Atari collapsed, too. That's the one thing about Silicon Valley. Damn! When that bomb goes off it all happens in a hurry. Everyone talks about how fast Silicon Valley grows, this whole a-rising-tide-lifts-all-boats thing, but people forget the opposite is equally true: When the tide goes down, everybody goes down.

Brad Handler: Seeing a company like Exodus go to zero...that was a little sobering.

Ray Sidney: Exodus is the data center where Google had its first server farm. It was also hosting machines from Inktomi, which was another early search company, and from Yahoo—all the big names.

Chris Caen: It's "Oh my God! The bubble's burst! Silicon Valley is ending!" Everyone forgets that 2000 is the second time that the bomb went off in Silicon Valley. The first time was '83–'84.

Marc Andreessen: Then 9/11 happened, which was obviously tragic, and in addition to being tragic, had the effect of shutting down any business that was hanging on in our industry. Basically, 9/11 was it. Business just shut down.

Mike Slade: So 9/11 happens on a Tuesday morning, right, and the world's on fire, right? I went to Apple, and we were all sitting around and Steve looked at me and he goes, "There's going to be a war." People were like, "Should we even bother with this meeting?" People weren't really sure that we would function as a business. I'm serious! What the hell, right? It was weird.

Fred Davis: So, after 9/11, the market opened and every freaking man, woman, on the street dumped all their tech stocks. I had to go into the office and fire thirty people the next day because everything was gone.

Yves Béhar: Life was easy and beautiful and happening, and recent college graduates from the Midwest and East Coast had gotten hired at great salaries, and then suddenly there was no work in San Francisco. People's dreams were shattered.

Patty Beron: When the towers fell, my boyfriend—who's now my husband—and I were in Europe for a month, traveling around. I was like, "Let's go on a trip and when we come back we'll get new jobs, and we'll just keep making money and doing whatever." But when we came back he could not get a job to save his life.

John Markoff: It was like a neutron bomb went off. It all happened incredibly quickly. One day the sky was the limit, and then the sky wasn't the limit, and it just came crashing down so very quickly.

Jamie Zawinski: That's all nonsense. It's all blown out of proportion. There were a bunch of people who counted their chickens before they were hatched, who found out that they didn't make as much money as the lying liars told them they would. I knew a few people who bought their expensive condo and filled it with ridiculous furniture by leveraging money that didn't show up. Then the tax bill comes due, and they're fucked.

Fred Davis: I needed a couple more million bucks to lend my company because the bubble burst in 2000. And I had a huge number of AskJeeves shares, so I go to my broker and he says, "Don't sell the shares, just borrow on margin and lend it to the company. It's a small amount." And he said, "You're totally protected because Jeeves was trading around $14. It has $7 of cash in the bank, so it'll never go below $7." Okay, great! But after 9/11 Jeeves went to eighty cents because people were panicking. Two years later, it sold to Barry Diller for $35 a share. But because I had such a large position in it, I couldn't last until the buyout happened.

Jeff Rothschild: So, you know, a lot of people got greedy, and maybe they can blame their brokers, but a lot of people did stupid things. They would exercise their stock options, and then hold before they would sell—because they were counting on a higher price. So they might exercise a stock option for $100 a share, at which point they owe $50 a share in taxes—but if the stock price falls to $10 a share when the tax bill of $50 a share comes due…you sell your house, you sell everything you've ever owned, and you will still be working for the IRS for the next ten years. This happened to a lot of people I know.

Fred Davis: I saw tens of millions of dollars disappear in one day.

Jamie Zawinski: Okay, well, some very bad money decisions there. But I'm sorry, there was no "collapse." There were inflated promises that didn't come true. It's not like property prices have ever gone down in San Francisco, ever once. Maybe for a minute after the last earthquake.

Yves Béhar: Maybe they were stupid dreams: stock options that looked like they were worth millions; recent college graduates thinking that they had made it—and it was real easy!

Patty Beron: We ended up leaving because my husband got a job in Utah. We were like, "Oh God, we don't want to go to Utah, that sounds horrible!" But you're going to go to where you're going to get a job. And that was kind of our exit from San Francisco.

Yves Béhar: People had to go home. One hundred thousand people left the Bay Area at that time.

John Markoff: It really cleared out.

Patty Beron: The world was a changed place. The mood was totally different. It was very somber, very quiet: no e-mailing, no people calling. I did keep SFgirl.com going until we left, but the site was basically about parties, and the mood was not the mood to be throwing parties of any kind. It had run its cycle.

John Markoff: The best thing about the collapse was all of a sudden you could drive on 101 and 280.

Yves Béhar: Traffic went from being absolutely completely impossible to being a breeze. It suddenly was a pleasure to be on the freeway between San Francisco to Silicon Valley.

Aaron Sittig: You'd go to dinner at a restaurant and you'd get amazing service, and so you'd strike up a conversation with the server and find out that they were like an unemployed physics PhD who had just gotten a degree from Stanford. It was really rough. Everyone was just scraping by.

Sean Parker: It went from being the center of the universe and all these bright-eyed, bushy-tailed people showing up from all over the country and reading *Wired* magazine cover stories about all of the young people who were emigrating to Silicon Valley to seek their fortune—to nuclear winter. Only the cockroaches survive, and you're one of the cockroaches.

Mark Pincus: San Francisco was a shell of a city, and all these people had come in and they left and all these companies had collapsed. It was very drastic.

Po Bronson: All this money was gone and then there were investor lawsuits and then there wasn't even the money to fund the investor lawsuits; it just all kind of became barren.

Sean Parker: And then after the dot-com bust, there was a feeling everything that there was to be done with the internet had already been done, that there were no new ideas, that nothing new could ever come out of the consumer internet.

Mark Pincus: I'm sure it was only in my head, but I felt that there were about six people that I knew who were interested in doing anything in the consumer internet, and we all kind of went to the same two coffee shops in the Mission.

Sean Parker: So you're just sort of like hanging out with the other cockroaches and it's terrible. There's no sense of enthusiasm, it's the most unglamorous, unsexy, miserable thing you could be doing. No one understands why you're doing it; certainly investors didn't, nobody my own age did.

Mark Pincus: No one in San Francisco had any interest or even understood what you were talking about. If anything, they might be a little annoyed.

Ev Williams: The thing that people missed was that it was the financial bubble that burst, not the internet. This whole time usage of the internet continued to climb dramatically.

Aaron Sittig: Everyone who had come to pursue a quick fortune left, and the only people still around were people who believed in the real promise of what the internet could provide, what technology was capable of doing for people. A lot of the cynicism that happens during flush times went away, and the attitude changed. There was a lot of mutual support and looking out for each other, hoping that someone would find a way to break through and succeed.

Sean Parker: So you have this really narrow population of people who believed. The only people who got through that period were the ones who were truly passionate or the ones who had incredible capability.

Andy Grignon: The pioneers got the arrows. And collectively, the hive watched. They take the lessons learned and start to make profitable companies based on them.

Ev Williams: I was a coder at the time, so I was like, *Well, I'll just build shit*, and that was kind of fun for a while. I just didn't leave the basement.

Sean Parker: Everybody else got annihilated and left.

The Return of the King

iCame, iSaw, iConquered

*T*he dot-com boom was cold comfort to the Valley's traditional hardware business. While NASDAQ was on a steep climb, Apple was facing bankruptcy. By 1997 Macintosh had sunk to a pathetic 2 percent share of the market—and Microsoft had the rest. That was the year Jobs finagled a return to the company which he had founded two decades before but had never actually controlled. This time he would be firmly in charge. The first thing that the older and wiser Jobs did was launch an advertising campaign, "Think Different," which flattered the Apple faithful with comparisons to Albert Einstein, Buckminster Fuller, Pablo Picasso, and the like. Then he launched a new computer, the iMac, which piggybacked on the vogue for all things internet. The third launch was a clever new product aimed at the Napster generation. The iPod was a way to put an MP3 in your pocket. Apple's update of Sony's venerable Walkman idea proved to be a game changer, a completely new direction for the company. Last in the computer marketplace, Jobs was the first to realize that the whole of the consumer electronics industry was ripe for plunder. Apple was back, Steve Jobs was calling the shots, and this time it really would be different.*

Larry Ellison: Back in mid-'95 Steve was finishing up *Toy Story* at Pixar and running NeXT, the computer company he founded after he left Apple. Apple was in severe distress. It had gone steadily downhill during the ten years of Steve's absence. The problems were now so serious that people were wondering if Apple would survive. It was all too painful to watch and stand by and do nothing.

Trip Hawkins: At that period of time in the nineties, they really seemed like they were up the creek without a paddle.

Michael Dhuey: What was keeping Apple alive was the Apple II. Luckily the Apple, because of its education market, kept selling year after year. That's what propped up the company until Steve Jobs came back.

Larry Ellison: We used to go for these long walks in Castle Rock State Park over by the coast. It cost $5 billion to buy Apple in those days, it was run by a fellow named Gil Amelio. I had raised the money. I was going to give Steve 25 percent of the company, and we were going to take it private and he was going to run it. And Steve said, "You know, Larry, I think I figured out a way I can get control of Apple without you having to buy it." I'll never forget this conversation as long I live. I said "Okay..." and then he said, "Yeah, I think I can get them to buy NeXT Computer and I'll go on the board, and eventually the board will recognize that I'm a better CEO than Gil Amelio and I'll become CEO."

Steve Wozniak: We bought NeXT because they had a real operating system and we desperately needed an operating system. Steve didn't understand what an operating system was, or what its importance was, when he built the Mac.

Jon Rubinstein: So I got to Apple in early 1997, right? And Steve hadn't come back yet. I came in as part of the restructuring that Gil Amelio was doing. They acquired NeXT as part of that. So I kind of came into Apple as part of the acquisition.

Larry Ellison: And then I went on the board.

Jon Rubinstein: About seven months later they fired Gil. Fred Anderson was interim CEO for a couple months, and then Steve came back as interim CEO.

Tom Suiter: At the time, there was a *Wired* magazine cover with the Apple logo on it and one word. It said, "Pray." That was the dire straits that Apple was in. Everybody said, "Oh, they have like ninety days of cash left: They're going to die."

Andy Hertzfeld: This is a little subtle and hard to understand, but Steve couldn't have come back to Apple without the Pixar success. That enabled it, because it gave him the ability to not put everything on the line with Apple. Remember: When he came back he wouldn't take the CEO job.

Wayne Goodrich: Steve was the "iCEO."

Trip Hawkins: When Steve came back he was a different Steve. He had grown up a lot—but he could still bring it on in terms of insanely great creative thinking.

Mike Slade: The first thing he did was a classic judo move. The "Think Different" campaign was brilliant, because it gave people permission to feel different and to feel good about the 2 percent market share that spring.

Michael Dhuey: Jobs spent a month going around to see what were all the projects running at Apple. He decided to go back to a simple product line, because Apple at that time had a ridiculous number of products. The Macintosh had like twenty slightly different versions or something, it was just nuts.

Jon Rubinstein: We convinced Steve that we should do a network computer.

Michael Dhuey: The idea was, *So what if we gave you a box and all you do is plug it into the internet and you're on the internet and you don't have to know anything about how computers work or the internet but you just use it?* That was a really good idea, because the internet was the big thing then.

Jon Rubinstein: The network computer evolved into the iMac, which was a very, very successful product for us.

Tom Suiter: I'll never forget the day he grabbed us and said, "I've got a little something I want to show you guys," and so we leave the Mariani building and bounce across the street and go over to where Jony Ive's studio was, and there are these things with shrouds on them and in a typical P. T. Barnum way, he pulls off the coverings and it's the iMac in five Lifesaver color versions.

Andy Grignon: That got Apple on people's radars again, because computers at the time were boring beige boxes. Even at Apple in the midnineties we made boring beige boxes.

Larry Ellison: Steve Jobs picked the colors of the original iMacs: "Sorry! No beige."

Tom Suiter: And they're so cool and translucent and it's like something from outer space. I remember saying, "I just want to lick them."

Michael Dhuey: It was sort of back-to-the-future. The iMac was kind of the replacement for the Apple II.

Jon Rubinstein: That was the beginning of the turnaround for Apple.

Mike Slade: And by late '98 or early '99 they had $4 billion in the bank. But it wasn't iMacs, okay? The iMac didn't make any money. It was mostly the expensive towers. The next campaign after "Think Different" was this thing where they make fun of Intel, and that's because Steve was trying to appeal to the Photoshop weenies and get them to buy overpriced G3 and

G4 towers. Okay, it was also PowerBooks, but mostly towers. It's really simple, right? Sell the high-gross-margin stuff. It might as well be a clothing store.

It took eighteen months for Apple to get back on its feet. Meanwhile, Tony Fadell, the young engineer who had cut his teeth at General Magic, was trying to get his own company off the ground. The idea was to do a line of MP3 players.

Guy Bar-Nahum: I remember Tony walking around, like a loser in the Valley with a secret in his pocket. He's the guy that had in his brain these two ideas: MP3s plus storage. There was a time when he was alone in the world with this. He was going around the Valley looking for people who would tolerate him, who would give him space, physical space, to sit in and draw.

Yves Béhar: Tony had just come out of General Magic, and we had a meeting in South Park, and Tony got really, really animated and passionate and excited about digital music and moving forward. Essentially he was founding a company, and he wanted to go investigate the hardware and the software that could live around digital media.

Tony Fadell: Hardware plus software plus services: General Magic was the first link that I ever saw like that. That influenced me for everything I did in my career since.

Yves Béhar: And he raised money from Samsung, and we took a space on Broadway and Columbus and started sketching and drawing different hardware and essentially a whole line of products: everything from home stereo systems to portable systems to DVRs. The whole product line would have a minimum amount of connectors and a very simple interface. The way Tony saw it was, "Let's create a whole ecosystem of products that will replace the entirety of your home audio and video components." And one of those things was a small portable MP3 music player.

Guy Bar-Nahum: Tony was like a guy with a secret in his pocket walking around, like, "Look at this! Look at this! I have the iPod." He didn't call it the iPod, and there were a lot of missing pieces, right? But that basic enabling idea? He had it first.

Yves Béhar: We had designed the whole system of products to be square with a circular interface that you sort of navigate around: basically a circle within a square. It was a design that we had scaled down to a small portable player, and up to a nice little cube stereo system.

Tony Fadell: I never went off to the internet. When everyone went off to the internet I stayed doing devices.

Yves Béhar: We had made a lot of progress: We had a great product lineup, we had some hardware designed, we had some mock-ups, the technology was aligning...

Tony Fadell: And then the internet crunch happened.

Yves Béhar: Just as it was for everyone else, it was impossible to find money to fund the next round for the company.

Tony Fadell: No one wanted to fund it. They were like, "You are crazy, you should only be doing software, hardware is dead," blah blah blah.

Yves Béhar: There was a belief back then that hardware would disappear. Hardware was going to be totally annihilated by web pages, the internet, e-commerce.

Silicon Valley was focused on the internet, but Apple had hardware coded in its DNA—and Jobs, like Fadell, awoke to the opportunity that Napster had created.

Sean Parker: At the time, there may have been as many as four hundred million people actively trading MP3s and listening to music on their computers, and that meant there was a market for this thing called an iPod.

Jon Rubinstein: Battery technology as well as display technology, driven by cell phones, was maturing. And so I kept going to Steve and going, "You know, Steve, I can't do a better music player yet." The trade-off was this: You get a large hard drive and a lot of songs, and it was huge and heavy and delicate, or it was really tiny and held five or ten songs. So neither of those products made sense. And so in February of 2001 I'm over in Japan for Macworld Tokyo, and every time I'm in Japan I go visit all the vendors. There was a handful of us that went on these supplier visits. We visited battery manufacturers because we need batteries for the notebook, and visited display manufacturers because we use displays for a lot of things, and we visited hard drive manufacturers because we use hard drives everywhere, and so we got to Toshiba. We had a meeting and we went through the hard drives for all the different products, and at the end of the meeting they go, "Oh, by the way, we've got this other thing we're working on, but we don't know what to do with it," and that was the small form-factor hard drive.

Guy Bar-Nahum: Now it's still a mystery to me why Toshiba made a 1.8-inch diameter drive with five gigs capacity. Only the Japanese, God bless them, do things like that sometimes. They make a piece of technology that doesn't have any justification in the market. At the time there was no industry to justify that illogical decision. The whole arc of development of the hard drive was for laptops—not handhelds.

Jon Rubinstein: So I went back to Steve and said, "Steve, I think we can do this: The drives won't be ready until late in 2001, but this is now doable, we can build a really cool product. It will be about the size of a pack of cards, and I need $10 million to do it, because I need to assemble a team and go build this out." Steve said, "No problem, I'll write you a check." So I call Fred Anderson, the CTO, and made sure the check wouldn't bounce, because you always had to double-check with Fred, and Fred said, "The check is good," and so we started assembling a team to do it.

Tony Fadell: So I am getting on a ski lift in Vail, and I get this call: "Hi, this is Jon Rubinstein from Apple," blah blah blah. The very next week I had my first meeting with Jon.

Jon Rubinstein: I said, "Hey, I've got this project I need some help on." I didn't tell him what it was but I said, "It's a small consumer form-factor product."

Tony Fadell: I thought they wanted me to restart Newton or something. So I did not know what it was and they would not tell me. Then after I signed the NDA, it was, "We have iTunes, now we want iTunes-on-the-go. The MP3 players out there today suck, and we think we can make a better one. We know you make these kinds of devices, so can you come in and sketch what you would build?"

Yves Béhar: And then one day Tony said, "You know, I'm going to take a contract, I'm going to work with Steve Jobs for six weeks. And if I can deliver this project for Steve in six weeks, we'll see what happens."

Jon Rubinstein: I hired him as a consultant and it was, "Here's what we've been looking at, and so now go thrash out what's available to make all this happen."

Tony Fadell: I said, "Okay," and then I worked nights, weekends, all day, and I let my start-up company just run and I went off and dove into trying to design what would become the iPod. So I had to learn about new chips and all the different things and put it together. What I remembered from my General Magic days is that you have to make sure that you do a lot of due diligence before going in, because if you commit to something they are going to expect that that is the way it turns out. So I did a ton of due diligence. I did as much as I could in that short amount of time. Six weeks later I pitch it to Steve.

Jon Rubinstein: There was a key meeting. It's me and Phil Schiller and Jony—there was a group of us, but Tony is the lead.

Tony Fadell: Steve comes in the room, does not say a word, takes the checklist, throws it on the desk, and gives the vision for it.

Ron Johnson: It was, "I want to make a music device and I want it to hold all your music, I want it to be digital, with great software so you could take your music everywhere." I don't think he had thought of "a thousand songs in your pocket" yet, but that was what he was imagining.

Larry Ellison: He was obsessive about making an MP3 player easy to use, and affordable, and beautiful...all the things you'd think Sony would do.

Tony Fadell: Then we flip to the next page of the checklist, and there was something about Sony. "And Sony, yes they are number one, but we are going to beat them up!" Next page...

Larry Ellison: Steve had decided, rather brilliantly, "Why should I compete with Microsoft and bang my head up against the wall when I can compete with Sony instead?" We'd go on these long walks and said, "Who do you want to compete with: Microsoft or Sony? I'll take Sony every day of the week, because I'm a software company, and Sony is not."

Jon Rubinstein: We knew that someone was going to replace the Walkman. Either we were, or they were.

Tony Fadell: And he would just skip around, so it was just this back-and-forth; he was riding the tide of momentum.

Larry Ellison: Apple is a software company, Microsoft is a software company, but Sony was a hardware company. You can kill a hardware company, because most hardware products, like an iPod, are 95 percent software.

Tony Fadell: And then at the end I had all these little mock-ups. I had bought these modern-day calculators and all the MP3 players, tore them apart, and made little pieces, and I put the little bits all into the box and I said, "Okay, here are the different battery technologies we could use, here are the different storage types. Here are the different displays. Here are the different interfaces, boards, connectors," and basically assembled the different options in front of him. I did three different options of "Here is what they could look like. Here is what the cost could be," those kinds of things.

Jon Rubinstein: Tony was coached: "Here's the choices we're going to present to Steve. Here's the order you present them in. Here's how you present it so that when we come to the end, we get the right answer. Right?"

Tony Fadell: For the final option, which a bunch of us thought was the best option, I made a Styrofoam model, with laser printouts of what I thought would be the right interface and everything, and then weighted the model so it felt about the right size and everything...I then had all the little blocks that would go inside, because I wanted to show that this was real enough. So I showed it to him. He picked it up and he was like, "Okay," but then he said, "But I do not think this is the right user interface for it." And then Phil goes, "Can we tell him?" and Steve looks at me and goes, "Yeah, we can trust him." Because they were all Apple employees except me. I was a consultant. So Phil Schiller runs into Steve's office, picks out a device, and has the device in his hand and says, "See that? Look at that wheel. We want something like that on it. Do you think you can do that?"

Mike Slade: Phil had a Bang & Olufsen remote, because Phil was one of those guys with Bang & Olufsen stuff all over his house, and he brought in the remote and showed us the scroll wheel, which Bang & Olufsen never patented, by the way, and I remember being there going, "You're right, this is cool, this is a really cool user interface."

Tony Fadell: I looked at it and figured out in my brain how they did it. I was like, "Sure. We can add that." And Steve is like, "All right. We are going to commit to this project. We are going to do it, and we want you to be the guy." And Jon looks at Steve and he goes, "Yeah, I will take care of it." And I go, "Look, I have got a start-up company," and Steve is like, "Do not worry about that. We will take care of it." I was like, "Okay..." I did not know what that meant.

Jon Rubinstein: I went back to Tony. In front of the whole team I said, "Okay, Tony, if you want to do this, you've got to join the company, I'm not going to do this with you as a consultant." So I kind of twisted his arm to take the job, basically.

Guy Bar-Nahum: Ruby is always cursing. He's a very ruthless and smart and disgusting person, and basically, according to the legend, he said to Tony,

"It's time to shit or get off the pot!" And he did that to him in a very theatrical way, during a staff meeting: in front of maybe twenty, twenty-five people. And Ruby enjoyed putting Tony on the spot in a very, very public way. He told Tony, "Look, man, I'm not continuing this meeting. I've had enough. You either join, or we just kill your project."

Jon Rubinstein: It wasn't really true, but…

Guy Bar-Nahum: Tony was really put on the spot, and he really didn't know what to do; he kind of hesitated.

Tony Fadell: I am like, *Wait a second! I have a company and there are people over there working on this other thing. How am I going to do this?* I was just like "Whoa!" So I just got in my car and I started driving through the hills of Saratoga and Los Gatos. So I go up to Skyline Drive. I wind up those roads. And I am just sitting there going, *What am I going to do? What am I doing?* Apple had… they liked to say, "single-digit market share." It was literally 1 to 1.5 percent of computer sales in the US, nowhere else in the world, just in the US. One and a half percent! And the stock had crashed, and I am like, *Why am I going to join Apple? I love the place, but why am I joining?*

Guy Bar-Nahum: Why would he sell for cheap the iPod to Apple? Was it fear? "I can't do it myself—I need a full company behind me to actually make it." Was it a discussion with Steve? Some reassurance?

Tony Fadell: I had a very pointed conversation with Steve, going, "Look, let me tell you, I was at General Magic and I watched them come out with a great product and not be able to market it and sell it. I am going to pour my life into this. I am going to build a team. But at the end of the day when we are going to be going up against Sony, how do I know you are going to be able to sell and market it?" And Steve just said, "Look, I will devote the entire Mac advertising budget to this if you can pull this off." I said, "Really?" and he goes, "Yes."

Jon Rubinstein: Now what start-up is going to ever be able to do that? Right? The Mac had a large installed base, a tremendous fan base that was very very attached to Apple.

Ron Johnson: This is about when the stores launched: May 2001. The stores had a good start, in terms of the Mac faithful.

Jon Rubinstein: We had a captive market, right? So if Apple released the iPod, and if we could get 10 or 20 percent of the installed base to buy iPods, it would be dramatically successful.

Ron Johnson: But the challenge for the stores was not to serve the faithful. It was to reach new customers. And so the iPod was our chance to really make a difference.

Jon Rubinstein: Our demographics were terrible, and we hoped the iPod would bring more younger people in.

Tony Fadell: Now, everything is half-truths so I was like, *Okay, Steve is saying the right things but will he really do it?* Then in the back of my head I am like, *If we do not run like hell we are going to get canceled.* Because I saw so many projects get canceled. I am like, *I've got to build the team, and we've got to ship this by Christmas, because if we do not ship by Christmas we are dead as a product.* It was 24/7 working for that six to eight months, depending on how you count.

Michael Dhuey: He hardly ever slept. He would send e-mails at two or three in the morning. It gave the appearance that he was a vampire.

Jon Rubinstein: The fear was Sony or some other big consumer electronics company was going to come after us and actually kick our butt. And so we kept the pace of innovation such that there was no way anyone was going to get in front of us.

Guy Bar-Nahum: We basically hired twenty or thirty people, some of them from inside Apple—all the lost souls within the company, people who had fought with their managers and needed something new to do.

Michael Dhuey: We were put in a separate building, and we had security badges so only people in our group could open the doors to our area. It was pretty high security. You couldn't tell anyone what you were working on.

Jon Rubinstein: Why give a heads-up to your competitors? Why let Sony know you're working on an iPod?

Guy Bar-Nahum: It was: You show up, you have computers you need to set up, networks, and you can't really use the company, because of secrecy, and you just don't know who to talk to, so you just do it yourself. It was all asses and elbows, and that's where we shined, really.

Michael Dhuey: The funny thing is that almost nothing on the iPod was actually done by Apple. Underneath the hood it was a lot of stuff grabbed from other parts of the industry.

Guy Bar-Nahum: By luck, Tony found this one company that actually made platforms for MP3 players. I don't know why they did that, because there was no market for it. And that company was Portal Player in Kirkland, Washington.

Sanjeev Kumar: We were from the PC industry. We had not done anything in consumer electronics and frankly did not know much about MP3 music at the time—except that Napster was a big thing and lots of people were downloading digital music.

Guy Bar-Nahum: So they had a solution: software and hardware—basically all of the under-the-hood stuff that you need in order to create an MP3 player.

Sanjeev Kumar: We had the main processor and the hardware interfaces for all your storage devices, displays, and input devices. In addition, we had all the software to run the processor. It was something we'd been working on for two years.

Guy Bar-Nahum: The other part was Pixo. Pixo was a bunch of geezers that left Apple to create a software framework for cell phones. And we needed this kind of technology for iPod.

Andy Grignon: So some friends of ours from Apple reached out. They said, "Hey, we're making this new thing, and we'd like to see if we could use your software. Can you make us an example application?" And we were like, "All right, what do you guys want?" They were like, "Just a sample MP3 player." I'm like, "Great!" So we whipped up a sample MP3 player app. But what was funny was we didn't know what we were building! We never saw the form factor. We never saw anything. It was just "make your stuff work" and that was it. Obviously it was an iPod, right? But we just made a sample app. And to this day in every iPod that's ever shipped, "TMP3SampleApp" is what gets launched when you turn on your iPod.

Guy Bar-Nahum: We took all these code bases, and we just very hackily kind of threw them together with bubble gum and masking tape. We put them together in a very haphazard way, because if you have a finite amount of time, right, a finite amount of resources, you really lower the bar on the software, right? You bring it in.

Sanjeev Kumar: Apple was very pragmatic. They had their date to launch and they had absolutely no problem in dropping a feature if that feature was going to be a roadblock to shipping on that date. Obviously you can't ship something that doesn't work, but the iPod specs in the beginning and what we launched were two very different things.

Guy Bar-Nahum: And it worked, but if you challenged it a little bit, it was going to crash. So we shipped software that was really alpha level.

Sanjeev Kumar: That first iPod was really a science project.

Jon Rubinstein: We had to get it done.

Wayne Goodrich: The evening before the launch, Steve was looking at it and playing with it and Jon Rubinstein was walking up the aisle and just as Jon got there, Steve pushed his glasses up on his head and looked at the screen and he goes, "You know, I really wish we'd been able to get color on this thing." And then Jon just patted him on the back and said, "Next version, Steve, next version..."

Mike Slade: The introduction was in that little tiny Town Hall war room, and it wasn't even very crowded.

Steven Levy: I was in New York, working at *Newsweek*. We were consumed by 9/11, so I wasn't going to go to California for, you know, what seemed at the time like not a major product announcement.

Mike Slade: 9/11 happened a month before the iPod came out.

Steven Levy: Apple asked me if I wanted one, and of course I did. So they hand-delivered it to me the day it was released. It came with CDs because they didn't want us to steal music to put on it. So they figured that in case we didn't own CDs they would give us a stack of them, all Steve Jobs's favorite music: *Bob Dylan Live 1966*, Glenn Gould's *Goldberg Variations*. That night or the night after I was going to a Microsoft event in New York. The latest version of Windows was going to come out. So I brought my iPod and sat next to Bill Gates at the dinner and pulled it out of my pocket and said, "Have you seen this?" And he said no, and I handed it to him. And he fiddled with the dials and played with it, trying to understand it. It was like he was an alien that had the power to suck up all the information on a piece of technology. So there's this little tunnel between him and the iPod, and he got all the information, and he handed it back to me and said, "This is only for Macintosh?"

Mike Slade: People forget that it was designed as a Mac-only niche-y add-on: no power cord, assume FireWire, all these things that no self-respecting MBA would ever green-light because it limited the market size. Steve didn't care. He just wanted it to be good.

Ron Johnson: So it really wasn't a product for Windows people. If they wanted an iPod, they'd have to buy a Mac.

Jon Rubinstein: The goal of the iPod was not to change Apple at the time. The goal was to sell more Macs. You bought your iPod, so you'd spend money on that. And that then locked you into the Mac ecosystem.

Steven Levy: I liked it from the beginning. To me the most exciting thing about the iPod was the shuffle thing. You could take all your music and shuffle it, and it would all be yours but there would be the element of surprise, too. It was like a radio station where every single selection was a song you liked. To me it was a metaphor for the digital age. A symbol of the promise of the digital age.

Jon Rubinstein: We only sold a few hundred thousand of that first product.

Ron Johnson: So the iPod was not successful. It was a good product, but a low-value product from 2001 to 2004.

Steven Levy: In 2004, I noticed the thing was taking off in a cultural sense. The iPod itself was a thing: You would walk through the streets and you would see someone else with these white ear buds, and it was like a secret society—you were part of this cult. The iPod was a fetish object, an object of lust for the people who owned it. It could deliver music, better than anything before it. The fact that it looked beautiful just amplified how much people loved it.

Sanjeev Kumar: Even though the initial volume for the first few years of iPod were not that big, and the revenues were not that big, it created the halo effect. Apple products became cool again.

Ron Johnson: But the iPod wouldn't have changed the music world without the iTunes Music Store.

Mike Slade: The iTunes Music Store comes in late '03.

Jon Rubinstein: The iTunes Music Store allowed you to actually buy a track at a time, for ninety-nine cents, and download the music directly in iTunes. So you didn't have to buy CDs or records or anything else anymore. You could download directly from the store and get the latest music. So it was very cool, and pretty exciting.

Mike Slade: Steve is the one who made it happen, because at first the labels just said no to everything. They were just like, "Yes, we're doing our own thing, fuck off," over and over. They were the worst people I've ever dealt with in the entire galaxy, but Steve convinced them to let him do it on the Mac.

Jon Rubinstein: We put together a pitch to the music companies: "Look, we're at 2 percent market share with the Mac—we are not successful, so you have no downside to licensing us the music. You can do this experiment and it will cost you nothing." If we'd been wildly successful in the market, there's no way they would have licensed it to us. We didn't matter, so they could license to us and who cares, right? "Run the experiment, what do you have to lose? If you don't do this, it's just going to get stolen anyway." And so we kind of got away with it, because it didn't matter. Now, they probably ended up doing deals they probably regretted later, but that's how it all kind of played out.

Nolan Bushnell: I tried to do a digital jukebox in 1992 or '93, and to get the rights from the record companies was just impossible! Fast-forward ten years: Steve got all four of the major record labels to let him sell tracks for ninety-nine cents. That was a negotiation! And I really appreciated that he was able to do that, because I had gone down that same path and failed. So you gotta say, "Attaboy, Steve!"

Sean Parker: It turns out that it's a lot easier to get a deal done with the record labels after their business has collapsed. The music business went from a $45 billion global industry to a $12 billion global industry. It's a much easier conversation at that point. There is no hypothetical. You don't have to get them to believe that this catastrophic event is going to occur. It's already occurred.

Ali Ayar: The iPod took advantage of the fact that Napster existed. The fact that you could put "a thousand songs in your pocket"? Where are you getting a thousand songs?

Jon Rubinstein: People were stealing music, yes, no question about it.

Steve Wozniak: Then Steve did a strange thing.

Ron Johnson: Several members of the executive team, most notably Phil Schiller, and probably Jon Rubinstein, convinced Steve to open up the product to Windows. That was like heresy to Steve.

Steve Wozniak: He went open.

Ron Johnson: Mac was a closed ecosystem, right? But he made the decision to open it up to Windows.

Steve Wozniak: He okayed it. He said, "You don't have to have a Macintosh to use an iPod. We'll write iTunes for Windows."

Mike Slade: People didn't really think that the iPod was going to be a viable product for Windows, because it was a pain in the butt to hook it up to FireWire. It was kind of a hack for a Windows user until the USB 2.0 iPod came out. That's one of the reasons why the music companies were willing to give him the Windows option.

Jon Rubinstein: By then it was, "You're kind of pregnant, so it is too late to back out now."

Ron Johnson: It was a huge economic impact for the music industry, because before if you used to have a great song, you'd get $15 because you sold a record. Now that song was worth ninety-nine cents. It disaggregated the album.

Guy Bar-Nahum: iPod basically freed Apple from the PC wars. Apple was entrenched in this losing war. It was like World War One: suffering and bleeding and losing territory and losing cultural relevance until iPod came and opened up the big sky of the universal app that is music. Today it looks small, but back then it was infinitely bigger. Music crosses cultures, and operating systems, and everything that was limiting the growth of Apple. It was a big thing.

Steve Wozniak: Apple didn't grow in size ever over the Apple II days until the iPod. And it didn't grow in size when he introduced the iMac. It all started with the iPod—and it was the openness.

Sanjeev Kumar: It brought Apple back from the brink.

Andy Hertzfeld: They were making literally billions every week from the handheld music players.

Ron Johnson: The iPod gave Steve confidence that when we're not competing with a monopoly, Apple could win, because we got up to almost 90 percent market share on music players.

Steve Wozniak: The openness made our revenues double, our profits double, our stock double—and the board gave him billions of dollars in stock and eight jet airplanes.

Ron Johnson: So the feeling was, *Wow, we can do this!*

Yves Béhar: Before the iPod you couldn't really go to any meeting anywhere, and say, "Look, design is important and central to a business." We couldn't go do that as designers, because the minute you would say this to Microsoft, Compaq, HP, anybody in the business, they would laugh you out of the room, because you were basically taking the example of a loser: Apple. So the iPod really is what put design front and center. It really gave the hardware guys a show-off moment, a win.

Jordan Ritter: Of course I had wanted it to be Napster, but I didn't really care. I knew that someone would break through all these asinine people, companies, and industries that were saying no. And there it was. There it was. *Finally!*

The more philosophically minded had a different reaction. They felt that Jobs had sold out. He saved Apple, yes, but that big win came at a big cost. Jobs betrayed the power-to-the-people ideals that had animated his youth, early Apple, and the personal computer movement as a whole.

John Markoff: I was super disappointed. For me iTunes meant that the PC was becoming an entertainment device. I was still back in the bicycle-of-the-mind world. For me it was about augmentation: "Steve, where are the better tools for writers and thinkers?" We actually had that conversation, and he blew me off.

Alan Kay: When Steve came back to Apple he was not the same Steve. He was a very effective Steve, but he had lost his zeal for what computers are really good for socially—like education. Completely lost it. And you know I would try to get him back interested in wheels for the mind—"Remember that, Steve?"—instead of dumbed-down user interfaces.

John Markoff: Steve didn't look backward very much. NeXT and the Mac were really targeted at this notion of augmentation that Engelbart had thought up years before: Computers were these tools that could really enhance your ability to do intellectual work. Steve was turning the computer into a new entertainment device—first with audio, but later with video.

Jon Rubinstein: The early Steve was the bicycle-for-the-mind thing: "How do you enhance human creativity, human capabilities, by using computers?" With the iPod that got extended into: "How do you make people's lives better?" The side effect, of course, was that people tuned out.

Steve Wozniak: To me, it looks like, "Oh my gosh, they could walk out in front of a car!" You think a whole bunch of negative thoughts because you've grown up another way, but it's really not bad, it's just different. They have their way, we have ways from the past, and they can be very, totally different, and yet that doesn't make one of the ways bad compared to the other.

Alan Kay: I used to get after Steve and he didn't care, because he was selling it to people who were just happy to continue the dumbing down. It's
bad, really bad. Most people can't see it. People don't have any idea what a
computer is. They are just using it to play movies on. This is the corruption of consumer electronics—using a computer basically for convenience
rather than for actually doing primary needs. If you think about it, this is
completely fucked up.

I'm Feeling Lucky

Google cracks the code

*G*oogle was a dot-com poster child, but unlike the others it managed to escape immolation in the dot bomb. Brin and Page had the good fortune to secure $25 million of venture money in 1999, just before the fever broke. They also had the good sense to bank most of those millions. It took them a few years to figure out how to make search profitable, but again luck was on their side. The money sustained the young company through the launch of AdWords, a self-service ad-buying tool, in 2001. Two years later, Google perfected their advertising system with AdSense, which turned the entire internet into a billboard that Google could then sell. The timing was, again, perfect. Google became enormously profitable at the very moment that the rest of the Valley was at its most desperate. Google was able to expand on the cheap. With ad money pouring in, Brin and Page went on a buying spree: sucking whole companies, buildings by the dozen, and thousands and thousands of PhDs into their ever-expanding headquarters: the Googleplex. The much-anticipated Google IPO in 2004 marks the start of what Silicon Valley calls Web 2.0—a reboot, a leveling up, a phase shift. Post-Google, the Valley began to abandon the notion of the internet as a free "cyberspace" that one could "surf," and instead started to regard the web as a vast machine possessed of a native intelligence—which it could direct, program, even own.

Douglas Edwards: If you read Larry and Sergey's original paper that they wrote at Stanford where they talked about creating a search engine, they specifically said that advertising was wrong and bad and it would inherently corrupt the search engine if you sold advertising. So they were adamantly opposed to the notion of having advertising on Google.

Ray Sidney: So when I originally met Larry and Sergey in 1999, the plan was to license search—no advertising. We were going to have the best search out there and portals and companies and whatnot, and we were going to license it.

Marissa Mayer: The first revenue came on June 24, 1999, and it was licensing. Netscape was basically paying Google to power their search.

Larry Page: OEM-ing a search: You can get revenue from that, but it is limited. People aren't used to paying huge amounts of money for that.

Ray Sidney: Then people started reading about how much money was being brought in to various other companies by search advertising, and it was kind of decided that we were leaving money on the table.

Douglas Edwards: There was a lot of pressure to generate revenue, and so Larry and Sergey decided that advertising doesn't *have* to be evil—if it's actually useful and relevant.

Ray Sidney: We had this great website and that was really getting popular so maybe we should think about putting some ads up there? "Let's try it out!" Thus was born the first Google ad server. They had to call companies: "Hey, would you like to advertise on Google?" It was a manual process. The first ads were July or August of '99.

David Cheriton: The thing that many people don't recall from that first web era was that a lot of the companies that started out as search companies introduced advertising and then compromised the search in favor of the advertising, so nobody trusted the search.

Paul Buchheit: Companies would just mix the ads in with the regular search results so people would think it was a search result. It's kind of like fake news or something.

Brad Templeton: In most of the publishing business, there is a history of having a wall between editorial and advertising. Google really took that to heart. They tried to keep their wall between the advertisement and the actual editorial content pretty strictly divided. But all the other websites were basically tearing down the wall. Google said, "Let's not do that."

Douglas Edwards: That was something Larry and Sergey were adamantly opposed to. They felt that it was so corrupt and so disingenuous. It was like a religious thing with them, so they wanted to make sure that nobody would mistake Google ads for search results.

Paul Buchheit: Sometime in early 2000 there was a meeting to decide on the company's values. They invited a collection of people who had been there for a while. I was sitting there trying to think of something that would be really different and not one of these usual "strive for excellence" sort of statements. I also wanted something that, once you get it in there, would be hard to take out.

Brad Templeton: "Don't be evil" was the phrase.

Paul Buchheit: It just sort occurred to me.

Sergey Brin: We have tried to define precisely what it means to be a force for good—always do the right, ethical thing. Ultimately, "Don't be evil" seems the easiest way to summarize it.

Paul Buchheit: It's also a bit of a jab at the other companies, especially our competitors, who at the time were, in our opinion, kind of exploiting the users to some extent. They were tricking them by selling search results—which we considered a questionable thing to do, because people didn't realize that they were ads.

Sergey Brin: We think that's a slippery slope.

Brad Templeton: That's what "being evil" would be, actually.

Scott Hassan: Then Larry came up with a great idea: having *self-serve* ads, where you could just go to a website and click a couple things, and then ads would be shown on the Google website, without talking to a salesperson at all.

Marissa Mayer: AdWords was "pay by credit card and we're going to run your ad."

Ray Sidney: That's definitely important. You can call the big advertisers and say, "Hey, how would you like to start advertising with us?" But you can't call up all the little people who are interested in spending a small amount of money to advertise whatever small things they're selling—what you would call the long tail. So that was definitely key.

Douglas Edwards: One of the key innovations was that the ads were going to be created by the advertisers themselves and not in any way be vetted by Google. I thought that that was absolutely insane. I was sure that was suicide. And I told Larry as much. And so we did it anyway. We put it up on the home page as a link. And about two hours after we did that we got our first AdWords customer: Lively Lobsters in Kingston, Rhode Island. Ryan Bartholomew was the owner and sole employee.

Ryan Bartholomew: I didn't even know what I was doing, I just thought it was cool to buy lobsters for five bucks off the boat and try to ship them to buyers. And I was just working around the clock trying to figure out ways to get my listing out there. This was back in the days of Yahoo and

AltaVista and I don't even remember the others. But Google was around and very new, and I was up at two in the morning playing around on Google looking for something or other.

David Cheriton: The web had grown from this very early stage where there was a small amount of content and you could sort of find things with the help of Yahoo, to the case where the content had grown so rapidly, and had attracted all of these compromisers and game players, that it became harder and harder to find things. And so that was the environment that Google was in.

Ryan Bartholomew: And there was this little box to the side that was looking for advertisers for Google. I had never seen that before and I ended up clicking on it and running an ad for "live lobsters." I later found out that I was the first advertiser on Google, ever. It was a self-serve, all-text display ad. You bid on a keyword and you make the ad and you pay a certain amount for it.

Ray Sidney: So in a very real sense, every time you do a search query on Google, an auction happens behind the scenes, between the various advertisers that are interested in key words that are in the query.

Ryan Bartholomew: Because you are bidding competitively against other people for keywords. It was priced in CPM.

Brad Templeton: That's cost per thousand views. Ten dollars is a very common CPM in the marketplace. If you prefer to think of it as a cost per view, you can think of that as a penny per view. But usually people use the cost per thousand. *M* is the Latin for "one thousand."

Ryan Bartholomew: I spent eighty-three bucks, and as far as I could tell, I didn't get a single order! I was pretty pissed off. I ended up calling Google and I got somebody and I said, "This sucks!"

Brad Templeton: For many years nobody was able to really get advertisers to pay very much. So, basically, the philosophy of sites at the time was "Yes, what can we do for you, sir?" And so you had web pages and they'd have pop-ups and pop-unders and dancing bears and everything else. And whatever the advertisers would pay for, you would do.

Ryan Bartholomew: I asked them, "Why aren't you doing pay-for-click? Why aren't you charging per click rather than just on the number of impressions? I didn't get anything for my ads!" So they took my feedback and that was that.

Brad Templeton: So "I want to pay per click."

Ryan Bartholomew: And after a while they did move to a pay-per-click model.

Marissa Mayer: It was "we're going to let you pay per click."

Brad Templeton: And again the other sites are doing it.

Ryan Bartholomew: GoTo.com was the first pay-per-click engine, there was a company called FindWhat.com, there was this whole pay-per-click ad model around prior to Google, and I had already been doing pretty well with it.

Marissa Mayer: But we're going to rank your ad based on a quality score *and* how much you are willing to pay.

Brad Templeton: The brilliant thing that Google said was, "Okay. Pay per click, but if people do not click on your ad, we're not running it." Google charges you a fee if someone clicks on your ad. But if your ad is not performing enough, they take you out of the way, so they end up getting a minimum for every thousand impressions they do. Therefore they won't run your ad unless it's paying them the CPM figure they want. That was the genius! Google found a way that the advertiser could think that they were paying per click—which is what advertisers like—but they, the publisher of the website, could get paid per thousand views, which is what publishers like. They found this marriage which gave them what they wanted, and gave the advertisers the illusion that they were getting what they wanted, and of course who wins? Google wins.

Ryan Bartholomew: It was called AdWords Select.

Brad Templeton: And this was—by that standard of that day—a complete shock. "What do you mean, you won't take my money?" That was just a completely foreign idea to web advertising at the time—and it forced all the advertisers to write good ads that people would actually click on.

Douglas Edwards: And it wasn't just advertising for lobsters. Ryan quickly understood the opportunity for arbitrage. Amazon had an affiliate program where if you sent someone to Amazon and they bought a book, Amazon would pay you. So Ryan began taking ads out on Google not just for lobsters, but for books that would link to his Amazon affiliate page, where he would collect commissions. And of course in the same way he was using Amazon to generate revenue, he began linking to adult sites that were paying for traffic.

Ryan Bartholomew: I just couldn't make much money on the lobsters, so in 2001 I ended up selling that company and ended up focusing on porn basically. Google allowed you to bid on adult keywords and run traffic to the porn affiliate programs, which get people to give their credit card and sign up for a free trial and all that stuff. It was a really simple business model: I would bid on any sex-related keyword that I could get, traffic was cheap—especially for the obscure stuff—and I would send it to a page that was nothing but text links. It would break down whatever their interest was, any kind of fetish, and then they would click it and it would take them to a porn site targeted to that interest. At one point there must have been forty or fifty affiliate programs for everything under the sun.

Brad Templeton: I remember the first party they had after they moved to downtown Palo Alto where they had actually put a projector doing search queries onto the sidewalk of University Avenue and that's how you knew where it was.

David Cheriton: And one of them showed me his laptop or some display, where you could see the searches that were being filtered out, and of course they were all sort of porn type searches. Without the filter, we would've all been arrested. And it just sort of struck me. I hadn't really thought about the magnitude of the use of the search engine for searching for porn.

Ryan Bartholomew: The beauty of it is that it didn't matter whether it was lobsters, porn, Aerosmith CDs, or snowblowers! It was all just arbitrage on the value of the traffic. This is what Google was built upon. This is ultimately what Google is. You take a general search and filter it by what people are interested in. The more accurately you can figure out what people are searching for, the more likely they are to buy that product or service. Amazon is kind of doing that, too; it's a very sophisticated machine for the same idea—narrowing you down to exactly what you want.

Young Harvill: It's still not really talked about a lot in Silicon Valley, but a lot of the killer app for some of the first technologies is porn.

Ryan Bartholomew: I remember spending $13,000 in one day on all adult-oriented traffic. And my margins on that traffic were something like 3X. So I must have made thirty or forty thousand dollars in one day. That was right at the peak.

Brad Templeton: This turned into Google's great advertising success.

Ryan Bartholomew: And when the holidays rolled around I got a lava lamp from Google as a thank-you gift for sending them millions of dollars in business.

Scott Hassan: But AdWords was just about putting ads on the Google search engine. The real advance was AdSense in 2003.

Marissa Mayer: Which was "because you are paying per click or per lead, now we're going to take your ad and put it on different websites."

Paul Buchheit: AdSense, the content-targeted ads, was actually something that, if I recall, I did on a Friday. It was an idea that we had talked about for a long time, but there was this belief that somehow it wouldn't work. But it seemed like an interesting problem, so one evening I implemented this content-targeting system, just as sort of a side project, not because I was supposed to. And it turned out to work.

Susan Wojcicki: It was a really novel idea at the time to serve ads that were targeted dynamically. People were saying, "This is a sports site, so we'll serve a sports ad." And we were saying, "No. We can actually look at the page in real time and figure out what this page is about."

Paul Buchheit: What I wrote was just a throwaway prototype, but it got people thinking because it proved that it was possible, and that it wasn't too hard because I was able to do it in less than a day. After that, other people took over and did all the hard work of making it into a real product.

Ryan Bartholomew: AdSense is just the inverse of AdWords. With AdWords you are paying for when the ad is clicked on, then you get that traffic and then it's up to you how you make money on that traffic. AdSense is just the program where, if you are the publisher of a web page, you get a cut of what those AdWords buyers are spending. It's like owning a billboard. You run AdSense if you owned a billboard and the person who is buying the ad on the billboard would be an AdWords customer—and Google is the middleman.

Douglas Edwards: Google manages the system: They collect both the advertisers and the sites that run the ads. And so there's a viral effect that goes into play when you build a critical mass. The advertisers say, "Where are all the publishers?" "Well, they're at Google." And the publishers say, "Where are the most advertisers?" "Oh, they're at Google." So it feeds on itself until you become the predominant player in that space.

Ev Williams: That's when Google came knocking. It was a really hard decision to sell Blogger to Google because Google was still private, no one knew what was going to happen. This was pre-Gmail, pre-anything but search and advertising. So that was confusing: "Why would they want to buy Blogger? We're not a search engine."

Douglas Edwards: It was really simple: Once we had AdSense, we wanted more places to put ads, and we thought Blogger was going to generate millions of pages of content on which we could put ads.

Scott Hassan: It expanded the reach that their ad network had and allowed them to have a lot of inventory cheaply. It was a win-win for everybody.

Biz Stone: After Google acquired Blogger, I wrote an e-mail to Ev. And I said to him, "I feel like I'm the missing member of your team. And so, if you ever need to hire more people for the Blogger team—I'm your guy." I was actually hired to be on the AdSense team—or whatever it was called in 2003—but I really worked on Blogger under Ev.

Ev Williams: And we had to go down to Mountain View, because that's where the real Silicon Valley was.

Biz Stone: We were in "Building Pi," of course. Nerd factory, right?

Brad Templeton: And by that point, they had become a pretty big company.

Larry Page: We finally moved into a big corporate park in Mountain View, where we are now.

Heather Cairns: We were moving into the old Silicon Graphics campus, where there's still old Silicon Graphics people working there, and they weren't too happy to see us.

Marissa Mayer: At that point SGI was doing poorly, so there were about fifty people on that whole campus.

Jim Clark: It was a sinking ship.

Heather Cairns: We are like, "Yay! Here are our pool tables and our candy! Yay! We're Google!" And they're looking out the windows at us playing volleyball and they're like, "Fuck you!"

Jim Clark: They were upset that they had missed out on being part of Netscape.

Marissa Mayer: We'd been so disrespectful—loud and annoying.

Heather Cairns: We weren't trying to be disrespectful. We were just stupid. We didn't have a sensitivity to the fact that those people probably wouldn't have jobs in a few months. They were clear on that. And they're just watching the new blood come in—happy, eager, just bouncing off the walls.

Biz Stone: Google was a weird place—like a weird kid land. Adults worked there, but there were all these big, colorful, bouncy balls. Eric Schmidt had a twisty playground slide that he could exit his office from—which now seems almost perverted.

Heather Cairns: I did the employee manual and I modeled our culture after Stanford—because that's where most of our people were coming from.

Sean Parker: Google really did set themselves up to get great engineers by trying to make their environment feel as similar to grad school as possible. Google could make the case: "Oh, don't worry, this is going to feel a lot like when you were a researcher. This isn't like selling out and going into the corporate world, you're still an academic, you just work at Google now." They ended up getting a lot of really smart people because of that.

Ev Williams: And then we show up and we are these weirdos: bloggers from the city. So I always felt like an outsider, but I was still impressed by Google.

Douglas Edwards: It was a difficult integration, because we had somewhat different cultures. So when there were issues of technical integration there was friction. The expectations of the engineers at Google were unreasonably high, perhaps, and Blogger was a little bit more laid-back.

Mark Pincus: Evan Williams and the people at Blogger were these superhip sort of creative people that wanted to do this kind of fringy thing that was cool—and they were buried inside Google.

Biz Stone: Google was not a normal place at all. There was just all kinds of weird stuff going on. I would just walk around and check stuff out like the kid in *Willy Wonka* going around the chocolate factory.

Heather Cairns: Larry and Sergey would be doing crap with Legos, building stuff with Legos.

Larry Page: Lego Mindstorms. They're little Lego kits that have a computer built in. They're like robots with sensors.

Heather Cairns: I remember them making a rubber wheel and moving it over paper. I was like, "What are you doing?" "Well, we want to scan every book and publication and put it on the internet." I'd say, "Are you crazy?" And they'd say, "The only thing that holds us back is turning pages. Right now it costs a lot of money to pay a human to do this with a scanner 24/7, so we are trying to find a way to walk the pages under this wheel to automate it."

Biz Stone: Like one day I walk into a room and there was just a whole bunch of people dazed out on these automated contraptions with lights and foot pedals and books. And I was like, "What are you guys doing?" They said, "We are scanning every book published in the world." And I was like, "Okay. Carry on."

Heather Cairns: Part of me even wondered: *Are they losing interest in the search engine?* Sometimes it seemed like it but no, they just always had about seven things going at once—because that was their interest.

Biz Stone: And then I distinctly remember going into what I thought was a closet, and there was this Indian dude on the ground with no shoes on and he had a screwdriver and he was taking apart all these DVRs. He looked like he had been up all night or something. And I said, "What are you working on in here?" And he said, "I'm recording all broadcast TV." And I was like, "Okay. Carry on."

Marissa Mayer: I was there the day we did the first Street View experiments. It was a Saturday and we just wanted to blow off some steam. We rented an $8,000 camera from Wolf Camera that was considerably less when rented per day. We drove around in a little blue Volkswagen Bug with the camera on a tripod in the passenger seat. We just started driving around Palo Alto taking a photo every fifteen seconds, and then, at the end of the day, we took photo stitching software to see if we could stitch the pictures together.

Heather Cairns: Larry and Sergey were first and foremost, and probably still are, inventors. That was their true love.

Douglas Edwards: Sometimes it was hard to separate what was fantasy from what was a real proposal. From the very beginning Sergey and Larry— well, in particular, Larry—was talking about space tethers, and how that's the business we should really be in. You never knew whether to take them seriously or not.

Marissa Mayer: I hosted weekly brainstorming sessions because we wanted people to think big. One week I started the session with the space tether. We started brainstorming about building it out of carbon nanotubes, and could we use it to do pizza delivery to the moon?

Douglas Edwards: Sergey would just throw out these marketing ideas. He wanted to project our logo on the Moon. He wanted to take the entire marketing budget and use it to help Chechen refugees. He wanted to make Google-branded condoms that we would give out to high schools. There were a lot of ideas that were floated and most of them never became full-fledged projects. But if Larry and Sergey suggested something you pretty much had to take it at face value for a while.

Marissa Mayer: Some things we actually did go out and build—like driverless cars. We brainstormed that.

Biz Stone: It was just strange, it was really weird, but it was awesome, too.

Ev Williams: So I was not happy at Google, which today I totally blame myself for. It was eight hundred people when I got there. I considered it a big company, but it was not a big company. It had just grown really fast and everything was broken and immature and chaotic, and I let myself think that was wrong when, in fact, it was exactly as it should have been.

Charlie Ayers: The whole climate of the company was a focus on growth, growth, growth.

Ev Williams: Their perspective was that search and advertising were on fire, and we're going to IPO next year—and there's this Blogger thing, where people are posting their cat pictures.

Heather Cairns: I would say that by 2003 it's a very different place than when we started. We're like two thousand people, and people were talking about going public.

Ev Stone: We were making a lot of money, but before going public, no one knew how much money Google was making. But the management did, they were very excited about it.

Heather Cairns: *Going public. Being rich. Going public. Going public.* That was really on the forefront of so many people's minds.

Charlie Ayers: At that point there were a lot of us who had been there forever, who pretty much would just come to hang out. They were waiting, not even working anymore. You were seeing that happen with a lot of people.

Ray Sidney: I got burnt-out. I was not feeling very productive. I thought, *You know what? I need to get away.*

Charlie Ayers: A lot of the early-timers were looking at, like, *How much does this island cost?* There was a lot of distraction.

Ray Sidney: Originally I thought, *You know what? I just need to take a month or two off, and then I'll kind of get that fire back in my belly.* And that never happened. I left in March of 2003.

Charlie Ayers: So it was smart to have a whole new wave coming in, because they were the ones that were focusing on making the business happen.

Biz Stone: I got there in 2003, worked there for a year, and in 2004 they go public. It was perfect timing.

Charlie Ayers: As the IPO got closer, the level of distraction got greater and greater and greater. Their eyelids were too heavy with dollar signs.

John Battelle: With the benefit of hindsight, Google's IPO in 2004 was as important as the Netscape IPO in 1995. Everyone got excited about the internet in the late nineties, but the truth was a very small percentage of the world used it. Google went public after the dot-com crash and reestablished the web as a medium. Web 1.0 was a low-bandwidth, underdeveloped toy. Web 2.0 is a robust broadband medium with three billion people using it for everything from conducting business to communicating with your friends and family.

Biz Stone: I just got there in just the right time to get a feel for the before and the after.

Douglas Edwards: After the IPO it became more buttoned up, more metrics driven—which was good for the company, probably. But it was not the culture that I was used to and had enjoyed when I was there.

Charlie Ayers: They're like, "we're publicly traded now." So 2004 was not the best year at Google, morale-wise. They started sending more of us to Dale Carnegie classes.

Heather Cairns: Larry and Sergey used to hold their forks and knives in a fist, scooping. They used to scoop food into their mouths, which would be a couple of inches from the plate and I'd be like, *I can't even watch this. I can't. I'm going to be sick.* They had to be taught not to do that.

Charlie Ayers: There were a handful of us that would go to public speaking classes, media training classes, leadership classes.

Heather Cairns: Nobody has superbad, disgusting behavior anymore. It's really depressing. The personality has been coached out of them—all of them.

I'm CEO...Bitch

Zuck moves to Silicon Valley to "dominate" (and does)

*I*n Silicon Valley during the mid-aughts, the conventional wisdom was that the internet gold rush was largely over. The land had been grabbed. The frontier had been settled. The web had been won. Yet nobody ever bothered to send the memo to Mark Zuckerberg—because at the time, Zuck was a nobody: an ambitious teenaged college student obsessed with the computer underground. He knew his way around computers, but other than that, he was pretty clueless—someone had to explain to him that internet sites like Napster were actually businesses, built by corporations. Zuckerberg could hack, however, and he knew exactly what he wanted to build: TheFacebook.com—a college-specific version of Friendster, a pioneering social network based in Silicon Valley. Zuckerberg's knockoff site was a hit on his Harvard campus, and in 2004 he decided to spend a summer in the Valley rolling it out to other colleges nationwide. The whole enterprise began as something of a lark. Indeed, Zuckerberg's first business cards read, "I'm CEO...bitch." The brogrammer 'tude was a joke...or was it?

Sean Parker: The dot-com era sort of ended with Napster, then there's the dot-com bust, which leads to the social media era.

Steven Johnson: At the time, the web was fundamentally a literary metaphor: "pages"—and then these hypertext links between pages. There was no concept of the user; that was not part of the metaphor at all.

Mark Pincus: I mark Napster as the beginning of the social web—people, not pages. For me that was the breakthrough moment, because I saw that the internet could be this completely distributed peer-to-peer network. We could disintermediate those big media companies and all be connected to each other.

Steven Johnson: To me it really started with blogging in the early 2000s. You started to have these sites that were oriented around a single person's point of view. It suddenly became possible to imagine, *Oh, maybe there's another element here that the web could also be organized around? Like I trust these five people, I'd like to see what they are suggesting.* And that's kind of what early blogging was like.

Ev Williams: Blogs then were link heavy and mostly about the internet. "We're on the internet writing about the internet, and then linking to more of the internet, and isn't that fun?"

Steven Johnson: You would pull together a bunch of different voices that would basically recommend links to you, and so there was a personal filter.

Mark Pincus: In 2002 Reid Hoffman and I started brainstorming: What if the web could be like a great cocktail party? Where you can walk away with these amazing leads, right? And what's a good lead? A good lead is a job, an interview, a date, an apartment, a house, a couch…And so Reid and I started saying, "Wow, this people web could actually generate something more valuable than Google, because you're in this very, very highly vetted community that has some affinity to each other, and everyone is there for a reason, so you have trust." The signal-to-noise ratio could be be very high. We called it Web 2.0, but nobody wanted to hear about it, because this was in the nuclear winter of the consumer internet.

Sean Parker: So during the period between 2000 and 2004, kind of leading up to Facebook, there is this feeling that everything that there was to be done with the internet has already been done. The absolute bottom is probably around 2002. PayPal goes public in 2002, and it's the only consumer internet IPO. So there's this weird interim period where there's a total of only six companies funded or something like that. Plaxo was one of them. Plaxo was a proto–social network. It's this in-between thing: some kind of weird fish with legs.

Aaron Sittig: Plaxo is the missing link. Plaxo was the first viral growth company to really succeed intentionally. This is when we really started to understand viral growth.

Sean Parker: The most important thing I ever worked on was developing algorithms for optimizing virality at Plaxo.

Aaron Sittig: Viral growth is when people using the product spreads the product to other people—that's it. It's not people deciding to spread the product because they like it. It's just people in the natural course of using the software to do what they want to do, naturally spreading it to other people.

Sean Parker: There was an evolution that took place from the sort of earliest proto–social network, which is probably Napster, to Plaxo, which only sort of resembled a social network but had many of the characteristics of one, then to LinkedIn, MySpace, and Friendster, then to this modern network which is Facebook.

Ezra Callahan: In the early 2000s, Friendster gets all the early adopters, has a really dense network, has a lot of activity, and then just hits this breaking point.

Aaron Sittig: There was this big race going on and Friendster had really taken off, and it really seemed like Friendster had invented this new thing called "social networking," and they were the winner, the clear winner. And it's not entirely clear what happened, but the site just started getting slower and slower and at some point it just stopped working.

Ezra Callahan: And that opens the door for MySpace.

Ev Williams: MySpace was a big deal at the time.

Sean Parker: It was a complicated time. MySpace had very quickly taken over the world from Friendster. They'd seized the mantle. So Friendster was declining, MySpace was ascending.

Scott Marlette: MySpace was really popular, but then MySpace had scaling trouble, too.

Aaron Sittig: Then pretty much unheralded and not talked about much, Facebook launched in February of 2004.

Dustin Moskovitz: Back then there was a really common problem that now seems trivial. It was basically impossible to think of a person by name and go and look up their picture. All of the dorms at Harvard had individual directories called "face-books"—some were printed, some were online, and most were only available to the students of that particular dorm. So we decided to create a unified version online and we dubbed it "The Facebook" to differentiate it from the individual ones.

Mark Zuckerberg: And within a couple weeks, a few thousand people had signed up. And we started getting e-mails from people at other colleges asking for us to launch it at their schools.

Ezra Callahan: Facebook launched at the Ivy Leagues originally, and it wasn't because they were snooty, stuck-up kids who only wanted to give things to the Ivy Leagues. It was because they had this intuition that people who go to the Ivy Leagues are more likely to be friends with kids at other Ivy League schools.

Aaron Sittig: When Facebook launched at Berkeley, the rules of socializing just totally transformed. When I started at Berkeley, the way you found out about parties was you spent all week talking to people figuring out what was interesting, and then you'd have to constantly be in contact. With Facebook there, knowing what was going on on the weekend was trivial. It was just all laid out for you.

Facebook came to the Stanford campus—in the heart of Silicon Valley—quite early: March 2004.

Sean Parker: My roommates in Portola Valley were all going to Stanford.

Ezra Callahan: So I was a year out of Stanford, I graduated Stanford in 2003, and me and four of my college friends rented a house for that year just near the campus, and we had an extra bedroom available, and so we advertised around on a few Stanford e-mail lists to find a roommate to move into that house with us. We got a reply from this guy named "Sean Parker." He ended up moving in with us pretty randomly, and we discovered that while Napster had been a cultural phenomenon, it didn't make him any money.

Sean Parker: And so the girlfriend of one of my roommates was using a product, and I was like, "You know, that looks a lot like Friendster or MySpace." She's like, "Oh yes, well, nobody in college uses MySpace." There was something a little rough about MySpace.

Mark Zuckerberg: So MySpace had almost a third of their staff monitoring the pictures that got uploaded for pornography. We hardly ever have any pornography uploaded. The reason is that people use their real names on Facebook.

Adam D'Angelo: Real names are really important.

Aaron Sittig: We got this clear early on because of something that was established as a community principle at The Well: You own your own words. And we took it farther than The Well. We always had everything be traceable back to a specific real person.

Stewart Brand: The Well could have gone that route, but we did not. That was one of the mistakes we made.

Mark Zuckerberg: And I think that that's a really simple social solution to a possibly complex technical issue.

Ezra Callahan: In this early period, it's a fairly hacked-together simple website: just basic web forms, because that's what Facebook profiles are.

Ruchi Sanghvi: There was a little profile pic, and it said things like, "This is my profile" and "See my friends," and there were three or four links and one or two other boxes below that.

Aaron Sittig: But I was really impressed by how focused and clear their product was. Small details—like when you went to your profile, it really clearly said, "This is you," because social networking at the time was really, really hard to understand. So there was a maturity in the product that you don't typically see until a product has been out there for a couple of years and been refined.

Sean Parker: So I see this thing, and I e-mailed some e-mail address at Facebook, and I basically said, "I've been working with Friendster for a while, and I'd just like to meet you guys and see if maybe there's anything to talk about." And so we set up this meeting in New York—I have no idea why it was in New York—and Mark and I just started talking about product design and what I thought the product needed.

Aaron Sittig: I got a call from Sean Parker and he said, "Hey, I'm in New York. I just met with this kid Mark Zuckerberg, who is very smart, and he's the guy building Facebook, and they say they have a 'secret feature' that's going to launch that's going to change everything! But he won't tell me what it is. It's driving me crazy. I can't figure out what it is. Do you know anything about this? Can you figure it out? What do you think it could be?" And so we spent a little time talking about it, and we couldn't really figure out what their "secret feature" that was going to change everything was. We got kind of obsessed about it.

Two months after meeting Sean Parker, Mark Zuckerberg moved to Silicon Valley with the idea of turning his dorm-room project into a real business. Accompanying him were his cofounder and consigliere, Dustin Moskovitz, and a couple of interns.

Mark Zuckerberg: Palo Alto was kind of like this mythical place where all the tech used to come from. So I was like, *I want to check that out.*

Ruchi Sanghvi: I was pretty surprised when I heard Facebook moved to the Bay Area, I thought they were still at Harvard working out of the dorms.

Ezra Callahan: Summer of 2004 is when that fateful series of events took place: that legendary story of Sean running into the Facebook cofounders on the street, having met them a couple months earlier on the East Coast. That meeting happened a week after we all moved out of the house we had been living in together. Sean was crashing with his girlfriend's parents.

Sean Parker: I was walking outside the house, and there was this group of kids walking toward me—they were all wearing hoodies and they looked like they were probably pot-smoking high school kids just out making trouble, and I hear my name. I'm like, *Oh, it's coincidence*, and I hear my name again and I turn around and it's like, "Sean, what are you doing here?" It took me about thirty seconds to figure out what was going on, and I finally realize that it's Mark and Dustin and a couple of other people, too. So I'm like, "What are *you* guys doing here?" And they're like, "We live right there." And I'm like, "That's really weird, I live right here!" This is just superweird.

Aaron Sittig: I get a call from Sean and he's telling me, "Hey, you won't believe what's just happened." And Sean said, "You've got to come over and meet these guys. Just leave right now. Just come over and meet them!"

Sean Parker: And so I don't even know what happened from there, other than that it just became very convenient for me to go swing by the house. It wasn't even a particularly formal relationship.

Aaron Sittig: So I went over and met them, and I was really impressed by how focused they were as a group. They'd occasionally relax and go do their thing, but for the most part they spent all their time sitting at a kitchen table with their laptops open. I would go visit their place a couple times a week, and that was always where I'd find them, just sitting around the kitchen table working, constantly, to keep their product growing. All Mark wanted to do was either make the product better, or take a break and relax so that you could get enough energy to go work on the product more. That's it. They never left that house except to go watch a movie.

Ezra Callahan: The early company culture was very, very loose. It felt like a project that's gotten out of control and has this amazing business potential. Imagine your freshman dorm running a business, that's really what it felt like.

Mark Zuckerberg: Most businesses aren't like a bunch of kids living in a house, doing whatever they want, not waking up at a normal time, not going into an office, hiring people by, like, bringing them into your house and letting them chill with you for a while and party with you and smoke with you.

Ezra Callahan: The living room was the office with all these monitors and workstations set up everywhere and just whiteboards as far as the eye can see.

At the time Mark Zuckerberg was obsessed with file sharing, and the grand plan for his Silicon Valley summer was to resurrect Napster. It would rise again, but this time as a feature inside of Facebook. The name of Zuckerberg's pet project? Wirehog.

Aaron Sittig: Wirehog was the secret feature that Mark had promised was going to change everything. Mark had gotten convinced that what would make Facebook really popular and just sort of cement its position at schools was a way to send files around to other people—mostly just to trade music.

Mark Pincus: They built in this little thing that looked like Napster—you could see what music files someone had on their computer.

Ezra Callahan: This is at a time when we have just watched Napster get completely terminated by the courts and the entertainment industry is starting to sue random individuals for sharing files. The days of the Wild West were clearly ending.

Aaron Sittig: It's important to remember that Wirehog was happening at a time where you couldn't even share photos on your Facebook page. Wirehog was going to be the solution for sharing photos with other people. You could have a box on your profile and people could go there to get access to all your photos that you were sharing—or whatever files you were sharing. It might be audio files, it might be video files, it might be photos of their vacation.

Ezra Callahan: But at the end of the day it's just a file-sharing service. When I joined Facebook, most people had already kind of come around to the idea that unless some new use comes up for Wirehog that we haven't thought of, it's just a liability. "We're going to get sued someday, so what's the point?" That was the mentality.

Mark Pincus: I was kind of wondering why Sean wanted to go anywhere near music again.

Aaron Sittig: My understanding was that some of Facebook's lawyers advised that it would be a bad idea. And that work on Wirehog was kind of abandoned just as Facebook user growth started to grow really quickly.

Ezra Callahan: They had this insane demand to join. It's still only at a hundred schools, but everyone in college has already heard of this, at all schools across the country. The usage numbers were already insane. Everything on the whiteboards was just all stuff related to what schools were going to launch next. The problem was very singular. It was simply, "How do we scale?"

Aaron Sittig: Facebook would launch at a school, and within one day they would have 70 percent of undergrads signed up. At the time, nothing had ever grown as fast as Facebook.

Ezra Callahan: It did not seem inevitable that we were going to succeed, but the scope of what success looked like was becoming clear. Dustin was already talking about being a billion-dollar company. They had that ambition from the very beginning. They were very confident: two nineteen-year-old cocky kids.

Mark Zuckerberg: We just all kind of sat around one day and were like, "We're not going back to school, are we?" *Nahhhh.*

Ezra Callahan: The hubris seemed pretty remarkable.

David Choe: And Sean is a skinny, nerdy kid and he's like, "I'm going to go raise money for Facebook. I'm going to bend these fuckers' minds." And I'm like, "How are you going to do that?" And he transformed himself into an alpha male. He got like a fucking supersharp haircut. He started working out every day, got a tan, got a nice suit. And he goes in these meetings and he got the money!

Mark Pincus: So it's probably like September or October of 2004, and I'm at Tribe's offices in this dusty converted brick building in Potrero Hill—the idea of Tribe.net was like Friendster meets Craigslist—and we're in our conference room, and Sean says he's bringing the Facebook guy in. And he brings Zuck in, and Zuck is in a pair of sweatpants, and these Adidas flip-flops that he wore, and he's so young looking and he's sitting there with his feet up on the table, and Sean is talking really fast about all the things Facebook is going to do and grow and everything else, and I was mesmerized. Because I'm doing Tribe, and we are not succeeding, we've plateaued and we're hitting our head against the wall trying to figure out how to grow, and here's this kid, who has this simple idea, and he's just taking off! I was kind of in awe already of what they had accomplished, and maybe a little annoyed by it. Because they did something simpler and quicker and with less, and then I remember Sean got on the computer in my office, and he pulled up The Facebook, and he starts showing it to me, and I had never been able to be on it, because it's college kids only, and it was amazing. People are putting up their phone numbers and home addresses and everything about themselves and I was like, *I can't believe it!* But it was because they had all this trust. And then Sean put together an investment round quickly, and he had advised Zuck to, I think, take $500,000 from Peter Thiel, and then $38,000 each from me and Reid Hoffman. Because we were basically the only other people doing anything in social networking. It was a very, very small little club at the time.

Ezra Callahan: By December it's—I wouldn't say it's like a more professional atmosphere, but all the kids that Mark and Dustin were hanging out with are either back at school back East or back at Stanford, and work has gotten a little more serious for them. They are working more than they were that first summer. We don't move into an office until February of 2005. And right as we were signing the lease, Sean just randomly starts saying, "Dude! I know this street artist guy. We're going to come in and have him totally do it up."

David Choe: I was like, "If you want me to paint the entire building it's going to be $60,000." Sean's like, "Do you want cash or do you want stock?"

Ezra Callahan: He pays David Choe in Facebook shares.

David Choe: I didn't give a shit about Facebook or even know what it was. You had to have a college e-mail to get on there. But I like to gamble, you know? I believed in Sean. I'm like, *This kid knows something and I am going to bet my money on him.*

Ezra Callahan: So then we move in, and when you first saw this graffiti it was like, "Holy shit, what did this guy do to the office?" The office was on the second floor, so as you walk in you immediately have to walk up some stairs, and on the big ten-foot-high wall facing you is just this huge buxom woman with enormous breasts wearing this Mad Max–style costume riding a bull dog. It's the most intimidating, totally inappropriate thing. It's not so much that we set out to paint that, because that was the culture. It was more that Sean just hired this guy, and that set a tone for us. A huge-breasted warrior woman riding a bull dog is the first thing you see as you come in the office, so like, get ready for that!

Ruchi Sanghvi: Yes, the graffiti was a little racy, but it was different, it was vibrant, it was alive. The energy was just so tangible.

Katie Geminder: I liked it, but it was really intense. There was certain imagery in there that was very sexually charged, which I didn't really care about, but that could be considered a little bit hostile, and I think we took care of some of the more provocative ones.

Ezra Callahan: I don't think it was David Choe, I think it was Sean's girlfriend who painted this explicit, intimate lesbian scene in the woman's restroom of two completely naked women intertwined and cuddling with each other—not graphic, but certainly far more suggestive than what one would normally see in a women's bathroom in an office. That one only actually lasted a few weeks.

Max Kelly: There was a four-inch-by-four-inch drawing of someone getting fucked. One of the customer service people complained that it was "sexual in nature," which, given what they were seeing every day, I'm not sure why they would complain about this. But I ended up going to a local store and buying a gold paint pen and defacing the graffiti—just a random design—so it didn't show someone getting fucked.

Jeff Rothschild: It was wild, but I thought that it was pretty cool. It looked a lot more like a college dorm or fraternity than it did a company.

Katie Geminder: There were blankets shoved in the corner and video games everywhere, and Nerf toys and Legos, and it was kind of a mess.

Jeff Rothschild: There's a PlayStation. There's a couple of old couches. It was clear people were sleeping there.

Karel Baloun: I'd probably stay there two or three nights a week. I won an award for "most likely to be found under your desk" at one of the employee gatherings.

Jeff Rothschild: They had a bar, a whole shelf with liquor, and after a long day people might have a drink.

Ezra Callahan: There's a lot of drinking in the office. There would be mornings when I would walk in and hear beer cans move as I opened the door, and the office smells of stale beer and is just trashed.

Ruchi Sanghvi: They had a keg. There was some camera technology built on top of the keg. It basically detected presence and posted about who was present at the keg—so it would take your picture when you were at the keg, and post some sort of thing saying, "So-and-so is at the keg." The keg is patented.

Ezra Callahan: When we first moved in, the office door had this lock we couldn't figure out, but the door would automatically unlock at nine a.m. every morning. I was the guy that had to get to the office by nine to make sure nobody walked in and just stole everything, because no one else was going to get there before noon. All the Facebook guys are basically nocturnal.

Katie Geminder: These kids would come in—and I mean kids, literally they were kids—they'd come into work at eleven or twelve.

Ruchi Sanghvi: Sometimes I would walk to work in my pajamas and that would be totally fine. It felt like an extension of college; all of us were going through the same life experiences at the same time. Work was fantastic. It was so interesting. It didn't feel like work. It felt like we were having fun all the time.

Ezra Callahan: You're hanging out. You're drinking with your coworkers. People start dating within the office...

Ruchi Sanghvi: We found our significant others while we were at Facebook. All of us eventually got married. Now we're in this phase where we're having children.

Katie Geminder: If you look at the adults that worked at Facebook during those first few years—like, anyone over the age of thirty that was married—and you do a survey, I tell you that probably 75 percent of them are divorced.

Max Kelly: So, lunch would happen. The caterer we had was mentally unbalanced and you never knew what the fuck was going to show up in the food. There were worms in the fish one time. It was all terrible. Usually, I would work until about three in the afternoon and then I'd do a circuit through the office to try and figure out what the fuck was going to happen that night. Who was going to launch what? Who was ready? What rumors were going on? What was happening?

Steve Perlman: We shared a break room with Facebook. We were building hardware: a facial capture technology. The Facebook guys were doing some HTML thing. They would come in late in the morning. They'd have a catered lunch. Then they leave usually by midafternoon. I'm like, *Man, that is the life! I need a start-up like that.* You know? And the only thing any of us could think about Facebook was: *Really nice people, but never going to go anywhere.*

Max Kelly: Around four I'd have a meeting with my team, saying, "Here's how we're going to get fucked tonight." And then we'd go to the bar. Between like five and eight-ish people would break off and go to different bars up and down University Avenue, have dinner, whatever.

Ruchi Sanghvi: And we would all sit together and have these intellectual conversations: "Hypothetically, if this network was a graph, how would you weight the relationship between two people? How would you weight the relationship between a person and a photo? What does that look like? What would this network eventually look like? What could we do with this network if we actually had it?"

Sean Parker: The "social graph" is a math concept from graph theory, but it was a way of trying to explain to people who were kind of academic and mathematically inclined that what we were building was not a product so much as it was a network composed of nodes with a lot of information flowing between those nodes. That's graph theory. Therefore we're building a social graph. It was never meant to be talked about publicly. It was a way of articulating to somebody with a math background what we were building.

Ruchi Sanghvi: In retrospect, I can't believe we had those conversations back then. It seems like such a mature thing to be doing. We would sit around and have these conversations and they weren't restricted to certain members of the team; they weren't tied to any definite outcome. It was purely intellectual and was open to everyone.

Max Kelly: People were still drinking the whole time, like all night, but starting around nine, it really starts solidifying: "What are we going to release tonight? Who's ready to go? Who's not ready to go?" By about eleven-ish we'd know what we were going to do that night.

Katie Geminder: There was an absence of process that was mind-blowing. There would be engineers working stealthily on something that they were passionate about. And then they'd ship it in the middle of the night. No testing—they would just ship it.

Ezra Callahan: Most websites have these very robust testing platforms so that they can test changes. That's not how we did it.

Ruchi Sanghvi: With the push of a button you could push out code to the live site, because we truly believed in this philosophy of "move fast and break things." So you shouldn't have to wait to do it once a week, and you shouldn't have to wait to do it once a day. If your code was ready you should be able to push it out live to users. And that was obviously a nightmare.

Katie Geminder: Can our servers stand up to something? Or security: How about testing a feature for security holes? It really was just shove it out there and see what happens.

Jeff Rothschild: That's the hacker mentality: You just get it done. And it worked great when you had ten people. By the time we got to twenty, or thirty, or forty, I was spending a lot of time trying to keep the site up. And so we had to develop some level of discipline.

Ruchi Sanghvi: So then we would only push out code in the middle of the night, and that's because if we broke things it wouldn't impact that many people. But it was terrible because we were up until like three or four a.m. every night, because the act of pushing just took everybody who had committed any code to be present in case anything broke.

Max Kelly: Around one a.m., we'd know either we're fucked or we're good. If we were good, everyone would be like "whoopee" and might be able to sleep for a little while. If we were fucked then we were like, "Okay, now we've got to try and claw this thing back or fix it."

Katie Geminder: Two a.m.: That was when shit happened.

Ruchi Sanghvi: Then another push, and this would just go on and on and on and on and on until like three or four or five a.m. in the night.

Max Kelly: If four a.m. rolled around and we couldn't fix it, I'd be like, "We're going to try and revert it." Which meant basically my team would be up till six a.m. So, go to bed somewhere between four and six, and then repeat every day for like nine months. It was crazy.

Jeff Rothschild: It was seven days a week. I was on all the time. I would drink a large glass of water before I went to sleep to assure that I'd wake up in two hours so I could go check everything and make sure that we hadn't broken it in the meantime. It was all day, all night.

Katie Geminder: That was very challenging for someone who was trying to actually live an adult life with like a husband. There was definitely a feeling that because you were older and married and had a life outside of work that you weren't committed.

Mark Zuckerberg: Why are most chess masters under thirty? Young people just have simpler lives. We may not own a car. We may not have family... I only own a mattress.

Kate Geminder: Imagine being under thirty and hearing your boss say that!

Mark Zuckerberg: Young people are just smarter.

Ruchi Sanghvi: We were so young back then. We definitely had tons of energy and we could do it, but we weren't necessarily the most efficient team by any means whatsoever. It was definitely frustrating for senior leadership, because a lot of the conversations happened at night when they weren't around, and then the next morning they would come in to all of these changes that happened at night. But it was fun when we did it.

Ezra Callahan: For the first few hundred employees, almost all of them were already friends with someone working at the company, both within the engineering circle and also the user support people. It's a lot of recent grads. When we move into the office was when the dorm room culture starts to really stick out and also starts to break a little bit. It has a dorm room feeling, but it's not completely dominated by college kids. The adults are coming in.

Jeff Rothschild: I joined in May 2005. On the sidewalk outside the office was the menu board from a pizza parlor. It was a caricature of a chef with a blackboard below it, and the blackboard had a list of jobs. This was the recruiting effort.

Sean Parker: At the time there was a giant sucking sound in the universe, and it was called Google. All the great engineers were going to Google.

Kate Losse: I don't think I could have stood working at Google. To me Facebook seemed much cooler than Google, not because Facebook was necessarily like the coolest. It's just that Google at that point already seemed nerdy in an uninteresting way, whereas like Facebook had a lot of people who didn't actually want to come off as nerds. Facebook was a social network, so it has to have some social components that are like really normal American social activities—like beer pong.

Kate Geminder: There was a house down the street from the office where five or six of the engineers lived that was one ongoing beer pong party. It was like a boys' club—although it wasn't just boys.

Terry Winograd: The way I would put it is that Facebook is more of an undergraduate culture and Google is more of a graduate student culture.

Jeff Rothschild: Before I walked in the door at Facebook, I thought these guys had created a dating site. It took me probably a week or two before I really understood what it was about. Mark, he used to tell us that we are not a social network. He would insist: "This is not a social network. We're a social utility for people you actually know." MySpace was about building an online community among people who had similar interests. We might look the same because at some level it has the same shape, but what it accomplishes for the individual is solving a different problem. We were trying to improve the efficiency of communication among friends.

Max Kelly: Mark sat down with me and described to me what he saw Facebook being. He said, "It's about connecting people and building a system where everyone who makes a connection to your life that has any value is preserved for as long as you want it to be preserved. And it doesn't matter where you are, or who you're with, or how your life changes: because you're always in connection with the people that matter the most to you, and you're always able to share with them." I heard that, and I thought, *I want to be a part of this. I want to make this happen.* Back in the nineties all of us were utopian about the internet. This was almost a harkening back to the beautiful internet where everyone would be connected and everyone could share and there was no friction to doing that. Facebook sounded to me like the same thing. Mark was too young to know that time, but I think he intrinsically understood what the internet was supposed to be in the eighties and in the nineties. And here I was hearing the same story again and conceivably having the ability to help pull it off. That was very attractive.

Aaron Sittig: So in the summer of 2005 Mark sat us all down and he said, "We're going to do five things this summer." He said, "We're redesigning the site. We're doing a thing called News Feed, which is going to tell you everything your friends are doing on the site. We're going to launch Photos, we're going to redo Parties and turn it into Events, and we're going to do a local-businesses product." And we got one of those things done, we redesigned the site. Photos was my next project.

Ezra Callahan: The product at Facebook at the time is dead simple: profiles. There is no News Feed, there was a very weak messaging system. They had a very rudimentary events product you could use to organize parties. And almost no other functions to speak of. There's no photos on the website, other than your profile photo. There's nothing that tells you when anything on the site has changed. You find out somebody changed their profile picture by obsessively going to their profile and noticing *Oh, the picture changed.*

Aaron Sittig: We had some people that were changing their profile picture once an hour, just as a way of sharing photos of themselves.

Scott Marlette: At the time photos was the number one most requested feature. So, Aaron and I go into a room and whiteboard up some wireframes for some pages and decide on what data needs to get stored. In a month we had a nearly fully functioning prototype internally to play with. It was very simple. It was: you post a photo, it goes in an album, you have a set of albums, and then you can tag people in the photos.

Jeff Rothschild: Aaron had the insight to do tagging, which was a tremendously valuable insight. It was really a game changer.

Aaron Sittig: We thought the key feature is going to be saying who is in the photo. We weren't sure if this was really going to be that successful, we just felt good about it.

Facebook Photos went live in October 2005. There were about five million users, virtually all of them college students.

Scott Marlette: We launched it at Harvard and Stanford first, because that's where our friends were.

Aaron Sittig: We had built this program that would fill up a TV screen and show us everything that was being uploaded to the service, and then we flicked it on and waited for photos to start coming in. And the first photos that came in were Windows wallpapers: Someone had just uploaded all their wallpaper files from their Windows directory, which was a big disappointment, like, *Oh no, maybe people don't get it? Maybe this is not going to work?* But the next photos were of a guy hanging out with his friends, and then the next photos after that were a bunch of girls in different arrangements: three girls together, these four girls together, two of them together, just photos of them hanging out at parties, and then it just didn't stop.

Max Kelly: You were at every wedding, you were at every bar mitzvah, you were seeing all this awesome stuff, and then there's a dick. So, it was kind of awesome and shitty at the same time.

Aaron Sittig: Within the first day someone had uploaded and tagged themselves in seven hundred photos and it just sort of took off from there.

Jeff Rothschild: Inside of three months, we were delivering more photos than any other website on the internet. Now you have to ask yourself, *Why?* And the answer was tagging. There isn't anyone who could get an e-mail message that said, "Someone has uploaded a photo of you to the internet"—and *not* go take a look. It's just human nature.

Ezra Callahan: The single greatest growth mechanism ever was photo tagging. It shaped all of the rest of the product decisions that got made. It was the first time that there was a real fundamental change to how people used Facebook, the pivotal moment when the mind-set of Facebook changes and the idea for News Feed starts to germinate and there is now a reason to see how this expands beyond college.

Jeff Rothschild: The News Feed project was started in the fall of 2005 and delivered in the fall of 2006.

Dustin Moskovitz: News Feed is the concept of viral distribution, incarnate.

Ezra Callahan: News Feed is what Facebook fundamentally is today.

Sean Parker: Originally it was called "What's New," and it was just a feed of all of the things that were happening in the network—really just a collection of status updates and profile changes that were occurring.

Katie Geminder: It was an aggregation, a collection of all those stories, with some logic built into it because we couldn't show you everything that was going on. There were sort of two streams: things you were doing, and things the rest of your network was doing.

Ezra Callahan: So News Feed is the first time where now your homepage, rather than being static and boring and useless, is now going to be this constantly updating "newspaper," so to speak, of stuff happening on Facebook around you that we think you'll care about.

Ruchi Sanghvi: And it was a fascinating idea, because normally when you think of newspapers, they have this editorialized content where they decide what they want to say, what they want to print, and they do it the previous night, and then they send these papers out to thousands if not hundreds of thousands of people. But in the case of Facebook, we were building ten million different newspapers, because each person had a personalized version of it.

Ezra Callahan: It really was the first monumental product-engineering feat. The amount of data it had to deal with: all these changes and how to propagate that on an individual level...

Ruchi Sanghvi: We were working on it off and on for a year and a half.

Ezra Callahan:...and then the intelligence side of all this stuff: How do we surface the things that you'll care about most? These are very hard problems engineering-wise.

Ruchi Sanghvi: Without realizing it, we ended up building one of the largest distributed systems in software at that point in time. It was pretty cutting-edge.

Ezra Callahan: We have it in-house and we play with it for weeks and weeks—which is really unusual.

Katie Geminder: So I remember being like, "Okay, you guys, we have to do some level of user research," and I finally convinced Zuck that we should bring users into a lab and sit behind the glass and watch our users using the product. And it took so much effort for me to get Zuck and other people to go and actually watch this. They thought this was a waste of time. They were like, "No, our users are stupid." Literally those words came out of somebody's mouth.

Ezra Callahan: It's the very first time we actually bring in outside people to test something for us, and their reaction, their initial reaction is clear. People are just like, "Holy shit, like, I shouldn't be seeing this, like this doesn't feel right," because immediately you see this person changed their profile picture, this person did this, this person did that, and your first instinct is *Oh my God! Everybody can see this about me! Everyone knows everything I'm doing on Facebook.*

Max Kelly: But News Feed made perfect sense to all of us, internally. We all loved it.

Ezra Callahan: So in-house we have this idea that this isn't going to go right: This is too jarring a change, it needs to be rolled out slowly, we need to warm people up to this—and Mark is just firmly committed. "We're just going to do this. We're just going to launch. It's like ripping off a Band-Aid."

Ruchi Sanghvi: We pushed the product in the dead of the night, we were really excited, we were celebrating, and then the next morning we woke up to all this pushback. I had written this blog post, "Facebook Gets a Facelift."

Katie Geminder: We wrote a little letter, and at the bottom of it we put a button. And the button said, "Awesome!" Not like, "Okay." It was, "Awesome!" That's just rude. I wish I had a screenshot of that. Oh man! And that was it. You landed on Facebook and you got the feature. We gave you no choice and not a great explanation and it scared people.

Jeff Rothschild: People were rattled because it just seemed like it was exposing information that hadn't been visible before. In fact, that wasn't the case. Everything shown in News Feed was something people put on the site that would have been visible to everyone if they had gone and visited that profile.

Ruchi Sanghvi: Users were revolting. They were threatening to boycott the product. They felt that they had been violated, and that their privacy had been violated. There were students organizing petitions. People had lined up outside the office. We hired a security guard.

Katie Geminder: There were camera crews outside. There were protests: "Bring back the old Facebook!" Everyone hated it.

Jeff Rothschild: There was such a violent reaction to it. We had people marching on the office. A Facebook group was organized protesting News Feed and inside of two days, a million people joined.

Ruchi Sanghvi: There was another group that was about how "Ruchi is the devil," because I had written that blog post.

Max Kelly: The user base fought it every step of the way and would pound us, pound Customer Service, and say, "This is fucked up! This is terrible!"

Ezra Callahan: We're getting e-mails from relatives and friends. They're like, "What did you do? This is terrible! Change it back."

Katie Geminder: We were sitting in the office and the protests were going on outside and it was, "Do we roll it back? Do we roll it back!?"

Ruchi Sanghvi: Now under usual circumstances if about 10 percent of your user base starts to boycott the product, you would shut it down. But we saw a very unusual pattern emerge.

Max Kelly: Even the same people who were telling us that this is terrible, we'd look at their user stream and be like: *You're fucking using it constantly! What are you talking about?*

Ruchi Sanghvi: Despite the fact that there were these revolts and these petitions and people were lined up outside the office, they were digging the product. They were actually using it, and they were using it twice as much as before News Feed.

Ezra Callahan: It was just an emotionally devastating few days for everyone at the company. Especially for the set of people who had been waving their arms saying, "Don't do this! Don't do this!" because they feel like, "This is exactly what we told you was going to happen!"

Ruchi Sanghvi: Mark was on his very first press tour on the East Coast, and the rest of us were in the Palo Alto office dealing with this and looking at these logs and seeing the engagement and trying to communicate that "It's actually working!," and to just try a few things before we chose to shut it down.

Katie Geminder: We had to push some privacy features right away to quell the storm.

Ruchi Sanghvi: We asked everyone to give us twenty-four hours.

Katie Geminder: We built this janky privacy "audio mixer" with these little slider bars where you could turn things on and off. It was beautifully designed—it looked gorgeous—but it was irrelevant.

Jeff Rothschild: I don't think anyone ever used it.

Ezra Callahan: But it gets added and eventually the immediate reaction subsides and people realize that the News Feed is exactly what they wanted, this feature is exactly right, this just made Facebook a thousand times more useful.

Katie Geminder: Like Photos, News Feed was just—*boom!*—a major change in the product and one of those sea changes that just leveled it up.

Jeff Rothschild: Our usage just skyrocketed on the launch of News Feed. About the same time we also opened the site up to people who didn't have a .edu address.

Ezra Callahan: Once it opens to the public, it's becoming clear that Facebook is on its way to becoming the directory of all the people in the world.

Jeff Rothschild: Those two things together—that was the inflection point where Facebook became a massively used product. Prior to that we were a niche product for high school and college students.

Mark Zuckerberg: Domination!

Ruchi Sanghvi: "Domination" was a big mantra of Facebook back in the day.

Max Kelly: I remember company meetings where we were chanting "dominate."

Ezra Callahan: We had company parties all the time, and for a period in 2005, all Mark's toasts at the company parties would end with "Domination!"

Mark Zuckerberg: Domination!!

Max Kelly: I especially remember the meeting where we tore up the Yahoo offer.

Mark Pincus: In 2006 Yahoo offered Facebook $1.2 billion I think it was, and it seemed like a breathtaking offer at the time, and it was difficult to imagine them not taking it. Everyone had seen Napster flame out, Friendster flame out, MySpace flame out, so to be a company with no revenues, and a credible company offers a billion-two, and to say no to that? You have to have a lot of respect to founders that say no to these offers.

Dustin Moskovitz: I was sure the product would suffer in a big way if Yahoo bought us. And Sean was telling me that 90 percent of all mergers end in failure.

Mark Pincus: Luckily, for Zuck, and history, Yahoo's stock went down, and they wouldn't change the offer. They said that the offer is a fixed number of shares, and so the offer dropped to like $800 million, and I think probably emotionally Zuck didn't want to do it and it gave him a clear out. If Yahoo had said, "No problem, we'll back that up with cash or stock to make it $1.2 billion," it might have been a lot harder for Zuck to say no, and maybe Facebook would be a little division of Yahoo today.

Max Kelly: We literally tore the Yahoo offer up and stomped on it as a company! We were like, "Fuck those guys, we are going to own them!" That was some malice-ass bullshit.

Mark Zuckerberg: Domination!!!

Kate Losse: He had kind of an ironic way of saying it. It wasn't a totally flat, scary "domination." It was funny. It's only when you think about a much bigger scale of things that you're like, *Hmmmm: Are people aware that their interactions are being architected by a group of people who have a certain set of ideas about how the world works and what's good?*

Ezra Callahan: "How much was the direction of the internet influenced by the perspective of nineteen-, twenty-, twenty-one-year-old well-off white boys?" That's a real question that sociologists will be studying forever.

Kate Losse: I don't think most people really think about the impact that the values of a few people now have on everyone.

Steven Johnson: I think there's legitimate debate about this. Facebook has certainly contributed to some echo chamber problems and political polarization problems, but I spent a lot of time arguing that the internet is less responsible for that than people think.

Mark Pincus: Maybe I'm too close to it all, but I think that when you pull the camera back, none of us really matter that much. I think the internet is following a path to where the internet wants to go. We're all trying to figure out what consumers want, and if what people want is this massive echo chamber and this vain world of likes, someone is going to give it to them, and they're going to be the one who wins, and the ones who don't, won't.

Steve Jobs: I don't see anybody other than Facebook out there—they're dominating.

Mark Pincus: So I don't exactly think that a bunch of college boys shaped the internet, I just think they got there first.

Mark Zuckerberg: Domination!!!!

Ezra Callahan: So, it's not until we have a full-time general council on board who finally says, "Mark, for the love of God: you cannot use the word *domination* anymore," that he stops.

Sean Parker: Once you are dominant, then suddenly it becomes an anticompetitive term.

Steven Johnson: It took the internet thirty years to get to one billion users. It took Facebook ten years to get to one billion users. The crucial thing about Facebook is that it's not a service or an app—it's a fundamental platform, on the same scale as the internet itself.

Steve Jobs: I admire Mark Zuckerberg. I only know him a little bit, but I admire him for not selling out—for wanting to make a company. I admire that a lot.

Purple People Eater

Apple, the company that cannibalizes itself

*O*nly *three years after reviving Apple on the back of the iPod, Steve Jobs decides to make it obsolete. As an Atari alumnus, Jobs knows firsthand the danger of trying to milk a hot product for all it is worth. He knows that somebody at some point will unseat Apple's iPod monopoly with an MP3 player that can also place a phone call. Thus, Jobs concludes, Apple has to act first. So he sets up a skunkworks: Project Purple. Then, to make sure that the phone-of-the-future project has the proper urgency, Jobs divides Purple into two groups—and pits them against each other. P1 is quarterbacked by Tony Fadell, the hardware hero who had changed Apple's fortunes with the iPod. P2 is led by Scott Forstall, a software wiz from Apple's Macintosh team. Will the new Apple phone be a tiny Macintosh or a supersized iPod? In the end, Jobs chooses Forstall's version—a tiny Macintosh—and uses the iPhone's debut to send a message to Fadell. Jobs disses Fadell from the stage, surreptitiously but before an audience of millions. Fadell's days at Apple are numbered.*

Phil Schiller: Apple had been known for years for being the creator of the Mac computer. And it was great but it had a small market share. And then we had a big hit called the iPod, and this really changed everybody's view both inside and outside the company. People started asking, "Well, if you can have a big hit with the iPod, what else can you do?" And so we were searching for what to do after the iPod that would make sense.

Scott Forstall: And one thought was a tablet.

Phil Schiller: People were suggesting every idea: make a camera, make a car, crazy stuff!

Scott Forstall: And we settled on this idea of a beautiful tablet without a keyboard, without a hinge.

Jon Rubinstein: But the technology just wasn't there yet. Different combinations of horsepower, battery life, display technology: None of those worked for a tablet yet. But we'd acquired the technology for doing multitouch, so we could develop that. Multitouch is the ability to directly interact with your screen and be able to pinch and zoom and drag and all those kind of things.

Steve Jobs: I had this idea of being able to get rid of the keyboard and type on a multitouch glass display. So I asked our folks if we could come up with a multitouch display that I could rest my hands on and actually type on.

Phil Schiller: We thought that if you wanted to type really fast, you are going to be touching that surface with more than one finger. At the time no one had really done anything with multitouch, and we knew that was really important.

Scott Forstall: So we started building prototypes and demos to see if we could build a tablet. That was 2003.

Steve Jobs: And about six months later they called me in and showed me this prototype display and it was amazing.

Nitin Ganatra: There was just this enormous contraption that was created partly by the Mac hardware team. So what we actually had was this tablet-looking device with a big fat cable that connected it to a Macintosh.

Scott Forstall: We were asking ourselves, "Could we use the technology that we were prototyping for this tablet to build a phone?"

Tony Fadell: Mobile phones were coming on strong...

Nitin Ganatra: But all of the technical advancements around phones were happening somewhere other than Silicon Valley. All the interesting work was either in Finland or in Japan or in Canada. There was nothing in the US, much less in Silicon Valley.

Scott Forstall: We all hated our phones. We had flip phones at the time.

Nitin Ganatra: And every single one of them to a T is just dogshit. We'd hold up our phones and show each other the latest and greatest and look at them. "This is the latest and greatest because it has 256 colors on it?" We're talking about the early 2000s. "This is considered state of the art? Even though, in computing, this problem was tackled twenty years before? How backward is that?" I would talk to Scott about this a fair bit, saying, "When are we going to do a phone?"

Scott Forstall: And so we prototyped a thing which I will never forget. We took that tablet and built a small scrolling list. Now on a tablet we were doing pinch and zoom and scrolls and all these things and we wanted it to fit in the pocket, so we built a small corner of it as a list of contacts.

Nitin Ganatra: All the pixels and everything were sort of shoved down into a corner to more accurately reflect what a phone's form factor would actually look like.

Scott Forstall: And you'd sit there and you'd scroll this list of contacts and you could tap on that contact. It would slide over and show you the contact information and you could tap on the phone number and it would say, CALLING. It wasn't calling but it would say it was calling, and it was just amazing.

Steve Jobs: And when I saw the rubber-band inertial scrolling and a few of the other things, I thought, *My God, we can build a phone out of this!*

Jon Rubinstein: Because the key to a phone is "Can you get through the names quickly?" Right? One of the big issues with the iPod was, "How the hell do you find the song that you want to play?" So if you are not going to have a scroll wheel on the iPhone, then you needed an equivalent user interface paradigm to get to who you wanted to call.

Nitin Ganatra: Even though it was this big, clunky contraption and things didn't quite work right, you could see that it was far and away better than the piece of trash that you had in your pocket.

Scott Forstall: We realized that a touch screen that could fit into your pocket would work perfectly as one of these phones.

Phil Schiller: So the combination of "What should come after the iPod?" and having this multitouch glass technology came together around the same time to give us this vision.

Scott Forstall: The whole illusion that we were trying to create was that your fingers were literally reaching through the screen to the content that you saw behind it.

Nitin Ganatra: If you looked at that UI, it was like comparing heaven and earth. That's exactly how I felt and how everyone on our team felt too.

Scott Forstall: So we shelved the idea of the tablet for years and in 2004 we switched over to building what became the iPhone.

Ron Johnson: The phone is a product everyone uses—like music—right? Steve always wanted to do things that had a big market.

Nitin Ganatra: Purple was the code name for the software stack that would become the iPhone. And to get the kind of UI we wanted, it was understood that this project was going to be really ambitious, but there was also an understanding that it was going to be quite a long period of time before we could actually have it out there.

Scott Forstall: Steve knew that there were going to be a lot of people that were going to have to be involved, a lot of groups, so he put different people in charge of each of the groups. I was put in charge of all of the software, so I started and built a team.

Andy Grignon: And since Tony was making hardware, it landed in Fadell's kind of purview.

Nitin Ganatra: The iPod team felt like it was their right to develop the next embedded hardware product that Apple has. They already had a phenomenal success with the iPod. They had delivered this thing in a very short amount of time, and it was already sort of seen as the future of Apple, even at that point. So of course the group that is going to develop the next thing that fits in your pocket is going to be the iPod group.

Tony Fadell: So first it was the iPod, then it was the iPod with music syncing, then it became the iPod with music and photo syncing, then it became the iPod with the iTunes Music Store so you did not have to rip anymore, then it worked with videos, then it became the iPod with games and videos and all the other stuff, so that was great, and then it was the iPod with games and downloadable apps or downloadable games, so now let's make it a phone.

Nitin Ganatra: In early 2005 there was a meeting. Steve Jobs did most of the talking. Phil Schiller was there, and Tony Fadell, and then a lot of leadership from the Purple side, and Jon Rubinstein as well. So we all pile into this room, and Steve starts talking about how we have got this fantastic plan and we've got this great UI, but we don't have time to wait.

Andy Hertzfeld: It was clear that iPods were going to disappear within five years because smartphones were going to subsume them.

Guy Bar-Nahum: It was a panic by Steve Jobs that Nokia and Motorola would kill iPod.

Phil Schiller: Because we realized that if anything were ever to challenge the idea that you have all your entertainment in your pocket—your movies, your photos, your music—it would be the cell phone.

Nitin Ganatra: At this point, almost on cue, Tony, who is sitting at the center of a sort of longish table, starts talking about a contingency plan for something that's going to bridge between now and when we can actually ship Purple. That is when the designations P1 and P2 were created.

Tony Fadell: P1 was the iPod plus phone, the iPod-phone. It had a small screen and a wheel and a cell phone built into it.

Nitin Ganatra: P1 was ultimately a way to sort of unseat what Tony saw as being an unrealistic plan to begin with, which was to fit an OSX-based software stack onto iPhone hardware. P1 was Tony's attempt to basically own the iPhone.

Andy Grignon: P2 was this purely experimental concept: Can we ship OSX on a phone? Is it possible? If we strip it back? P2 was the more radical design.

Nitin Ganatra: Tony thought the P2 team was never really capable of shipping the iPhone software to begin with, and once the P1 came out in record time it was going to be understood that P2 was a science project.

Andy Grignon: P2 was bloated, it was huge, it didn't work. MacOS was built for heavy iron. P2 required an entirely new piece of silicon because of MacOS. This was Forstall's attempt to insert himself into the phone project, right?

Nitin Ganatra: My whole goal at the time was to make it so that any functionality was demoed on P2 before it demoed in P1, because I want to start chipping away at this perception that this thing is a science project, and that it's never going to ship. We definitely felt like we were the underdogs. One of the things that internally we used as a rallying point was "We need to demonstrate this functionality before the P1 team can," and goddamn it if that didn't work to motivate us.

Scott Forstall: They were there at night. They were there on weekends.

Nitin Ganatra: We lost Christmas vacation time, Thanksgiving time, holidays, weekends, nights, things like that. The first breakthrough was just actually seeing the full stack running on something that was not a Macintosh. The first version of the iPhone hardware was actually this thing where there were three different logic boards that were all separate and all connected up with cables, and on one end there was a little phone display, but it still took up, literally, two-thirds of a conference table. But the thing that was running on the display was the actual iPhone software, and it was actually able to send and receive text messages: real SMS messages going through the software stack through this prototype hardware and then back up to another phone. For me that was sort of the first big oh-my-God we're-not-completely-off-in-the-weeds-here moment. We're actually going to be able to make this thing work!

Matt Rogers: We were working on something else: a click-wheel-based solution. We were basically fusing a phone with an iPod. I was deep in the lab making this thing happen, and Tony was selling it to Steve. Every other week Tony would take what we were doing and show it to Steve. And Steve would say, "This is shit, this is shit, this is shit, this is good, fix that."

Tony Fadell: It was about four or five months of research, but halfway through everybody knew it wasn't working. But Steve didn't want us to give up, so we kept working on it and getting more and more dejected.

Andy Grignon: We actually built a working iPod-phone, several hundred of them, actually. It was a little wider but still thinner than the iPods at the time. You would go into phone mode from player mode and the controls changed over to the rotary dial interface.

Ron Johnson: It was a familiar interface that was beloved. Everyone understood how to use the scroll wheel. And we could bring it out at, like, $400. To do the touch phone would be six or seven hundred dollars. So there was a long debate. There was a choice we had to make. We had to pick one, and I'm pretty sure everyone was leaning toward the iPod-phone. But Steve always leaned toward the future. Steve always leaned to the most innovative. I remember Steve saying, "If we don't do this all-display phone, someone else might do it first—and the all-display phone is the big idea."

Guy Bar-Nahum: I'm not sure if it's a real memory or not, but in my imagination, in my kind of mythical memory, there was a meeting where Steve said, "What is this thing that we're building?" And Steve basically was looking at people and asking a rhetorical question, and he answered it: "We're building software! This thing that we're building is software. Software is pixels, and because it's pixels, it has to be all screen."

Matt Rogers: We spent a good three or four months on that initial prototype that Tony presented to Steve, and we ended up killing it because it wasn't good enough. You can see why Steve made the call: "We can do better, guys." The iPod-phone thing didn't have apps. It was kind of music plus phone. OSX brings in all these other experiences.

Guy Bar-Naham: Before that, it was the land of confusion. It's all kinds of creatures that were born in sin, and there were several of them that we all want to forget. There's no value in going and digging and getting embarrassed by these weird things that we did before the decision to go all touch: weird amalgamations, incremental design away from iPod. Literally a phone with a wheel, right? Or a phone that was half touch surface, half screen. Those were exercises before the big insight, the big bombshell. The bombshell that Steve kind of laid on people was that "It has to be all screen." So once you say that it's all screen, it's obvious. The shape is obvious.

Phil Schiller: It's rounded corners, a rectangular shape, a full glass face with the black screen and the black area around the screen just seen as one.

Christopher Stringer: You want to create the simplest, purest manifestation of what the object can be.

Andy Grignon: So Tony, even though he delivered the hugely successful iPod business, had lost control. Tony was still on the hook to build the actual hardware, but Forstall had won the software war, and so now Tony was building Forstall's hardware—which chapped him to no end.

Tony Fadell: The confidentiality started driving a wedge between the teams. And then it was used as a tool . . .

Matt Rogers: I was part of the hardware team responsible for building the product. And there was a separate team at Apple led by Scott Forstall that was building the software. And we would meet at some kind of neutral territory to make sure things got together.

Andy Grignon: Scott knew that his path to world domination was holding that UI under lock and key.

Scott Forstall: On the front door of the Purple dorm, we put up a sign that said FIGHT CLUB, because the first rule of Fight Club is "You don't talk about Fight Club"; and the first rule about the Purple Project is "You do not talk about that outside of those doors."

Andy Grignon: This is when Steve's leadership and management style started to permeate the company. Tony and Forstall started to adopt Steve's mannerisms and persona because it was a pressure cooker—but also you emulate what works, right? And so everyone started screaming at each other. It became just like the thing to do: Fly off the fucking handle.

Tony Fadell: It became, "Your teams can't see the apps, your team can't touch any of this stuff." We had to create a whole other operating system for diagnostics and manufacture.

Andy Grignon: An app that my team had written called Skankphone is what most of the people involved on the project could actually use. It was like blue buttons on a red background. It was just awful to look at. We did it on purpose. It was supposed to be hideous.

Tony Fadell: Then Steve and Scott would say, "You can't even show the devices to your people." And my engineers were saying, "We need to use these things, we need to try them. Because we have to make phone calls in the wild, we have to start testing cell phone networks, we have to test all kinds of different things!" It was driving a wedge.

Matt Rogers: Imagine trying to get a product that has never been done before done in record time—and yet you were not working with the guys that were your counterparts. We went from there basically to the product we know today, the product that we ended up shipping. At the time it was two years out. It was a monumental effort around the company. It was nuts.

Andy Grignon: The iPhone is a product that by all accounts should never have succeeded. There wasn't a single stable proven element in the entire stack: We'd never made a phone before! Now granted, lots of people have been building phones, so what do you do? The very first thing is you start to go shop for people who know how to do this. Well, Steve being crazy, but also being the genius that he was, told us we were not allowed to hire people who knew how to build phones! It was not because he was being a dick. Actually he was a little bit—but he wanted us to invent the knowledge ourselves.

Scott Forstall: There were a number of challenges. One of which was that everything we dealt with before was based on a mouse and keyboard. So we had to rethink everything. Every single part of the device had to be rethought for doing touch. So we started with a brand-new user interface, instead of something that was existing.

Andy Grignon: Then product design—Jony Ive's guys—would come out with a model and say, "We're going to make it look like this." "Oh, it doesn't have any buttons? Yes, so we need to invent a keyboard? All right. We'll just add that to the list, right?"

Nitin Ganatra: One of the things that terrified me was *How the heck are we going to make a virtual keyboard work anywhere near as well as a physical keyboard?*

Scott Forstall: If you look back to 2005 when the engineering team started on this, smartphones all had physical keyboards. The most popular at the time was probably the BlackBerry, and it had a physical keyboard. And many people at the time thought we were actually crazy to try to build something without any form of a physical keyboard: not a slide-out one, not one on the front screen, nothing. And so it was a science project for us to be able to create an on-screen touch keyboard that could work really well and then get out of the way when you weren't using it.

Nitin Ganatra: We actually had sort of a hack contest that Scott Forstall put together, where anybody who was an engineer on the iPhone software development team could go off and create their own version of what a virtual keyboard should be and how it should work.

Andy Hertzfeld: Steve Jobs himself told me he thought a lot about General Magic while he was working on the iPhone. And, reading between the lines, I think he was mainly talking about the projected keyboard. That's the main breakthrough of the iPhone: to not have the keyboard. But because he had used the General Magic devices he knew that you can do the projected keyboard effectively. It turns out that the screen resolution of the original iPhone was identical to the General Magic devices.

Andy Grignon: But again, going back to why this thing shouldn't have succeeded...let's start from the bottom. We're using a chip that's never been tested before. It's a brand-new part, and Samsung has made lots of chips before, but they've never made one with the configuration and the random requirements that we had. So the chip is not stable. We don't have a reference design for a phone, so we had to make that. So now we've got a piece of hardware we think we can make work, but we're using a finger to interface with a product. We've never done that before, how do we do that? And by the way it's supposed to work with five fingers at a minimum. So now all of a sudden we were now increasing the complexity and craziness of this product. Now we get into the software space. We're now putting an operating system that to date has only run on a different class of processor, an Intel processor, and rebuilding the software for ARM, this whole other instruction set, which is completely different. That's crazy!

Nitin Ganatra: Seeing our software run on the very first iPhone, the M68 hardware. That was the second breakthrough.

Andy Grignon: Everything had code names, so P2 was the program name, and M68 was the name of the actual hardware, the actual thing.

Tony Fadell: The original iPod was called P68, and we were in the M-series now, so I thought, *Let's just name it M68, because hopefully it's as good as the regular P68.*

Andy Grignon: And we used to send people over to China with empty suitcases. They would take units off the factory line over at Foxconn and turn around: Just fly there, pick them up, and fly them back. That way nothing could get lost in shipping. We'd pay the border guards off, or whatever the fuck, but that's how we'd ensure that the units got back safely and quickly without going through all this mumbo jumbo. And we had gotten our very first units back. You're looking at it like, *This is my baby. This is why we've been working so fucking hard.* You're seeing a thing light up the first time and you're like, *Oh, that's fucking baller!*

Nitin Ganatra: It was another one of these holy-shit-we-actually-did-it moments. It actually works, and it actually works a little bit better in some ways than we were anticipating! Yes, it was just amazing.

Andy Grignon: And we were sitting around in the hallways: It was our very first time actually using a fully completed thing. And it was just a moment of silence. We were all sitting there playing with it. Right? It was the very first time you could hold the thing in your hand and it was like, "Oh, that's pretty fucking cool!" And of course you start doing the pinch. You start to do all this multitouch stuff, right?

Guy Bar-Nahum: My team was worried. They came to me saying, "The fucker doesn't dial well. It's the worst phone I've ever used!" I told them, "Look, guys, you're missing the whole point. We're not making a phone. We're making a laptop killer. That's what we're making here, right?" I told them, "Nokia is about connecting people. What do we do? We separate people. We're Americans. We want to be alone. We don't want to be connecting to other people!" Dialing people is not the killer app for this thing.

Andy Grignon: So we're sitting there. We're browsing. We're playing with Safari. And I was like, *I wonder what's the first thing that people are going to do? They're going to browse porn!* So I went to this porn site, foobies.com, which was just a boobie site, nothing big, but I was like, *I'm the first guy to surf porn on an iPhone!* I was like, *Yes!*

Guy Bar-Nahum: I told them, "The killer app for this thing is browsing the web and checking e-mail. And if someone is going to call me, I don't give a fuck about that." And after that they relaxed. They were like, "Okay, okay, I get it."

Andy Grignon: It still had a lot of flaws. You still look at your baby with a very critical eye, right? Like, *Oh, it doesn't do this. It's kind of fucked up. This shit is totally broke.* It's all that kind of stuff, right? None of us had really let the idea sink in that this was the next gigantic hit for Apple.

Nitin Ganatra: The big deadline that we had was the January 2007 announcement at Macworld. There was going to be a long and very extensive demo that was going to be done by Steve. We probably first learned about this in October 2006 and it was, "Holy crap! We are going to be demoing this thing in January!"

Matt Rogers: Apple is a unique place. Apple is tens of thousands of people and they work on five products. And those products are awesome. And every engineer, every designer, every product manager, knows that everything must be awesome. When you set a schedule, you don't miss a schedule. There are thousands of people depending on it.

Andy Grignon: And we had this impossible deadline, just to hit the announcement.

Nitin Ganatra: There was going to be this demo of the deep functionality of the iPhone and at the very end this grand finale where Steve then goes back and does this multistep task involving numerous applications to actually show both that, "Hey! This thing actually can have value!" and that this isn't just smoke and mirrors—that this is the real deal.

Christopher Stringer: This broke new ground. It was more than a phone. Smartphones existed, but they were more like tiny little computers.

Nitin Ganatra: So it was understood that Steve was actually going to try to use this thing like a normal user would, and he's going to do it in January, and he's going to do it up onstage in front of the whole world to see. We were just terrified, because it's one thing to write software that works in isolation, but to actually have something that works integrated with a lot of other components? That's when a whole other class of bugs make themselves known. *Holy crap!* Nobody was going home for vacation. Nobody is going home for Thanksgiving. In fact, what are you doing this weekend?

Matt Rogers: And the previous four weeks were insane. I cannot even describe it. Myself and a team of about twenty-five were over in China basically hand-building them from scratch. We literally were working 24/7 to make sure that those first two hundred or three hundred units were perfect.

Andy Grignon: These were the show phones, right? Jony and Steve literally graded, with a jeweler's loupe and white gloves, every single phone. They're looking for the most minute imperfections, and then they grade them. The best are AA: "If we were to make it today, that's what we would want." Now right before these phones go to Moscone Center, it's time to flash that new software that has all the fixes on them, right? And so we bless the build: Everything is good. And someone goes to install the software on our units—the show units—and he's got a gang programmer, so he can burn the software into eight phones at a time. He starts off with the AAs. And so when we put new software down, the phone thought that we were trying to do something subversive, it thought it was being tampered with, and it lit the fuse. The chip has a security feature which lights a piece of metal up deep in the processor for the baseband, and when that's gone the phone is dead. We turned these phones into bricks. We had burned up our best units for the show! On the weekend before we're starting to go through the run-through!! I should have gotten fired on the spot for that...So we ultimately fixed that problem. The phones are there. Then we did the launch.

Eddy Cue: It was the only event I took my wife and kids to because, as I told them, "In your lifetime, this might be the biggest thing ever." Because you could feel it. You just knew this was huge. The iPhone was the culmination of everything for Steve.

Andy Hertzfeld: It was kind of a privileged ticket to be there. And Steve really is a bit nuts about levels of privilege at events. He really cared about who he invited to that and who we didn't. And so I was excited to go.

Phil Schiller: People had been waiting so long and were so excited about this upcoming event at Macworld. The energy was amazing: Electric, I would call it.

Andy Grignon: Except there wasn't a completed anything, and so a couple things could happen. The phone could crash, hard crash, so you would see the Apple logo and it shits the bed and you know it's rebooting from scratch.

That's a bad one, that happened, but largely what happened is you would use all the memory and there was this part of the operating system which would shoot an app in the head to free up memory, right? Mail was a particularly poor performer in that respect. Or for web pages, same thing: The OS would be like, "You know, that one is using too much memory," and then it would just go away. You would see the app just disappear, right? Or more subtle things: The base band—the chip that makes the actual phone calls—would crash and the signal strength would come and go, or whatever, but that would be very obvious to the audience watching the projected screen. We were just on pins and needles. We're like, "What the fuck is going to go wrong?"

Nitin Ganatra: The most terrifying time is when your bits, the bits that you've actually worked on, are up on the big screen and Steve is interacting with them, because God help you if there's any kind of screwup at all.

Andy Grignon: You know, during the announcement Steve sent a message to Tony. He deleted Tony. And that was the end of Tony—onstage.

Nitin Ganatra: During the Macworld demo in front of the whole entire world, Steve went and deleted Tony Fadell's name from the list of favorites within that phone app.

Andy Grignon: He joked about it, like, "Let's say there's somebody here who I don't want to talk to anymore?" And then he scrolls through and he comes up to Tony's name. "And I can just do one swipe, tap Delete, and he's gone." People laughed about it, but everybody knew. Steve doesn't do things like that for fun. Steve was a complicated guy. Steve was in many ways diabolical, and Tony and Steve's relationship had grown increasingly rocky. Steve would just fly off the handle on random things and Tony just had to take it. And by this point Forstall and Tony had developed a full-on feud with each other.

Tony Fadell: That demo script was created by Scott Forstall.

Nitin Ganatra: Now you could innocently look at it as, "Look, I can add things to favorites easily and I can delete things that are no longer my favorites easily," but I think it was understood that Steve doesn't do these things by accident. Steve will run through these scripts dozens of times, just on his own. It looks so fluid and natural up there, but the amount of preparation that goes into it is just staggering.

Andy Grignon: We had outlined what was called the "golden path," right? The path that Steve had to do, each and every time, and if he didn't, the phone would crash because it was running out of memory or whatever. It was if you opened this e-mail before you opened that e-mail something would get corrupted, and we just couldn't figure out why. We'd eventually get to it, but we just don't know why then, right? But if Steve did it in this exact order, we're good—just nobody breathe on it.

Nitin Ganatra: We were sitting in the audience. We were in the third row and Andy pulled out a flask and there was this drinking game that happened where every single time somebody's piece was shown on-screen, and everything worked well, and Steve went on to the next thing, the flask got passed to the person who was largely responsible for those pieces working.

Andy Grignon: So after each chunk it was like, "Knock 'em back!" I ended up doing eight or nine shots and I was completely obliterated by the actual end. That's when it seems like a great idea to just finish the bottle, because you're like, "Oh fuck yes!" It was a flawless demo. So we drank the rest of that bottle, then we went out and smoked a bowl, and then just got messed up for the rest of that day.

Nitin Ganatra: Early on there was a whisper campaign by BlackBerry and by Nokia—these other companies who already had phones out there— that was going around: "Yes, the iPhone was really cool looking, but it's not a serious business device." But we understood that this is kind of how the business works.

Guy Bar-Nahum: I told my team, "Look, guys, look at Nokia. They're laughing at us. They're saying, 'That's not even a phone, that's a piece of crap,' but they don't know what just happened to them."

Nitin Ganatra: We were immediately drawing parallels to what happened to the Macintosh in the '80s. It was exactly what was being said about the Macintosh right when the Macintosh came out, too.

Guy Bar-Nahum: The first iPhone was just like the Mac: It was a fully owned, closed system. Steve was like, "I don't want third-party apps on my perfect thing."

Andy Grignon: Steve had made the edict. His reason was that the second we let some knucklehead app developer write some software, we take the risk that it crashes the phone. At the end of the day you don't know what the fuck some developer who is coked up and on a bender is going to cook up.

Guy Bar-Nahum: And then Google came along and said, "We're doing Android." So what's the difference? Android was going to allow people to download third-party apps.

Steve Wozniak: The first iPhone did not have the App Store.

Guy Bar-Nahum: Steve panicked, and literally all of a sudden everything clicks through his head and he realizes that he was like a mile behind Google!

Steve Wozniak: Openness is so important, and the App Store is a type of openness. It wasn't the original idea of the iPhone. Steve had to be convinced of it.

Guy Bar-Nahum: And so he went to Forstall and told him, "Hey, Forstall, we have to open up an ecosystem," and that's where Forstall shined the second time, after making the iPhone UI. He told him, "I was working on it the whole time." So Forstall had actually ignored Steve's verbatim order and he worked on it in the basement and made it happen. You have to have balls of steel to do something like that.

Guy Bar-Nahum: And then because of Google, Apple was pushed into putting apps on the iPhone. And then apps became the way people touched the web.

Steve Wozniak: People ask me, "What Apple product has changed your life the most?" There were times when I said, "The iPhone," but it's definitely the third-party App Store. Having that App Store let people create apps and everything we do became apps.

Andy Grignon: And then all of a sudden Facebook and Twitter come online and the phone started to tickle the part of our brain that wants constant stimulus. All of a sudden you're standing in line at Safeway and you tweet to this global audience of people. You can be constantly engaged.

Guy Bar-Nahum: Then something amazing happened. The social-mobile web of our time—the web that was about conversations—was born. This has absolutely changed the world.

Andy Grignon: It was the perfect marriage of services. I would never have predicted any of it. Not a single person did.

Guy Bar-Nahum: Apple is the most nonweb company in Silicon Valley. They are very product-centric in their DNA. It was really a mistake, almost.

Andy Grignon: We always joked that Steve was "user number one" and the iPhone was built for Steve. He saw it as a way to be productive, to carry less shit around.

Ron Johnson: Steve fundamentally guided every hardware design, every software design; his fingerprints were all over everything that was done. These were his products; we were his tools.

Steve Wozniak: He made sure every little detail was right for the phone that he himself used. He had a really good mind for elegance and simplicity. That means not having every junky idea any engineer can think of. Steve was really good at turning those things down.

Andy Grignon: And the irony here is that Steve never made his own phone calls!

Twttr

Nose-ring-wearing, tattooed, neck-bearded, long-haired
punk hippie misfits

*I*n Silicon Valley a fortunate few make it into orbit on their first rocket ride—
Nolan Bushnell, the two Steves, Jeff and Pierre, Larry and Sergey, Mark
Zuckerberg—and that's about it. The reality is that most moon shots fizzle. Take,
for example, the launchpad that is serial inventor Ev Williams's brain. During the
dot-com boom Williams built Blogger, an easy-to-use web publishing tool. Blogger
survived the bust—but not as an independent company. Blogger was swallowed up
by Google. Williams's next venture was Odeo, an easy-to-use podcasting tool. Odeo
laid an egg. Half of Williams's problem was timing. Odeo was simply years too early.
The other half was the people: Odeo featured a cast of characters so shaggy and anarchic as to be virtually unmanageable. Yet, in the summer of 2006, those very hairy
anarchists hatched a new idea: Twitter. In some ways it was the least promising idea
that Williams had ever backed. But it was exactly right for the time. Twitter hit
just as the web was starting to go mobile with the iPhone. And so while Twitter
is another of Silicon Valley's overnight success stories, it took Williams himself well
over a decade—an eon on the Valley's timescale—before he tasted the stratosphere.

Rabble: Odeo came about because Ev and Biz were commuting down to
Google. They were living in San Francisco and they had these commutes
and they were bored, so they listened to podcasts every day.

Biz Stone: We were in a car ride home, Ev and I, and Ev was driving his yel-
low Subaru and I was like, "Ev, I think I have an idea. I think I have a really
good idea!" And he was like, "What is it?" And I said, "Well, you know
how we can record our voices in Flash and on a browser and play them
back? Like Audioblogger."

Ev Williams: Noah Glass had this thing called Audioblogger; he was my neighbor.

Noah Glass: It allowed people to call up a telephone number and record a message, and it published it to your blog.

Ev Williams: We did a partnership with him, with Audioblogger, to allow people to call in on the phone, record a message, and get it posted to their blog.

Biz Stone: And Ev was like, "Yeah?" I said, "Couldn't we write software that would sync whatever you recorded to these new things called iPods because it seems like they're just getting more and more popular. And so, couldn't we just kind of democratize radio and let everybody have their own radio station?" And then Ev made his face that he makes when he thinks an idea is good. He like opened his eyes really wide and said, "Yeah, you could do all that!" And so that was Odeo.

Ev Williams: We didn't know that other people were talking about the same idea. It turns out that audio distributed to iPods was what podcasting was, and it was starting to get some traction. Like a lot of inventions, the adjacent possible reveals itself and everybody sees it at the same time and thinks about it. And we told Noah: "Noah—you should totally do this!" And he was like, "That's interesting..."

Adam Rugel: Odeo was interesting. The idea was to be a network where you could consume podcasts.

Dom Sagolla: Back in 2005, there weren't a lot of people podcasting. This is before Apple had even claimed the space.

Ev Williams: Unrelated to that, I quit Google fully intending to take some time off. And so I was home every day, and Noah was pursuing this podcasting idea, and he just came over every day because I had agreed to seed-fund it and I had agreed to advise, but it wasn't my start-up: It was his.

Adam Rugel: So Noah was sort of the true original founder of Odeo, and Rabble was his first engineering hire. And those are two of the most eccentric and interesting guys you will ever meet.

Biz Stone: Everybody loved Noah. He was great, big, a larger-than-life character. He'll give you a big bear hug and almost crush you to death. He will jump in the water and swim to Alcatraz. He's so strong! He's just a crazy guy, right? And literally sometimes he's an actual *crazy* guy.

Rabble: In about January of 2005 the guy from TED got ahold of Ev and said, "Hey, what are you doing?" And Ev really wanted to get another invite to go back to TED, and so he said, "Well, I'm doing this podcasting start-up."

Ev Williams: And he says, "Will you launch at TED?" And I was like, *Wow, what a huge opportunity. We can't not do that.* And by the way, he also lined up Markoff to come write a *New York Times* story about us. We weren't ready to launch, but we announced and we did a demo and it was like a three-minute talk at TED, but I presented and I remember getting oohs and aahs from the TED audience. I was like, "Wow, that's great."

Ray McClure: We're still working out of Ev's apartment, and the thing we are working on is getting talked about in the *New York Times*! Whenever Ev would be a part of something he would seem to really nail it and make a splash.

Ev Williams: It was a big deal, and I was kind of caught up in the excitement and podcasting was getting all this hype, and I thought, *Well, I feel committed now, and so I guess I'll work at this company and this will be my next company*. And then it all kind of snowballed uncontrollably. We set out to raise $1 million. It was an unproven market. We had no product. We barely had a team. And we were being offered $5 million.

Biz Stone: Silicon Valley is a crazy place: If you left Google after their IPO to start something you automatically could get money.

Dom Sagolla: Evan Williams had exited from Google through the IPO.

Biz Stone: It was like, "Oh! You were at Google? You must be a genius!" Even if you were like mopping the floors at Google you are a genius. "You didn't get the job at eBay in 2002 but you got the job at Google? You must be awesome! You must be some kind of brilliant genius, because Google's a good company."

Ev Williams: What we probably should have been at that point is like three people tinkering in a garage, but once we had the money we had to do something with the money.

Rabble: And so we started hacking on this thing and slowly assembled a team of kind of misfit-y people. Mostly because Noah really liked weird, freaky people.

Ev Williams: Roughly speaking, there are two engineering cultures in Silicon Valley, which you could describe as hackers and engineers. And obviously Google is engineers. Engineers generally have computer science degrees and higher. They study the fundamentals. Hackers just want to make stuff work, and it's not about doing it right necessarily. Facebook was kind of famously built by hackers, but they're not the hippie hackers. Hippie hackers are a particular strain of hackers, and Noah is a total hippie. And so we hired hippie hackers—I mean Noah really hired these people. They are his people, he got along with them.

And so Odeo, Williams's second start-up after Blogger, came together. But from the beginning there was a schism in the core group. There were the hackers—Ev Williams and those that were loyal to him—and the hippie hackers: Noah's crew.

Biz Stone: Ev became the CEO, but the culture was kind of based off of Noah's craziness.

Adam Rugel: It was a really tight group. Although Ev was off on his own. He was married before everybody else was. And Biz got married during that time.

Biz Stone: Jack and Noah, they just partied. They went out to clubs and partied and I don't know what they did. Like they went to Burning Man and Jack told me afterward, "Halfway there we had to go back because Noah forgot all his drugs."

Dom Sagolla: They were nose-ring-wearing, tattooed, neck-bearded, long-haired punk hippie misfits—dropouts.

Adam Rugel: The vibe of San Francisco was sort of embedded in that team.

Rabble: It was a cross between a commune and a really geeky frat-club-house-type thing. Company dinners were all vegan and that kind of thing. But it was very business-focused, too.

Blaine Cook: A lot of people involved in the start-up scene in San Francisco were definitely of that ilk: rebels and activists and artists and that sort of people.

Ray McClure: I remember realizing at a certain moment that Rabble could juggle. Then he throws the ball over to Blaine, and Blaine starts juggling, and pretty soon Noah is juggling, too. And I realize that all these people had taken it upon themselves, independently, to learn to juggle! That's the kind of crew it was. They're not clowns, but maybe they're travelers—people who have had some interesting experiences. It was an extremely liberal group all the way up to the top.

Rabble: So in 2005 we build the Odeo podcast production tool; we announce demos of it; we set up a beta program where we start letting people in; and we keep working on the product. Some of it works really well, and some of it doesn't work very well. But we implement these small teams and we kind of iterate on it.

Ev Williams: We did that for a few months and we were kind of just stuck, and then Eddy Cue invited us down to Apple. We thought, *Oh, maybe they want to buy us or partner with us or something?* We showed him everything we were working on, and he was like, "Oh, very interesting," and then like two weeks later Apple launched podcasts integrated into iTunes.

Rabble: Their directory was kind of crappy, but they did it quickly and it worked.

Ev Williams: It completely shocked us because podcasts were so new and Apple was so big, and it just seemed like such a weird thing for this super-mainstream company to do. Even the word *podcast* was still this weird geeky term.

Biz Stone: We thought that podcasting was a fringe-y nerd thing. Apple was putting it in iTunes?

Adam Rugel: When it happened there was a lot of self-justification, like, "We can still do this—this makes it seem like it's bigger!"

Ev Williams: At first it was kind of like the dot-com crash, a suspension of disbelief: "Oh well, that won't affect us too much."

Rabble: We kept working on it but by January or February of 2006 it was clear that there was this malaise in the office.

Ev Williams: We just weren't feeling it, basically, and I remember talking to Biz and Noah, and I was like, "I'm not loving this."

Rabble: So at some point over the spring in 2006 Ev goes to the board of Odeo and says, "I'm quitting."

Ev Williams: I said, "Look, I'm not feeling like this is going to work, really, and maybe we should stop." We had about $3 million of our $5 million in the bank, and the board said, "Well, you can stop and give the money back, or you could try something else," and so they introduced the whole idea of pivoting to me. I was like, "Oh! Well, I guess we could just do that."

Adam Rugel: It was like, "Hey, we've got a little bit of money; we've got a roof over our head; we've got network connections; we've got computers; we've got a smart team. Let's see what we can do!" Ev should be given a ton of credit for having the confidence to realize who he had in that room. Evan, Noah, Rabble, Ray, and then Jack later, and Blaine as well: They're all geniuses, all of them, or as close to it as you can get.

Rabble: We started having brainstorming sessions and coming up with things, and a lot of the ideas we came up with were around telephony and communication stuff. And so we had this idea of doing this hackathon thing.

Ray McClure: A hackathon was something that we knew other companies were doing. It was, "We're going to spend the next couple of days coming up with ideas."

Ev Williams: I was always proud of the fact that I have a million ideas.

Dom Sagolla: We all split into teams of three and I was on a team with Jack. I really wanted to hear what Jack was up to.

Biz Stone: Jack's really quiet and soft-spoken, but I could get him to laugh. And I liked him. So we picked each other and then we started brainstorming.

Dom Sagolla: We gathered at this children's slide in South Park, in San Francisco, which I thought would be a great place to be creative. So we grabbed our Mexican food and sat there. I don't know if we even ate it. We just talked about where we were going and what we wanted to work on.

Biz Stone: I was like, "How about this? How about just picture blogging? Like you just put a picture up and that's it." And Jack was like, "That's pretty good." I was like, "After all this complicated podcasting B.S. I just wanna do something really simple like that."

Jack Dorsey: We were on the playground and I said, "What if we just do this simple thing with SMS where we send it out and it goes out to all these people in real time? Let's just do that."

Biz Stone: He's like, "Do you think we can make a whole thing just out of that? Like a social network kind of thing?" And I was like, "I love the simplicity of it. Let's do it."

Dom Sagolla: Jack's original use case was identifying cool, happening places to be and sending out information about where the cool party is—at a club or a party that was so loud that you couldn't make a phone call. I'm like, "I don't like clubbing, so why do I need this thing?" He kept trying to describe use cases that would fit the SMS broadcast idea. And that's when Jack came at me and said, "Well, maybe there's a disaster? Maybe you're in trouble? There's a use case!" But it was still difficult to articulate this, even to ourselves, the three of us. So we spent a bunch of time, going back and forth.

Adam Rugel: It's a developers' world at this point. Can you communicate an idea in the form of a rough sketch of a product? Are you an engineer or a front-end designer who can produce a prototype in a day or two?

Biz Stone: We dreamed up how it would work and I made a totally fake, but seemed-real website. This was before the iPhone. And so, we demo. I was in charge of demoing the thing and I was like, "So, Jack and I have this thing and here is what it does. You say what you're doing and then people who want to follow you click a Follow button." And we demoed that and everyone was kind of like, "…And?" And we're like, "That's it!" And they're like, "All you do is get what they're doing? That's it? You can say what you're doing and you can get what people are doing?" And we're like, "That's it! Super simple. Isn't that great?" And they were like, "No, that sucks. It's not even something, it's nothing. What is that? It's nothing!"

Blaine Cook: We had a bunch of discussions among the team and there were a bunch of examples of services that had done things like this in the past.

Rabble: Back in 2004 I launched this software called TXTMob with some activists who came out of the MIT Media Lab.

Blaine Cook: TXTMob was a tool that could send text updates that would be distributed to a group of people who had signed up to receive those updates. It was a really simple system, but it was probably the first time that mobile phones had been hooked up in just that way.

Rabble: TXTMob was meant to be a social network for toppling governments and organizing protests and shutting down conventions and blockading streets.

Blaine Cook: Obviously TXTMob was a source of inspiration, but there were other group texting services that were sort of Twitter shaped, and so we talked about all these past projects: why had they failed, or in the cases of some of them that still existed, why had they not succeeded in a big way?

Ev Williams: It was clear that mobile messaging was going to be a thing. And we were playing with telephony which Rabble and Blaine knew a lot about—and Noah, because that is what Audioblogger was.

Biz Stone: Ev could see that we were really into it, and so he took us aside and was like, "You guys should keep working on this. There is something to this mixing the web with SMS—nobody else is doing that."

Ev Williams: I don't know at what point we started calling it Twitter. There was some generic name before that and then there was the name "Friend Stalker," which was always a joke.

Tony Stubblebine: The thing I don't remember is why we all thought Twitter was such a good idea. As I recall, we jumped on it really early, even before we had a prototype. I think part of the issue was that most of us weren't yet on Facebook, so we didn't have a way to keep in touch with our friends.

Ev Williams: We decided to pursue this idea. It seemed worth investing in. We had something that had a little bit of momentum, but not enough surface area. You don't want to put a ton of people on a nascent idea. It's like kindling that you are trying to get going: You don't want to overwhelm it.

Dom Sagolla: Four people split off from the main podcasting group to became Twitter. Jack Dorsey and Florian Weber are the ones who actually built the code. They were iterating on this: fixing the bugs and making it happen. And then sprinkle in what Biz would call un-design: taking things away, making it spare, making it metaphorically easy to understand. And then you give it all of the, like, wet energy from Noah. And it starts to flow.

Rabble: And the rest of us kept the podcasting thing moving forward. And so that continued for a few weeks and by March 21, 2006, enough code had been written and they had a little phone attached with a USB cable to a computer that could send the tweets back and forth.

Dom Sagolla: It began with a series of early tweets that were automated, saying, "just setting up my twttr." That was the way it worked: It automatically sent the first message. So everyone's messages, if you look back in the early days, say the same thing.

Adam Rugel: There was no *i* or *e*, it was T-W-T-T-R at the time. A lot of the start-ups had done vowel-less names.

Dom Sagolla: This was in the age of Flickr.

Adam Rugel: It was just a trendy thing, a sign of the times.

In March 2006, Jack Dorsey sent the first recorded (nonautomated) tweet: "inviting coworkers."

Adam Rugel: It was just to everybody in the room.

Dom Sagolla: The next messages were people just typing whatever they could into this web page prompt. We're all at this office at 164 South Park, and Jack and I were back-to-back, and I just typed something: "oooooooh," almost ironically. And Jack wrote, "waiting for Dom to update more." And my response was, "oh this is going to be addictive."

Rabble: Twitter is this little black box where you press the button for the reload and you don't know when you're going to get something awesome.

Did I get something new? Did I get something new? Did I get something new? Did I get something new? Oh wow, look at that! So it's incredibly addictive.

Dom Sagolla: Because it was almost too easy. You don't need to think about it. You just write.

Ray McClure: So we're tweeting about what we're eating and when we're going out: just really really personal but not very interesting content. Maybe by the end of the day there was a hundred and something people on Twitter and we're testing it: Is this a thing? And what can it be used for?

Adam Rugel: Everybody had a slightly different twist on what it was supposed to be.

Tony Stubbledine: When we had a version that was friends and family only we all had a magical experience of feeling connected to our friends.

Biz Stone: After we got it working I was getting on BART to go to Berkeley. And just as I was getting on I heard people talking: "Earthquake, earthquake…" And I'm about to go underwater—under the whole Bay—in a tube. I'm thinking, *I got to get off.* And my phone was like *zzzt, zzzt, zzzt*— lots of messages coming in at once. And I look and it said, "Earthquake on the Hayward fault: 2.3 on the Richter scale. No big deal." And so I stayed on BART. I was like, *Huh! That was useful.*

Rabble: Very quickly we started using Twitter for protests. And then lots of people sort of discovered that Twitter is a really good tool for activism.

Blaine Cook: Rabble was pretty bored of working at Odeo and distracted in all sorts of different ways.

Rabble: I'm an anarchist. I think we should have a revolution and we should restructure and we should have democratic control and decentralized worker cooperatives and everything else.

Ev Williams: It's hard to manage anarchists.

Biz Stone: They wore shirts that said ANARCHY. And then they would ask questions to all of Odeo, like, "What's a good hedge fund to invest in?"

Rabble: I was waiting around for my stock options and ready to move on and had enough of the dot-com thing. I felt like I wasn't changing the world enough.

Biz Stone: Those guys were a pain in the ass. They would specifically sit down during a stand-up meeting—on purpose! There was one stand-up meeting where they were sitting down and Ev was like, "Guys, please, I need everyone here by ten a.m." Because everyone was showing up at noon or whatever. And one of the guys raised his hands and said, "I have a question." And Ev was like, "Yes?" And he said, "What's our motivation?" And Ev just lost it. Ev just yelled out, "It's your fucking job!!"

Dom Sagolla: It was this weird intersection of I-don't-care versus I-care-too-much: "How much don't you care?" "I don't care so much that I'm going to wear a nose ring and I'm not going to cut my hair or take a shower! But if you get into a discussion with me, like, 'How is this going to be built?' I will sit here and argue with you over code until your eyes bleed."

Blaine Cook: Over the summer of 2006 it was just a pretty terrible place to work. We were depressed that Odeo wasn't working and weren't really sure what the heck this Twitter thing was going to do or what it was. We didn't have any users for Twitter, either, though we thought that there was some potential there.

Rabble: And so the next thing was: Let's figure out who really wants to be here. A bunch of people were fired.

Dom Sagolla: Rabble, me, Adam, and Tony all got laid off on the same day.

Rabble: It was this blurry thing. Like, Adam continued to work out of the office for the next couple of years. So it was like, "You're fired!" "Okay, I'll see you tomorrow at the office."

Ev Williams: So it was September or October of 2006. I went to the board meeting and said, "Well, here's the situation: We probably have like $2.5 million left, still no real growth with Odeo, and we have this Twitter thing going. We think it could be a thing, but there's no real evidence of that." And then they're like, "You tried to do something new. There's no evidence it's working. No one wants to buy the company. There is still a couple million dollars in the bank. You should just give that money back." That was the very clear direction from the board and I couldn't really argue with that. And then that's when I said, "Well, how about I just buy the company?" So, I bought Odeo, paid back the investors, put it in this new company called Obvious, and we were running Odeo and Twitter.

Biz Stone: Actually Ev was like, "Let's me and you say we bought it back in case things go wrong." And I was like, "Okay." And then things went right and he was like, "Yeah, I bought it back." Because it was all his money. So, he basically owned everything and then decided to dole out ownership based on what he thought was fair, which I thought was unfair for me, but whatever. It all came out in the wash.

Blaine Cook: Obvious was Ev's idea of a research lab or a prototyping lab like Xerox PARC or Atari Research, where we were going to work on a whole bunch of projects. And around that time there was a meeting: Do we kill Twitter? Do we kill Odeo? What do we do with these things? We were pretty sick of Odeo, so that got put into maintenance mode. And I think at the time we decided that we should probably keep doing this Twitter thing even though it wasn't an instant runaway success.

Adam Rugel: Then Noah was forced out.

Biz Stone: Noah had been acting really crazy. He was staying up at the office all night long with giant puffy hair making YouTube videos: "I don't know, man. I don't think we're gonna make it." That kind of stuff. And he was driving everybody crazy with weird little details that made no sense. And so Jack told me, "I can't do this anymore. I wanna go to fashion school and learn how to make jeans." And Florian was like, "I'm just gonna move back to Germany." And so, Ev was like, "Should I fire Noah?" I was like, "Well, yeah. If we don't have the engineers what are we gonna do? We can't do it by ourselves." And Ev was like, "Okay. I'm gonna do it right now." And he was really like kind of nervous about it, but he fired him.

Ray McClure: Noah did it to himself. A lot of it was because he was breaking up with his wife at the time and I think Noah became a little unhinged. It was a bummer.

Adam Rugel: He would have been the best billionaire of the bunch by far. He would have been a Howard Hughes–type billionaire. So it's a shame that we didn't get to see that, but I think it was inevitable that it wasn't going to work out.

Biz Stone: We all knew Noah was crazy. He was one of those guys that's fun at a party but working with him is just a nightmare.

Dom Sagolla: So a series of layoffs and then this pivot into Obvious Corporation; then a private launch to friends and family—not anyone from big corporations like Apple or Google; then the public launch in the summer, mostly to influencers and early adopters; then they purchased a couple of vowels and Twttr became Twitter. And now it's February 2007 and we see the iPhone demo and our eyes and minds are lighting up. Everyone is paying close attention to this potential for a mobile web, and then we have a demonstration in the halls of South by Southwest.

Blaine Cook: South by Southwest is a big conference with sort of three parts: music, film, and what they call "interactive" around web and online technologies. It's been going for years and years now. I guess we were probably nominated for an award that year, though we were still pretty small.

Ev Williams: Twitter was getting adoption from our set. The web geeks in San Francisco were using Twitter. So I thought we should show up at South by Southwest because all these people would be at South by Southwest—plus many more.

Biz Stone: Ev had this idea to do some marketing at South by Southwest. He was like, "We should do something different. We should do something that nobody else does." And I said, "Everyone hangs out in the regular hallways between the things. What if I designed a visualizer and we put plasma screens up there?"

Ev Williams: Ray coded this beautiful full-screen tweet visualizer and Jack and Biz went early to set the visualizer up. We had this cartoony aesthetic—birds and clouds.

Biz Stone: We were up all night trying to get them to work. Then as people started coming in the service went down and we were sitting there like schmucks. It's like ten, eleven a.m. and all the screens were black, and we were just like, "Come on!" Blaine was asleep and he refused to have a cell phone. So we had to get another guy to go to his house and wake him up. Then we finally got it to actually work.

Ev Williams: The people who were there at the conference: Their tweets would float by on-screen.

Adam Rugel: So they did a party and gathered everybody at the festival on Twitter, and people were using it as a back channel to the whole conference.

Blaine Cook: Twitter gave people a way to be able to sort of organize and communicate without texting directly.

Adam Rugel: I think all the people there saw the potential.

Ev Williams: There was enough momentum going into it, enough buzz. People were talking about it on the blogs, and then that same crew all showed up at once and they saw the tweets of a lot of people that they were excited about on the screen.

Ray McClure: Twitter just totally took over South by Southwest that year.

Biz Stone: A couple of things that happened. One is this story of the guy at the bar who tweeted, "This bar is too loud. If anyone wants to talk about projects let's go to this other bar that's quiet." And then in the eight minutes it took him to walk there eight hundred people had showed up! That was my realization that, *Oh, shit. We definitely invented a new thing. We definitely made something that the world had never seen.*

Ev Williams: And then because the tech press and the bloggers were all there, they just talked about it everywhere.

Biz Stone: Then we were bumbling into the award ceremony to watch the awards and I looked at Ev and Jack and I was like, "Wait a minute. What if we win something? What if we actually win? We might win!" And the guys were like, "Oh, shit." And I was like, "If we win we should have a speech." And Ev said, "Yes, we should: You write the speech; Jack says the speech." And I was like, "Great." So, I wrote the speech for Jack. And we did win.

Jack Dorsey (from the stage): I'd like to thank everyone in 140 characters— and I just did.

Biz Stone: We won for "best blog," which made no sense, but we won something.

Dom Sagolla: They had to be recognized for something. They had to win something.

Adam Rugel: Then it becomes an inside thing. So the tech community and the tech journalists are on it. They know that Evan is special and has this history, they saw how the product worked and functioned in that environment, and they wanted to be a part of something big and new: That was the energy that helped it grow.

Ev Williams: It was a moment. It was lightning in a bottle.

Biz Stone: We incorporated Twitter as a separate company two days after we got back.

Rabble: Twitter launched right before the iPhone, and the iPhone made the experience of sending and receiving text messages nice. But at first only Apple was allowed to build apps for the first year or two.

Blaine Cook: But iPhones had the web, so people could load Twitter.com on their iPhone and view the messages very easily.

Ev Williams: We had shifted it to be more of a web product from an SMS-focused product, and that was key.

Rabble: So, it was one of the few things that worked really well—uniquely well—on that first generation of iPhones.

Dom Sagolla: And as soon as we saw that, we realized that this was peanut butter and jelly. This is how Twitter was meant to be experienced:

not on the web, not over SMS, not on your desktop, but in your pocket wherever you were, talking about whatever's going on at that moment, to other people wherever they are, away from their desk. It was a watershed moment. And so you can actually follow the rate of adoption in Twitter's user base with the number of iPhone installs. There is a very solid correlation between the growth of iPhone and the growth of Twitter.

Rabble: Do you see the irony in this? Apple killed Odeo for the iPod, and then Twitter exists and is successful because of the iPhone.

Adam Rugel: It was just nuts from 2007 on.

Rabble: Twitter got really lucky.

Mark Zuckerberg: It's as if they drove a clown car into a gold mine—and fell in.

Ev Williams: It's a myth that Twitter was successful because we were lucky. I mean, *obviously* we were lucky, but it's naïve to think that we were *just* lucky.

Rabble: The other part of the luck was that Ev was really well connected with the digerati. You know: the cool folks who show up at parties and like to use each other's stuff.

Ev Williams: Our users were web geeks—bloggers. That's when it got the term *microblogging.*

Nick Bilton: The original concept was *What am I doing?* and the original concept was never realized.

Ev Williams: Twitter didn't find immediate product-market fit. Why did it then grow? Because we changed it! And—little-known fact—after that fateful South by Southwest, growth stalled again. And then we changed it more.

Nick Bilton: I can't remember the last time I saw someone tweet "Going to lunch with my friend Bob!" on Twitter—like you do on Facebook.

Ev Williams: Success at this magnitude is fucking hard—and unlikely—no matter how great the idea. Because it's not just *an* idea—it's dozens of ideas and hundreds of decisions. The myth is that ideas are stumbled onto fully baked. In reality, they have to be developed.

Nick Bilton: And at the end of the day, Twitter became a place for people to become more famous, and not a place for conversation.

@RealDonaldTrump: My daughter Ivanka thinks I should run for President. Maybe I should listen.

Nick Bilton: Twitter today is one person talking to a lot of people—not a conversation. And I don't think that we as human beings were designed to enable 320 million people to have a conversation together.

Steven Johnson: It's true, Twitter is not a conversational medium, and that's fine. Not everything has to be a conversation medium. So if you are judging it by that standard, then yes, it's a failure. But I have found Twitter, personally, to be just a great addition to my life.

@RealDonaldTrump: I love Twitter…It's like owning your own newspaper—without the losses.

Steven Johnson: What I find so beautiful about Twitter is that it's literally like the serendipity of the front page of the newspaper, times one hundred. So when people sit there and say, "Oh we're killing serendipity because Google has ruined all that. Now you just go to the web and search for what you want and you get it," I look at those people and think, *Have you never used Twitter?* Literally every time I hit Refresh there is an interesting hint of a take about something. And more often than not, a link to something longer that opens my mind in some way. Twitter is a serendipity engine.

Nick Bilton: I think we'll look back at Twitter in ten, twenty, thirty years—in the same way we look back at The Well today—and say, "Well, that didn't work out so well, did it?"

@RealDonaldTrump: The #1 trend on Twitter right now is #TrumpWon—thank you!

To Infinity…and Beyond!

Steve Jobs in memoriam

*S*ilicon Valley is a culture, a people, a point of view. But if the place had to be represented by just one person, it would have to be Steve Jobs. He was a native son who twice plunged into the wilderness. The first foray was in 1974, to India. The nineteen-year-old Jobs did not find the holy man he was seeking, Neem Karoli Baba, a living saint known to his followers as Maharaj-ji, but the trip stirred something within him. On his return to the Valley Jobs kickstarted the personal computer industry. The Apple II and the Macintosh were breakthrough products, but the Mac, at least initially, did not sell. Exiled from the company he cofounded, Jobs was a wash-up at the age of thirty. A dozen years later he was asked to come back and save the company from near-certain oblivion. Astonishingly, he did save the company—by moving beyond the computer-on-every-desk paradigm that he had established some three decades before with the Apple II. The iPhone ushered in an age of anywhere-anytime mobile computing. Thus the man who started the personal computer era also ended it. Jobs's premature death three years later permanently enshrined him in the Valley's firmament. His life was the stuff of myth—myths that Jobs often encouraged. But Jobs had no magical powers, no superhuman ability to see the future. The truth is both mundane and extraordinary: By sheer determination and cleverness, Jobs became the very man that he went looking for as a lad—the guru, the seer, the wizard. He died at home, in Palo Alto, on October 5, 2011.

Mike Slade: In October of 2003, Steve wanders into my office out of the blue on a Monday afternoon after the executive team staff meeting and he goes, "I need to talk to you."

Steve Jobs: I had a scan at seven thirty in the morning and it clearly showed a tumor on my pancreas. I didn't even know what a pancreas was.

The doctors told me this was almost certainly a type of cancer that is incurable, and that I should expect to live no longer than three to six months. My doctor advised me to go home and get my affairs in order, which is doctor's code for "prepare to die." It means to try to tell your kids everything you thought you'd have the next ten years to tell them in just a few months. It means to make sure everything is buttoned up so that it will be as easy as possible for your family. It means to say your good-byes.

Mike Slade: He goes, "I've got pancreatic cancer, I'm going to die" and started crying. And I'm like, *What the fuck?* And I'm freaking out and so he's like, "They're not 100 percent sure," but he's such a disaster and so I interact with him the rest of the day and it's kind of weird.

Larry Brilliant: The phone rang and you could always tell when it was Steve because the caller ID would always just say "Apple," and so I picked it up and he said, "Do you still believe in Neem Karoli Baba? Do you still believe in Maharaj-ji? Do you still believe in God? Do you still believe?" I said, "Hi, Steve, how are you?" And we talked, and he said, "I want to know if you still believe in Maharaj-ji." And I said, "Yes, sure. Why?" He said, "Because I've been diagnosed with pancreatic cancer."

Steve Jobs: I lived with that diagnosis all day. Later that evening I had a biopsy. They stuck an endoscope down my throat, through my stomach and into my intestines, put a needle into my pancreas and got a few cells from the tumor. I was sedated, but my wife, who was there, told me that when they viewed the cells under a microscope the doctors started crying, because it turned out to be a very rare form of pancreatic cancer that is curable with surgery.

Mike Slade: And then the next morning Steve takes me aside and he goes, "It's not pancreatic cancer, it's an islet cell tumor, so it's not as serious as I thought it was. It's still serious but I'm going to be fine..." And I'm like, "Whoa!"

Ron Johnson: Steve's strength throughout his life was his ability to think differently and to trust his intuition and to follow his convictions. He did that at work, and he did that in his personal life. And that's how he tackled his illness.

Mike Slade: He chose not to have surgery right away and did acupuncture, which was ultimately a fatal decision, although it took seven or eight years.

John Couch: Steve overcame so many obstacles that people said were insurmountable obstacles that he probably felt that he could beat his illness on his own as well.

Mike Slade: Steve was a hippie—and I don't mean that in a negative way. He was a pure vegan and he believed in alternative medicine, as demonstrated by not having the islet cell tumor surgery right away.

Tom Suiter: In 2005 I, like everybody else, saw the commencement speech at Stanford and kind of figured, *Great! He dodged the bullet. Thank God*.

Jon Rubinstein: He told everyone he was cured, right? So everyone went "Okay, he's cured: back to work." So there'd be stressful moments when it was first announced and everything, but after that everyone just got back to work.

Andy Hertzfeld: The cancer diagnosis was just after the iPod started up on that hockey-stick adoption curve—but before the iPhone. And it's very interesting to think about all that. Boy, did he accomplish a lot in a short time.

Wayne Goodrich: At the iPhone introduction in 2007 his stamina and his outlook was still very Steve—he was very adamant that it was business as usual.

Ron Johnson: There was never any bitterness about the challenges he faced. He didn't complain about it. I think he battled very gracefully.

Mike Slade: I visited him after he had his liver transplant in 2009 and a couple other times, and he always told me that he was okay. Knowing what I know now, he never actually told me the whole truth at any point in time, even though we were really close friends.

Mona Simpson: I remember my brother learning to walk again with a chair. After his liver transplant, once a day he would get up on legs that seemed too thin to bear him, arms pitched to the chair back. He'd push that chair down the Memphis hospital corridor toward the nursing station and then he'd sit down on the chair, rest, turn around, and walk back again. He counted his steps and each day pressed a little farther.

John Markoff: I remember him coming back from one of his sabbaticals and seeing him at the Stanford Apple Store. He was much thinner, but he looked pretty healthy.

Ron Johnson: He just became a lot leaner, physically. His body was wearing down, but his mind was sharp as ever when I saw him. His spirit was as strong as ever or even strengthening—which is not uncommon near the

end. It becomes this unique moment where people start to truly understand better how to think about life and the really important things in life.

Andy Hertzfeld: The cancer definitely changed his approach to life. He
got a little nicer or more straightforward. He had much more of a desire
to explain himself. Before, he always was very kind of imperious: "No!
We're not doing that!" But then after he got sick he started articulating his
thinking behind things.

John Markoff: He came back to Apple for the last time in the spring of 2011
and we knew he was in bad shape. I just sent him a note that said, "Hey!
Steve Levy and I would like to come visit." And we had a really nice two-
hour visit with him in May of 2011. He was drinking ginger ale the whole
time, so I knew he was on some kind of cancer drug or something like
that, but he didn't talk about his health, and we didn't ask about his health.
He told us at one point that he wished he had a little more energy and that
he wanted to take on the car industry. He really wanted to design a car! It
was a really nice conversation. And that was the last time I saw him. He
took his final leave from Apple two or three months later.

Mike Slade: Bill Gates goes, "Will you help me? I want to see Steve. I know he's
really sick: I want to go see him," and I go, "Just e-mail the guy," and he goes,
"Well, I did" and then it turned out that when Bill left Microsoft he changed
his e-mail address, and so it didn't get through to Steve. Bill thought Steve
was ignoring him, whatever. So Bill said, "Will you facilitate me meeting
him?" And I'm thinking to myself, *Why do I have to do this? You guys introduced
me to each other!* It's like your parents asking you to arrange dinner or something. So anyway, I set it up for May of 2011.

Bill Gates: I wrote Steve Jobs a letter as he was dying. He kept it by his bed.

Mike Slade: A phone call comes in and it was Steve, and I go, "Hi, Steve,"
and he was pretty sick and he goes, "I'm supposed to meet with Bill in like
an hour," and I go, "I know, I set it up—remember?" And he goes, "What
does he want?" Because he thought Bill wanted to give him the whole big
giving-pledge thing and Steve didn't want to do that shit, and I'm like,
"No, he just wants to hang out with you, really!" And he goes like, "Bill
Gates wants to hang out?" Because Bill is not a hang-out kind of a guy.
And I go, "Yes, he really does!" And he goes, "We're scheduled for ninety
minutes—that's not enough time to hang out." And I go, "Just go with it.
What's the worst that's going to happen? Give him a Diet Coke and he'll

be on his way! You should do this, you know?" And he goes, "Okay. Talk to you later. Bye." I think Bill stayed for three or four hours. It was a long time, and he was really happy he did it, let's put it that way. They talked about their kids and everything, and it's kind of like Ted Williams and Joe DiMaggio talking to each other. They're not pissed. It was sweet.

Bill Gates: We laughed about how fortunate it was that he met Laurene, and she's kept him semisane, and I met Melinda, and she's kept me semisane. We also discussed how it's challenging to be one of our children, and how do we mitigate that? It was pretty personal.

In August, Steve Jobs formally gives up the reins at Apple and retreats to his home in Palo Alto.

Larry Brilliant: One of the greatest pleasures I had in forty years of friendship with Steve was going for walks, frequently from his house to the frozen yogurt place, or the smoothie place.

Larry Ellison: We would always go for walks. And the walks just kept on getting shorter, until near the end we'd kind of walk around the block, and you would just watch him getting weaker.

Alan Kay: I told Steve before he died, I said, "Steve, you know the best thing you ever did was hanging on to Pixar for ten years." I hope he goes to heaven for just that alone.

Ron Johnson: I saw Steve the Sunday before he passed. The door was always open, so you could just kind of wander in and walk back and say, "Hi." He was in bed. And I sat next to his bed for quite a while. I had maybe two hours with Steve, just chatting. I wouldn't put myself in his inner circle, but I was in the next ring, and we had an incredible relationship and had many intimate, deep conversations about spiritual issues.

Steve Wozniak: Near the end of his life he seemed very changed and he really, he was going back in his mind thinking about those early times, before Apple even.

John Markoff: Jobs would get sentimental. Somehow that led into a discussion of how significant an event taking LSD had been for him. He said it was one of the two or three most significant things he'd done in his life.

Larry Brilliant: I think what's interesting is what Maharaj-ji said about it. He said, "LSD is yogi medicine." It's medicine for yogis, for people who are trying to become enlightened.

Dan Kottke: In the winter of 1972 we were reading all this stuff: We had read *Be Here Now*, we had read *Autobiography of a Yogi*, we had read *Ramakrishna and His Disciples*—that's all about enlightenment.

Larry Brilliant: But LSD only allows you into the presence of God, or the presence of Christ, for a minute. You can bow down and say, "Hello," but then you've got to get out of there when it wears off. So you can take LSD, and come into the presence for a minute, then you've got to get out of there like a thief in the night. It's better to earn your way in, and then you can stay forever.

John Markoff: What I found most striking was that he said that it was something that his wife didn't share with him and it was something that set him apart from the industry, because most of the CEOs of the companies he dealt with hadn't taken acid. And unless you've taken acid, you don't understand what that means. It is sort of a profoundly intellectually and psychologically disruptive thing. The world doesn't look the same after you've done it.

Larry Brilliant: In the ashram it was very, very rare for anyone to take psychedelics. It was already a place of such heightened feelings. But I never would have gone to India or found Maharaj-ji if I hadn't taken psychedelics, and I'm sure that's true for Steve as well. I have known him longer than anybody, I think. I met him in 1974, and we shared this experience of this ashram, and our conversations were about spiritual progress, and illness, and the meaning of life, and the end of life, and all those things.

Ron Johnson: The mood was reflective, unhurried: remembering a lot of things that we had done together. I just remember at the end saying, "Thank you." And he said the same thing. And I give him one last big hug. I had to climb into his bed to do it.

Larry Ellison: This is the strongest, most willful person I have ever met, and after seven years the cancer just wore him out. He was just tired of fighting, tired of the pain. And he decided—shocked Laurene, shocked everybody—that the medication was going to stop. He just pulled off the meds on Saturday or Sunday.

Mona Simpson: Tuesday morning, he called me to ask me to hurry up to Palo Alto. His tone was affectionate, dear, loving, but like someone whose luggage was already strapped onto the vehicle, who was already on the beginning of his journey, even as he was sorry—truly, deeply sorry—to be leaving us. So sorry we wouldn't be old together as we'd always planned, that he was going to a better place.

Dan Kottke: One of the books that we really liked was *Way of the White Clouds*, which is about Tibetan reincarnation. We just found that fascinating.

Larry Brilliant: Steve was in and out of the spiritual practice business, but he was a devotee of Neem Karoli Baba and had pictures of Maharaj-ji everywhere, until the day he died.

Mike Slade: He died on Wednesday, October 5, 2011. I was blown away when he died; I didn't expect it at all. I was shocked that he died. He told me he had a year. And then he was dead.

Mona Simpson: Steve's final words, hours earlier, were monosyllables, repeated three times: "Oh, Wow! Oh, Wow! Oh, Wow!"

Dan Kottke: It gives support for the idea that he was tripping at the moment of death. Aldous Huxley famously had himself injected with LSD when he was close to death, and Steve knew about that. But I doubt Steve did that. He would have wanted to be lucid. But DMT is endogenous, it is produced by the body.

Erik Davis: DMT is also known as the spirit molecule—it's a naturally occurring psychedelic compound widely found in nature.

Dan Kottke: And there is a theory that the pineal gland releases DMT at the moment of death, which is kind of interesting from a neuropharmacological point of view... that's my take on it.

Ron Jonson: Bill Campbell was in many ways Steve's best friend. He visited every day, took him for walks, spent time with him. Bill was coaching an eighth-grade football team over at Sacred Heart that day. He showed up at practice and brought his kids together into a huddle and said, "Guys, I want you to know that my best friend just died." And then a double rainbow appeared right over the field. Some people think those things happen by accident. Others think there is something deeper going on. It was an extraordinary moment.

Mike Slade: The next day I get an e-mail from Laurene's office saying, "We're having a small ceremony for fifty people, will you come?" So I went to it. It was a really hot, weird, Silicon Valley day. It was so close to when he died that everybody was still raw. It wasn't like one of those things where everybody really has their shit together. It was a really small, very eclectic group: Bob Iger was there, he spoke; Lee Clow read "The Crazy Ones"; Steve's biological sister Mona Simpson was there; all his children were there; Laurene was there; the only Apple people I recall were Jony Ives, Eddy Cue, and Tim Cook. George Riley was kind of running the show. Larry Brilliant was there.

Larry Brilliant: We read something from the *Bhagavad Gita*, but we're not going to talk about that.

Mike Slade: It was in the cemetery and his coffin was right there. We all formed a semicircle around the casket and then anybody who wanted to speak could speak. It was beautiful, and very visceral. It wasn't a religious funeral: There was no reverend telling you what to do. It was kind of cool that way. We all went to John Doerr's house in Woodside afterward and just sort of drank wine and shot the shit. I got to spend a bunch of time talking to Laurene's brothers, who are really funny guys from Jersey, talking about pranks they played on Steve and stuff like that. I'm really glad I went, but it was hard. It wasn't like the thing a week later at Stanford, which was sort of fun in a funny way.

Wayne Goodrich: For me the Stanford memorial was the moment that it all really, really became real.

Ron Johnson: Stanford was such a fitting choice for Steve. For years he admired the design. He loved the layout. He loved the stone. He told me that the quad—which was built in the late 1800s—was the single best piece of architecture in California. And it's in his backyard.

Andy Hertzfeld: It's walking distance. A long walk.

John Markoff: What I remember in walking up with Steven Levy was that the security was so intense that it felt like a presidential event. It wasn't just private security. It was governmental security: the Secret Service. We were certain that Obama was coming, but he didn't. Rahm Emanuel came, but no Obama.

Jon Rubinstein: Clinton was there—lots of billionaires, lots of famous people.

John Markoff: Joan Baez and George Lucas and Larry Ellison and Bill Gates and John Warnock, and, and, and...

John Couch: Everyone was there! There were competitors there. People that Steve had butted heads with. They were there.

Dan Kottke: I was not invited.

Alvy Ray Smith: I wasn't invited.

Steve Wozniak: I did not go.

John Markoff: The crowd was really kind of stunning. It was sort of an affair of state, a Silicon Valley affair of state.

Wayne Goodrich: Everything was just heightened. Everything was so introspective. Every moment you'd see something that brought back a flood of story and remembrance.

Ron Johnson: It's a majestic place. As you approach the quad there's this large, perfectly proportioned oval with the most beautiful green grass and some flowers in the center, and that leads you to the quad which was the original Stanford campus. And it's laid out such that you walk up these steps and then you enter the quad, and just right on center is among the most beautiful churches that you will ever see.

Mike Slade: It's like something out of the Renaissance.

Tom Suiter: It's just awesome. It's gorgeous. It looks like it was just dropped in from Rome.

Ron Johnson: And as you walk into that, the sunset is right behind the cathedral. It was just a beautiful evening.

Wayne Goodrich: Just walking through the last arch and into the actual quad, looking at the Memorial Church, and seeing so many people and faces of people that I had known.

Ron Johnson: Friends of Steve from Apple, friends of Steve from the Valley, people that you just knew because of their position in the world.

John Couch: It was almost like a physical rapture, everybody who is anybody walking silently toward the Stanford chapel.

Ron Johnson: People are in a very reflective frame of mind.

Tom Suiter: It was somber. It was like, *We're all part of this thing. This is a drag. This sucks.*

John Couch: There was this incredible, quiet respect and a sense of mortality for all of us.

Mike Slade: So anyway, we get escorted into the church: five hundred, seven hundred people?

John Markoff: They don't fill Memorial Chapel completely.

Jon Rubinstein: It was a beautiful service. Clearly it had been stage-managed by Steve from beyond the grave.

John Markoff: Yo-Yo Ma played first and wonderfully.

Andy Hertzfeld: It was really deep, just heartbreakingly beautiful, one of the most emotional pieces of music I've ever heard.

John Markoff: Afterward he briefly introduced the event and told a short, funny story about how Steve had wanted him to play at his funeral and how he had asked Steve to speak at his. As usual, he noted, "Steve had gotten his way."

Mike Slade: So then Laurene spoke and had written this beautiful speech about Steve that was surprisingly analytical.

Ron Johnson: She's just a very professional, poised, intelligent, articulate person who had a long time to prepare for this. She had a message to convey, and she delivered it very gracefully.

Mike Slade: She was like, "Look, if you guys think he was a dick, it's because he was in pursuit of beauty, and that sort of trumped everything, and most people didn't really get that about him." And so almost everything that was dickish—she didn't use those words—was because he was in pursuit of beauty. I've been to lots of funerals where the grieving widow was *the grieving widow*, and that's not what she did. So it was wonderful—really brilliant, and I actually learned a lot from it—but it was surprising, right?

John Markoff: Reed followed her and spoke about his dad. He told a sweet story about being a young child and sleeping in a crib in a corner of the house and being afraid and having Steve crawl into the crib with him to protect and stay there until he fell asleep—something that was frequently foiled when he tried to extract himself from the crib.

Wayne Goodrich: The empathy that I felt for them at that moment in time, knowing that they would never really know or interact with their father as true adults . . . I felt very bad, very bad, very empathic.

John Markoff: Evie read "The Crazy Ones" in a clear voice.

Eve Jobs (from the stage): "Here's to the crazy ones. The misfits. The rebels. The troublemakers. The round pegs in the square holes. The ones who see things differently. They're not fond of rules. And they have no respect for the status quo. You can quote them, disagree with them, glorify or vilify them. About the only thing you can't do is ignore them. Because they change things. They push the human race forward. And while some may see them as the crazy ones, we see genius. Because the people who are crazy enough to think they can change the world, are the ones who do."

John Markoff: I cried.

Wayne Goodrich: The wave of emotion in the church was such that it was a conscious effort to even keep my perception about me, instead of just breaking down in a pool of my own tears.

John Markoff: Mona Simpson spoke and told of how she met her brother and about their relationship. It was much closer than I realized. She, too, talked about Steve's search for beauty.

Mona Simpson: I remember when he phoned the day he met Laurene: "There's this beautiful woman and she's really smart and she has this dog and I'm going to marry her."

John Markoff: Joan Baez stood, and her guitar was brought out and she sang "Swing Low, Sweet Chariot."

Mike Slade: She sat down with the guitar and played it. She hit the high note and my spine shivered. I was just blown away.

John Markoff: Her voice is all still there.

Mike Slade: She was seventy and it just blew everybody's doors off.

John Markoff: Bono and Slash sang. First Dylan's "Every Grain of Sand" and then a U2 song, "One."

Mike Slade: Dylan was supposed to play the first song and he blew them off. They asked him to play and he said no.

John Markoff: To sing Dylan, Bono placed an iPad on a music stand to remember the words with.

Bono (singing Bob Dylan's song): "I hear the ancient footsteps like the motion of the sea / Sometimes I turn, there's someone there, other times it's only me / I am hanging in the balance of the reality of man / Like every sparrow falling, like every grain of sand."

Mike Slade: And then of course Larry Ellison gave this big eulogy, which was, you know, funny and charming and kind of all about Larry.

Larry Ellison: I wanted to talk about my friendship with Steve, what it was like, and a little bit about what Steve was really like.

Mike Slade: Bill and I were kind of eye-rolling each other about Larry because it had a lot of the pronoun *I* in it—shall we say—for a eulogy. But anyway, he meant well.

John Markoff: After the service everyone filed out and walked across the oval to the Rodin Sculpture Garden.

Ron Johnson: We get out of the church and the sun is starting to set, and we've got a long walk, probably half a mile. It was just like a procession because everyone followed the same path.

Wayne Goodrich: I don't remember who I walked over there with, but it was just very quiet, a lot of reflecting back, numbness. *Is this really happening? Did this just occur?* And the questions of why start to seep in: *Why now? Why here? Why?*

John Couch: It wasn't a boisterous drinking kind of thing. It was very sobering, just dealing with the fact of mortality of us as humans.

Wayne Goodrich: Everybody had a little bit of a story about their time with Steve, and some people even managed to talk about it. All the stories that all the people there had about him would have been an amazing thing to somehow capture. My hairs are standing up on my arm right now just thinking about it.

Larry Brilliant: I met Steve Jobs because all of us hippies who had gone to India to hang out with gurus were very poor, and we got very skinny, and we never had any money to buy any food. And when one of us—any one of us—got a gig, then everybody would go find that person and say, "Hey, I knew you from the ashram, I'm hungry," or something like that. And I got a gig, my gig was at the United Nations and so I was working at the WHO office in Delhi, and by the time Steve came in '74 the cafeteria had gotten a reputation for having good, safe lettuce. So Steve showed up looking for lettuce, and I took him to lunch, and he inhaled lettuce, he inhaled the salad. He was trying so hard to eat only living food, and in India, the best and the healthiest thing you can do with food is make sure it's dead, killed, boiled. So he was constantly getting sick, getting skinnier, and I just remember the first meeting we had was contentious because I at that time was eating liver. I had been a vegetarian for almost ten years and I didn't feel like I had much energy, which was fine when I was in the ashram but it wasn't so fine when I thought that if I didn't give an extra hour of work that day, then maybe babies would die of smallpox. I was skinny, vegetarian, and my boss said, "Oh, having a little trouble keeping up, eh?" She said, "Eat liver." Of course there was no cow liver, being India, so you eat buffalo liver. And so my first meeting with Steve was a debate between eating buffalo liver in order to get enough calories so that you could fight longer to stop more children from dying from smallpox, or eating living food because the planet would be better off in the long run if you only ate living food. And it was a great conversation.

Steve Wozniak: Steve Jobs was very different before we started Apple. Personality settles in around age twenty. So, going to India was before that. When we started Apple and he was the founder of a company that had money, it was, "I'm going to be in a suit. I'm going to be on the covers of magazines. I'm going to be the one that will get credit for it." Steve had changed so much.

John Couch: One of the stories that I tell is from when we were very close to each other in the early Lisa days. And you know Steve was a minimalist: He had a beautiful English Tudor home; he had a Maxfield Parrish painting; he had a Tiffany lamp; he had a Swedish sound system; and then he had a mattress on the floor and a dresser. And I walked up to the house one day and it was always impeccable, but there was a piece of paper on the front lawn, and I thought, *Well, that's really strange*, and I picked it up, and it was an Apple stock certificate: seven-point-five million shares of stock. And I knocked on the door and I said, "Steve, I think this belongs to you." And he goes, "Oh, oh yes," and he just kind of opened the dresser drawer and he put it in and he closed it and he said, "It must have blown out the window." And so I knew from that point on that it was never about money for Steve.

Steve Wozniak: He wanted to be one of those important people that really changed things—like from that Ayn Rand book—and that led him to exclude everything else. He wanted to do it his way. He was always fighting me.

Mike Slade: I'll tell you my last Steve Jobs story: Okay, so I love marketing, right? And one of the reasons I love it is that it's a lot of bullshit. And so I'm at Microsoft and Steve is trying to get me to come to work at NeXT and so we're sitting there talking and he's telling me, "You're going to rot up there in Seattle! It rains all the time. It's beautiful in Silicon Valley," and he was saying, "You know, Mike, Palo Alto is a special place." He goes, "Palo Alto is like Florence in the Renaissance. You walk down the street, and you meet an astronaut and a scholar," blah, blah, blah. I'm like, "Yes, that sounds cool!" So I come to work for NeXT. And so my wife and I, we didn't have kids yet, and so we'd eat a lot at Il Fornaio on University Avenue in Palo Alto. We were sitting there, in early '91, and I'm reading the menu, and on the back of the menu at Il Fornaio it says, "Palo Alto is like Florence in the Renaissance..." And it goes through the whole spiel! The fucking guy sold me a line from a menu! From a chain restaurant!! Bad ad copy from Il Fornaio, which was his favorite restaurant, right? Such a shameless bullshitter!

Jon Rubinstein: Steve was a very complicated individual—very complicated. There were so many aspects of him. I mean he was so passionate about product, not just our products, but products in general. He was really good at choosing a path: not a long-term path, this isn't a vision thing. This is, "You get to a fork in the road. What fork do you take?" He surrounded himself with good people and kept them motivated through both the good times and the bad times. He was a good marketing guy: He could sell ice cubes to Eskimos, right? When he was in front of an audience—it didn't matter if it was ten people or ten thousand people—everyone felt like Steve was speaking directly to them. And that's a remarkable talent.

Alvy Ray Smith: This will sound like a sick comparison, but Hitler had this articulate speech thing, too. Steve had this charisma that was just awesome, but to misuse that is evil—and I think he was evil.

Larry Brilliant: I saw the behavior that people described as mercurial and preemptory and harsh. But seventeen thousand people worked for him and would have killed to continue working for him, because he made them better.

Steve Wozniak: But people close to Steve had to deal with just wrong behavior.

Jon Rubinstein: Steve had all kinds of issues, but he had a bunch of really, really important gifts that allowed him to accomplish the things he wanted to accomplish.

Ron Johnson: I think even Steve recognized that the measure of life is not time, it's impact. And Steve's impact is as profound as anybody who has walked this planet.

Steve Wozniak: Look, I came up with the product that made Apple! If Steve Jobs had started without me, where would he have gone? Keep in mind, Steve tried to make four computers in his life—with millions of dollars—and they all failed: the Apple III, for marketing reasons; the Lisa, because Steve didn't understand costs; the Macintosh, which wasn't really a computer, just a program that looked like a computer and led to big problems later on; and the NeXT.

John Markoff: Still, Steve was emblematic of the Valley, rising to represent how it touched the popular culture. He is the vector for Alan Kay's insight. Kay was the person who understood that computing was a universal medium, that it wasn't a calculation tool, and as a universal medium it transformed any medium that it touched. Before computing, there had been paper, there had been music, there had been video, and computing sort of relentlessly transformed each one of them, and Steve was the vector. He was the messenger who sort of implemented Alan's vision. Steve was the one who shaped those products so that they were usable by mortals and so we could get beyond the original personal computers. And the modern world is different because of that.

Steve Wozniak: I swear he didn't have to be a bad person, he didn't even have to be a controlling person, even a closed person, to create all the good technology he did. So he was good and bad.

Wayne Goodrich: It got dark quite late.

Steven Levy: There was this beautiful moon hanging there and it was perfectly clear. I said to Laurene, "I bet Steve had a couple of earlier versions of the weather that he sent back until he came up with this." It was just a beautiful event.

Ron Johnson: The Rodin Sculpture Garden is a very peaceful place. It's one of my favorite places to go sit at Stanford and they had transformed it into almost like a nightclub. It was just really a lovely place to be. It was just this beautiful, beautiful setting, and the temperature is coming down, but there's a warmth and a freshness to the evening, and everyone just didn't want to leave.

Wayne Goodrich: Then it just started thinning out, and everybody just kind of disappeared.

Marc Benioff: And then we were all leaving, and on the way out they handed us a small brown box, and I received the box and I said, "This is going to be good." Because I knew that Steve made a decision that everyone was going to get this—and so whatever this was, it was the last thing he wanted us to all think about. So I waited until I got to my car and I opened the box. And what is the box? What is in this brown box? It was a copy of Yogananda's book.

Andy Hertzfeld: They gave *Autobiography of a Yogi* to every attendee as they were leaving. It was kind of an interesting thing to do.

Marc Benioff: Yogananda was a Hindu guru who had this book on self-realization, and that was the message: The message was "Actualize yourself."

Dan Kottke: I was actually surprised, because Steve was much more of a fan of *Zen Mind, Beginner's Mind*. That, I would say, is more his favorite book.

Marc Benioff: "Actualize yourself"—that was Yogananda's message. And if you look back at the history of Steve—that early trip where he went to India to the ashram of his maharishi. He had this incredible realization that his intuition was his greatest gift and he needed to look at the world from the inside out.

Larry Brilliant: You just wouldn't get a conversation with anybody else the way you would get it with Steve Jobs. You could be talking about politics, you could be talking about space travel, you could be talking about movies or music, technology, spiritual life. He was so smart, and his mind was so agile, and what he could do was he could take the experiences that he had with computers or the experiences that he had with a mystical quest and he could apply them to analyzing today's news or to political events, or geopolitical events even more so. I never had a conversation with him where I didn't learn something.

Marc Benioff: And here his last message to us was: "Look inside yourself, realize yourself, look to *The Autobiography of a Yogi*, which is a story of self-realization." I think that is so powerful. It gives a tremendous insight into who he was but also why he was successful. He was not afraid to take that key journey.

Ron Johnson: Steve was clearly a spiritual being.

Marc Benioff: He was the guru.

EPILOGUE

─────────────

The future is already here—it's just not very evenly distributed.
—WILLIAM GIBSON

The Endless Frontier

The future history of Silicon Valley

They come West seeking fame and fortune. They work their claims. Some even strike it rich. Swap the pickax for some coding chops and the six-shooter for a standing desk, and suddenly Silicon Valley becomes a much more familiar place. It is what it has always been: a frontier—the boundary between what is and what could be. It's a place where the future is dreamed up, prototyped, packaged, and ultimately sold. But what is the future of this place that makes the future? Where is Silicon Valley going? Where is its technology taking us? What are we to become? The best answers to those questions come from those who've already built the future: the one that we live in today.

Kevin Kelly: The biggest invention in Silicon Valley was not the transistor but the start-up model, the culture of the entrepreneurial start-up.

Marc Porat: It's the style of thinking and behaving that's called "being an entrepreneur."

Megan Smith: I grew up in it. It's extraordinary. An entrepreneurial culture of like, "Hey, how can we solve this?" And really caring about helping each other.

Carol Bartz: It really is just this need to change as fast as possible to enable the next great thing. We don't even have to imagine the next great thing yet. We just have to get the tools to do something and use trial and error until we have the next great thing.

Jim Clark: The venture capital business is a big part of that. They are in the business of betting on these people, and the good ones make money at it. Then it becomes kind of a self-fulfilling organism.

Tony Fadell: That's going to be Silicon Valley's legacy: It has cast a mold, cast the die, for this ever-evolving way of taking up risk, taking on new ideas, funding them, and making things happen.

Marc Porat: The elements that made Venice are the elements that made Palo Alto, are the elements that will make Silicon Valley, and it just grows and grows.

Ev Williams: Once it gets started then it just snowballs.

Scott Hassan: I try not to predict the future very much, but the one thing I know for certain is that in the future, there are going to be more computers, they're going to be faster, and they're going to do more things.

Tony Fadell: You're going to see every single industry, no matter how behind the times they are, adopting technology—deep technology.

Scott Hassan: Eventually computers are going to do everything. I don't think anything is safe. Nothing.

Tony Fadell: Change is going to be continual, and today is the slowest day society will ever move.

Kristina Woolsey: Technology is changing fundamental things. It changes where you can live and work; it changes who you know; it changes who you can collaborate with. Commerce has completely changed. Those things change the nature of society.

Scott Hassan: Never ever try to compete with a computer on doing something, because if you don't lose today, you'll lose tomorrow.

Tony Fadell: Tomorrow will get faster and every year after it's only going to continue to get faster in terms of the amount of change. That means all of these incumbents with these big businesses that have been around for a hundred or two hundred years can be unseated, because technology is the unseating element. Technology is the levelizer.

Carol Bartz: We are very arrogant out here that nothing can change unless technology is involved, and technology will drive any business out there to a disruption point.

Marc Porat: Technology, that's what we do here in Silicon Valley. We just push technology until someone figures out what to do with it.

Andy Hertzfeld: Right now the Valley is particularly excited about two things: One of them is machine learning; incredible progress has been made in machine learning the last three or four years. A broader way of saying it is artificial intelligence.

Kevin Kelly: The fundamental disruption, the central event of this coming century, is going to be artificial intelligence, which will be underpinning and augmenting everything that we do—it will be pervasive, cheap, and ubiquitous.

Marissa Mayer: I'm incredibly optimistic about what AI can do. I think right now we are just at the early stages, and a lot of fears are overblown. Technologists are terrible marketers. This notion of *artificial* intelligence, even the acronym itself, is scary.

Tiffany Shlain: There's all this hysteria about AI taking over. But here's the thing: The skills we need most in today's world—skills like empathy, creativity, taking initiative, and cross-disciplinary thinking—are all things that machines will never have. Those are the skills that will be most needed in the future, too.

Marissa Mayer: If we'd had better marketing, we would have said, "Wait, can we talk about *enhanced* intelligence or *computer-augmented* intelligence, where the human being isn't replaced in the equation?" The people who are working on artificial intelligence are looking at how they can take a repetitive menial task and make a computer do it faster and better. To me, that's a much less threatening notion than creating an artificially intelligent *being*.

Andy Hertzfeld: You know there wasn't so much a breakthrough algorithm that made it happen, although there was some improved algorithm work. Mainly it was just the ability to apply the algorithm to billions of data inputs—it was just a difference of scale. And so that scale just came naturally from Moore's law.

Alvy Ray Smith: Moore's law is astonishing, it's beyond belief, it's the dynamo of the modern age.

Andy Hertzfeld: It couldn't have happened no matter how brilliant your algorithm was twenty years ago, because billions were just out of reach. But what they found was that by just increasing the scale, suddenly things started working incredibly well. It kind of shocked the computer science world, that they started working so good.

Kevin Kelly: So just like the way we made artificial flying, we're going to make new types of thinking that don't occur in nature. It's not that it's going to be humanlike but faster. It's an alien intelligence—and that turns out to be its chief benefit. The reason why we employ an AI to drive our cars is because they're not driving it like humans.

Jim Clark: It's bound to happen, the robotic driving of cars.

Nolan Bushnell: Self-driving cars are going to change everything and help cities to literally become gardens because streets, in some ways, kind of go away.

Jaron Lanier: Once we have automated transportation you might stop thinking about home in the same way. The idea is that everybody would be in self-driving RVs forever, so there would just be like this constant streaming of people living a mobile lifestyle going from here to there all over the world. Why not? And indeed, there's something very attractive about that. I could see raising a family where you have a bunch of families with young kids of similar ages, all traveling together seeing different parts of the world, convening for their education and working remotely. That could actually be really nice, because what we do these days is we spend hours a day moving kids around from this lesson to that school to this soccer thing or whatever and it's kind of an insane way to live. That makes no sense to anybody. And so I could imagine something that's actually pretty nice coming together. I like that vision a lot.

Scott Hassan: I like the concept of self-driving cars, but I worry that our legal system can't really handle them. The problem with autonomous cars is that it's the manufacturer who is driving that car.

Kevin Kelly: Humans should not be allowed to drive! We're just terrible drivers. In the last twelve months humans killed one million other humans driving.

Scott Hassan: If we can bring that down to, like, three a day, that will be amazing, right? It will be a thousand times less, right? But the problem is those three deaths are going to be caused by an autonomous car killing them. And that could sink any company making autonomous cars, because the payout—three per day—that's a lot of lawsuits. So from a liability point of view, it actually might be easier to skip over the autonomous driving cars, and go direct to flying cars. It might actually be easier, because there's less liability involved. And so I think we're going to start seeing flying cars in the next five to ten years, and then they're going to get widespread in like twenty to thirty years. Because the nice thing about a flying car is that all you have to do is convince one agency that it's cool—the FAA—and there's really nothing anyone can do about it, because if they say, "It's okay," then it's okay.

Andy Hertzfeld: So the second thing Silicon Valley is particularly excited about right now is artificial reality, or you might say mixed reality or whatever you want to call it.

Kevin Kelly: That VR vision of the alternative world is still there, but the new thing is this other version of "augmented" or "mixed" reality, where artificial things are inserted into the real world, whether they be objects or characters or people.

Scott Hassan: VR blocks off your field of vision, and everything has to be reconstructed digitally. And so MR, which is mixed reality, is a technology that can selectively draw on any part of your vision. It can actually include all your vision, if that's what's required. So MR is, I believe, the next step in how we interface with computers and information and people. It's all going to be through mixed reality. And VR is a special case of mixed reality.

Nolan Bushnell: All of this is on a continuum, and right now augmented reality is a little bit harder than virtual reality, technically.

Steve Wozniak: Because of Moore's law, we always have more bits and more speed to handle those more bits on the screen. Well, we now have finally gotten to the point where we have enough computer power that you can put the screen on your head, and it's like you're living in a different world, and it fools you. It's enough to fool the brain.

Nolan Bushnell: I've seen how technology has moved from *Pong* to what we're playing today, and I expect the same kind of pathway to virtual reality. I think that twenty years from now we will be shocked at how good VR is. I like to say we are at the *"Pong* phase" of virtual reality. Twenty years from now, VR is going to be old hat. Everybody will be used to it by then. Maybe living there permanently.

Brenda Laurel: The only way I can see that happening is if we completely trash the planet.

Jim Clark: Nolan is a good friend, I know him well, and he can get hyperbolic. I do not think people are going to be living in virtual reality. That might be true in a hundred years—not in twenty.

Nolan Bushnell: So when is VR indistinguishable from reality? I've actually put a little plot together on that. I think we're about 70 percent of the way there visually. I think we're 100 percent of the way there with audio. I think we're 100 percent of the way there in smell. I think we're just scratching the surface on touch and fooling your inner ear, and acceleration, and the thing that I think will break the illusion will be food. I think that's going to be the hardest one to simulate in VR. So when you see the guy in *The Matrix* having a great bottle of wine and steak? That's going to be hard.

Jaron Lanier: Just on some spiritual level, it just seems terribly wrong to say, "Well, we know enough about reality that living in this simulation is just as good." Giving up that mystery of what the real world is just seems like a form of suicide or something.

Jim Clark: Plus I'd rather have real sex than virtual sex.

Nolan Bushnell: That's really a matter of haptics—a full haptic body suit, where the suit simulates temperature and pressure on your skin, and various things...

Brenda Laurel: You know what? If the boys can objectify software instead of people, then it's good for everyone—except the boys.

Scott Hassan: That same type of technology will be used in tele-operated robots; some people call them Waldos. Think of this device as a set of arms that rolls around, that's able to do stuff—two hands that can be manipulated from afar. Let's say it fits where your dishwasher used to be, and whenever you need it, it comes out of there and it unfolds and it's operated by somebody else in another location that has expertise that you want at that time. You want dinner made? Well, it's just remote-operated by a chef, in some type of rig, so that when they move their arms, the robot moves its arms, in the exact same way. And that person is wearing these gloves, so that whenever the robot touches something, they can feel that touch. So that that person can pick up things and chop things and go into the refrigerator, open the doors, as if they were there. Then that same Waldo, when that person is done with making dinner for you, instantly switches over to this other person who loves to clean up, and then they go and clean up the whole kitchen for you. It's the service economy, but it's in your own home, right? And so you basically have expertise on demand. So, well, something like that is going to be widespread in maybe ten years.

Nolan Bushnell: In twenty years, 80 percent of homes will have some kind of a robot.

Carol Bartz: Every inflection point really followed from the fact that you could make something affordable, so that the public or industry could do something with it. You could get this in the hands of more people, which meant it was a bigger market, and on and on, and off you went.

Scott Hassan: They'll probably be the same price as a refrigerator. It's going to be one of those things: You got your car, you got your house, and you got your Waldo. But the cool thing about that is, once that kind of stuff comes out, then people will write all these applications that help those people do certain tasks. So you would install an app so that you click on the potato, and then your Waldo takes over and does it for you automatically, really fast, right? So you would have all these application makers making little things that can make someone's job easier, and then eventually you get to a point where you're not just controlling one of these Waldos, you will be controlling maybe three or ten or one hundred of these simultaneously, and you're more managing these Waldos now, not controlling them individually. Does that make sense? So you've got this huge scaling effect.

Jim Clark: Yeah, I don't get excited about the virtual reality stuff, the car driving and robotics and stuff like that. It's just going to happen. The parts that really get my juices going are the human-computer interface, through the nervous system, and biology transformation. If I was a young man just getting a PhD, I would definitely do biology, because I think that's where it's going. A biologist armed with all this knowledge of computer science and technology can make a huge impact on humanity.

Adele Goldberg: If you were to predict the future based on seeing what is in the labs today and extrapolate, you would believe synthetic biology is the future, not electronics.

Andy Hertzfeld: Because the idea of bio being the next frontier is based on the silicon, really. There's about one hundred billion neurons estimated in most people's heads, and the world knew that thirty years ago and I remember thinking, *Boy, a hundred billion, that's enormous!* And now I think, *A hundred billion? Hey, that's not so much!* Right? If there was a byte per thing, that's not even a terabyte. I have thirty terabytes on my desktop computer upstairs! So it's just that Moore's law has gotten us to the point we're up to dealing with the biological scale of complexity.

Alvy Ray Smith: Moore's law means one order of magnitude every five years— that's the way I define it. And so what do you do with another two to three orders of magnitude increase in Moore's law? We humans can't answer that question. We don't know. An order of magnitude is sort of a natural barrier. Or another way to say it is, if you've got just enough vision to go beyond the order of magnitude, you would probably become a billionaire.

Jim Clark: I think that connecting humans to computers, having that interface, is increasingly going to be possible with a helmet that's measuring neurological signals from the brain and using that to control things. I'm pretty sure that twenty years from now we're going to be well into getting the human-computer interface wrapped around a direct kind of brain-fed interface.

Scott Hassan: We're going to tap right into the optic nerve, and insert things that you don't see, but your brain doesn't know that you don't see them. We're just going to insert it right into your optic nerve. We really don't understand how memory works and stuff like that, but we understand somewhat how the optic nerve works, because it's just a cable going back to your brain, and, you know, we know in theory how to insert things into it, so it's just a bunch of engineering work to make that happen.

Jim Clark: And as time goes on, I think we'll get more and more refined at being able to map and infer and project those signals, on the cortex, on the brain, and I feel as certain about that as I feel about anything.

Larry Page: Eventually we'll have the implant, where if you think about a fact, it will just tell you the answer.

Scott Hassan: It's maybe twenty years away. I mean it depends on how well the market takes up MR, mixed reality. If it really loves it, then it's going to be sooner, so if it's slow to pick up then it's going to be longer. But I think eventually it's going to be there.

Jim Clark: We will for sure be controlling computers with thoughts, and I think increasingly we're going to have kind of hybrid systems that are kind of biological- and computer-like, and they're going to be there to make humans more effective at whatever...

Tony Fadell: So then after that I think it's really going to be how we coevolve with AI. How do we as humans coevolve with the artificial intelligence machine-learning kinds of technologies we are creating now? So, look—chess champions, right? They got beat by Deep Blue back in the nineties, right?

Kevin Kelly: When Garry Kasparov lost to Deep Blue he complained to the organizers, saying, "Look, Deep Blue had access to this database of every single chess move that was ever done. If I had access to that same database in real time, I could have beat Deep Blue." So he said, "I want to make a whole new chess league where you can play as a human with access to that database." And it's kind of like free martial arts, where you could play anywhere, you could play as a human with access, you could play it as a human alone or you could play just as an AI alone. And he called that combination of the AI and the human, he called that a centaur. That team was a centaur, and in the last four years, the world's best chess player on the planet is not an AI, it's not a human, it's a centaur, it's a team.

Tony Fadell: And so the chess champions have now coevolved with the technology and they've gotten smarter and better. Too many times do we get stuck in our own way of thinking, and these things can give us a shot in the arm to think dramatically different. It's like an Einstein showed up to help us, right?

Kevin Kelly: Then I would say that in thirty years people will be beginning to get used to the idea that you can have artificial consciousness...And when you give a body to an AI you have a robot.

Nolan Bushnell: Now, when the robot manifests self-awareness, and then becomes aware of its existence and wants to self-preserve that, we've got some interesting issues that we're going to have to solve. What actions can it take in self-preservation? In a self-aware, self-programming, self-understanding, self-learning bot, can we truly control the limits on its actions? I think we're fifty years from that, but it's something that we're going to have to cross.

Steve Wozniak: But even if it got smarter than us, it would take us on as partners, because we're the creators, you know we're going to be its first friend. Right now humans are still in control, and I've never heard anyone talk like we're going to be out of control very soon.

Tony Fadell: And then I think there's going to be another big split when we decide that biological means locomotion, or biological means manipulation. Because robots are incredibly fragile, they're hard to repair, they don't self-heal, and—this is a crazy thing—we're going to figure out a way to actually turn the biological systems into the robots we want, that self-heal, that train more easily, that actually are much more energy efficient. Literally, the power efficiency of how much you eat versus how much energy you expend is so much more efficient in a biological system versus a mechanical system. So when we want superhuman things, how are they not going to be biological as opposed to a megatron? And then there's going to be all kinds of societal and philosophical and governmental and ethical issues that are going to arise, just like with AI.

Steve Wozniak: Hundreds of years downstream machines will be a superior species, but what will we humans be doing? Can a human be satisfied just being taken care of, like a family dog?

Kevin Kelly: I kind of reject the idea of this super-AI that becomes God-like as being unfeasible for a number of different reasons, but there will be aspects of it that I think may not be understandable, and that's one of the definitions of the singularity.

Steve Wozniak: I have started feeding my dog filet steaks, because if I'm going to be a pet some day, with all my needs, my comforts, my entertainment, my shopping, taken care of by computers, well—do unto others as you would have them do unto you. If they're going to do it, I want to be taken care of well.

Kevin Kelly: Yet we're already there in a sense; it's already begun and we don't even recognize it. The first part of it has already been completed in the sense that three billion people are online, so it has begun.

Steve Jobs: Humans are tool builders. And we build tools that can dramatically amplify our innate human abilities.

Kevin Kelly: We are making a set of tools that will allow us humans to collaborate and cooperate and to work together in making things and making things happen at a scale that was previously unimaginable, which is basically the scale of the planet.

Steve Jobs: And I believe that with every bone in my body—that of all the inventions of humans, the computer is going to rank near the top, if not at the top, as history unfolds and we look back. And it is the most awesome tool that we have ever invented. And I feel incredibly lucky to be at exactly the right place in Silicon Valley at exactly the right time historically where this invention has taken form.

Alvy Ray Smith: I was in the middle of Mountain View the other day, astonished at how this suburban town, where graduate students at Stanford used to live in funky apartments, is now this booming city! Silicon Valley is now a city: downtowns that go all night long, restaurants as far as you can see, you can't park anymore.

Marc Porat: Silicon Valley was Santa Clara and Sunnyvale, until it became Palo Alto, and then nobody lived in Menlo Park until they did. And Mountain View was a place you went to get some Asian food, until it wasn't. Redwood City was a place that basically was working-class until now—in some parts of Redwood City you can't afford it. Oakland was too dangerous to even contemplate going into, and now it's the new Brooklyn.

Alvy Ray Smith: In San Francisco, I've never seen skyscrapers go up so fast. It's completely changing the nature of San Francisco!

John Markoff: The Salesforce Tower—this vertical campus that they are building—has surpassed the Transamerica building, surpassed the Bank of America Center. It dominates the skyline in San Francisco in a really overwhelming way. That's probably the strongest statement about how the Valley has changed. It's certainly emblematic of a new era. We're getting these billion-dollar corporate statement headquarters now. It's not just Salesforce. It's Facebook, Apple, Nvidia—even Google.

Marc Porat: Eventually one could see this Silicon Valley thing spreading up to Mill Valley, across to Richmond, down to Oakland and Berkeley, and then down through Fremont and Hayward and then back down to San Jose. So the entire San Francisco Bay Area would be Silicon Valley. And that's a pretty good start.

Steve Jobs: And we've come only a very short distance. And it's still in its formation, but already we've seen enormous changes. I think that's nothing compared to what's coming in the next hundred years.

Scott Hassan: It's happening right now. In a twenty-mile radius the next hundred years is being invented.

Charlie Ayers: And if you're smart enough, you see the low-hanging fruit everywhere.

Scott Hassan: I know companies working on fusion power, that is going to be a thing. And when you have fusion power you can get water right out of the air. You can get food right out of the air. No problem. Fusion energy is going to make those things free, worldwide. Pretty much almost free. I think you're going to be able to live as long as you want, in the relatively near future. That's going to be solved.

Marc Porat: You take what you see and you extrapolate and you push incrementally on things. Size costs, capacity, functionality, whatever you push, until you push so hard there's a paradigm shift and things break.

Scott Hassan: Economics is based on a certain number of people dying every once in a while, right? Well, that's not going to happen, which is going to mess up everything. Fusion power is going to mess up a number of things.

Marc Porat: And there's a discontinuity or more calamity and disruption, and then unintended consequences follow, which then produces another period of push and then, and so on.

Kevin Kelly: So when I think of the future of Silicon Valley I see it as still being the center of the universe as defined by having the least resistance to new ideas, and that's just its cultural history of being tolerant of wild ideas.

Lee Felsenstein: Silicon Valley is a state of mind in a generalized physical area.

Charlie Ayers: Silicon Valley is an energy and a focus and a vibration that consumes you and takes you over, if you allow it to.

Marc Porat: So the IQ points come here, and IQ points are sort of radioactive in a sense that if you press them hard enough, they create fusion. We have the universities and we have the momentum and we have the fusion model, already working.

Charlie Ayers: And you could be part of it, or you can be against it.

Scott Hassan: If any one of the technologies that I know of that are being developed right now in Silicon Valley do really well, the world is going to be an amazing place. But the really amazing thing that I think is probably going to happen is that they *all* are going to do well. So, I'm a superoptimist.

Kevin Kelly: What we're really making here is something that is humanity plus: It's us, plus the machines, plus the planet.

Scott Hassan: I hope that these quantum entanglement systems will allow you to transmit information faster than the speed of light. If it happens, that's going to allow us to colonize the whole entire galaxy, and other galaxies. If that's actually, really possible, it's going to be phenomenal, and that's being developed right now.

Kevin Kelly: Not only will our tools tell us about ourselves, but they also will inform us as we invent new versions of ourselves. And it's going to have a huge effect on our own understanding and our own identity of who we are as humans.

Scott Hassan: The world is going to be an amazing place. But it's going to be a pretty weird world.

Kevin Kelly: And it won't ever end—we're just going to keep layering on more and more layers of connection and tools and abilities and intelligences to this thing that we are making. And that networked thing is the main event.

Scott Hassan: It's going to be like nothing you see today. Realistically? Nothing.

Kevin Kelly: That's the big story. Am I going big enough for you?

Cast of Characters

Chris Agarpao used to go to see Jackie Chan movies with his brother-in-law's best friend, Pierre Omidyar, who happened to mention that he might need a little help in this side business that he had started. So Agarpao became the first employee of eBay in 1996—where he still works, even today.

Al Alcorn was Atari's first engineer. In his first week on the job he produced *Pong*—the arcade video game that put Atari on the map. Atari was the first modern Silicon Valley company, and set the stage for virtually everything that came after.

Mitch Altman was an early employee of VPL, Jaron Lanier's pioneering virtual reality company. Today Altman travels the world visiting and promoting "hackerspaces"—community centers where the tools and expertise needed to make just about anything can be found.

Marc Andreessen and a group of young computer science students in Illinois hacked together Mosaic—the first browser to really take advantage of the just-developed hypertext transfer protocol—in 1993. A year and a half later he and his buddies were in Silicon Valley building an industrial-strength version: Netscape. Andreessen's browser was the most important piece of software to ever come out of Silicon Valley.

Don Andrews was a software engineer who worked with Doug Engelbart in SRI's Augmentation Research Center and was a participant in Engelbart's famous 1968 demo. He stayed with Engelbart until nearly the end, leaving in 1985 to work for Adobe.

Joey Anuff, aka the Duke of URL, is one of the founders of *Suck*, the first meme factory of the emergent web, and its most prolific poison pen. *Suck*'s specialty was exposing the pretensions of the day—which Anuff did with untrammeled glee. The sarcasm and snark that fills the web today is, essentially, Anuff's personality writ large.

Bill Atkinson was the software wizard responsible for the Macintosh's breakthrough look and feel. His program, HyperCard, was the first really successful example of hyperlinked "hypermedia" before the world wide web.

Ali Aydar was a twenty-three-year-old programmer and the guy that a fifteen-year-old Shawn Fanning looked up to and asked for advice. A few years later, Adyar found himself working for Fanning, as Napster's first employee.

Charlie Ayers is a Deadhead and a chef. Ayers even cooked for the band, on and off, after Jerry Garcia died. But by 1999 he had joined Google as their first chef. Ayers brought a lot of the Dead's distinctive culture to Google with him and it was a good fit—until the IPO in 2004.

Ralph Baer was awarded a patent for the "television gaming apparatus and method" in 1969 and throughout much of his life battled Nolan Bushnell for the honor of being called "the Father of Video Games." Baer died in 2014.

Karel Baloun was voted "most likely to be found under a desk" at one of the first Facebook awards gatherings, because he would sleep at the office two or three times a week just to avoid the hour-long commute back home.

Jim Barksdale was the first CEO of Netscape. He was recruited by Jim Clark, who put up the money and gathered the talent, but it was Barksdale who took Netscape public in 1995 and then kept the company alive while enduring constant attack from Microsoft, until Netscape's eventual sale to AOL in 1999.

John Perry Barlow wears many hats: rancher, activist, writer, lyricist. But if you ask him, he'll tell you that he's really a homesteader—in cyberspace. He was probably the best and certainly one of the most prolific writers on The Well, Stewart Brand's early experiment in social media. In the nineties he was instrumental in founding the Electronic Frontier Foundation and still serves on the board of directors.

Guy Bar-Nahum was one of the original engineers on the Apple's iPod project and, then again, on the iPhone. These were the key products that turned Apple around from a near-bankrupt also-ran in the personal computer market to the most valuable company in the world.

Hank Barry had been a rock drummer before he reinvented himself as, first, a lawyer; second, a Silicon Valley venture capitalist; and then finally the CEO of Napster. He took over from the first CEO, Eileen Richardson, when the legal challenge from the record industry started to heat up.

Ryan Bartholomew was the first person to buy an AdWord on Google. He bought the phrase "live lobsters." Bartholomew didn't manage to sell any lobsters, but he saw the potential of AdWords arbitrage. The strategic buying and selling of words (and their associated traffic)—was soon earning Bartholomew tens of thousands of dollars a day.

Carol Bartz is foul-mouthed, Republican, and the first woman to blast through Silicon Valley's glass ceiling. Though she retired in 2001, she is still revered as one of the Valley's most accomplished leaders—of either sex—having served as CEO at both Autodesk and Yahoo.

John Battelle was hired right out of Berkeley's graduate school of journalism to be *Wired*'s first managing editor. Soon after, he founded his own, more business-focused magazine, the *Industry Standard*. After that, it was a book on Google, an online advertising network, and a popular conference series. Today he's hard at work trying to reinvent traditional magazine journalism for a new, online age.

Michelle Battelle was *Wired* managing editor John Battelle's fiancée when she first walked into the magazine's offices in San Francisco. She soon found herself working there, helping to get the first issue of the magazine noticed by the media elite that she had left behind in New York City.

Andy Bechtolsheim was getting a PhD in electrical engineering at Stanford when the school gave him a mandate: Build a network-connected computer for every researcher and student at the university. The design he came up was called the SUN—short for Stanford University Network. The design was commercialized, and SUN swiftly became the brand of choice for scientists, engineers, and other power users. A decade or so later, Bechtolsheim paid it forward by giving some seed money to a few Stanford PhD students with a promising idea—Google, as it turned out.

Yves Béhar is an industrial designer who specializes in making high-tech beautiful. Béhar moved to San Francisco's South Park in 1993 but was only discovered by Silicon Valley after Steve Jobs demonstrated—with the iMac, iPod, iPhone, and iPad—what good product design could mean for a company's bottom line. Béhar's company, fuseproject, is one of the most important design firms in technology today.

Brian Behlendorf was *Wired*'s first webmaster. On the side, he created a free alternative to the software that Netscape was planning to sell. This project morphed into the open-source movement. Today Behlendorf is directing the open-source Hyperledger project, which is devoted to perfecting what may be the next revolutionary data-storage technology.

Marc Benioff is the cofounder and current CEO of Salesforce.com. The Salesforce Tower is the tallest skyscraper in San Francisco.

Patty Beron was the "it" girl of San Francisco's dot-com bubble. She always knew where the party was, and you did, too, if you checked her website, SFGirl.com.

Nick Bilton is the author of *Hatching Twitter*, a journalistic deep dive into the company and its turmoil that reads like a thriller. His latest book, *American Kingpin*, does the same for Silk Road, an illegal online marketplace specializing in drugs, guns, hacking software, forged passports, counterfeit cash, poisons, and the like.

David Boies made the government's case in *United States v. Microsoft Corp.*, and—rather famously—won. A few years later the celebrated trial attorney took on the defense of Napster—and lost.

Jack Boulware is a veteran of the little magazine explosion that came in the wake of desktop publishing in the late eighties and nineties. He had his own magazine, the *Nose*—a kind of West Coast take on *Spy*—and wrote for most of the others. Boulware is now the executive director of Liquake, San Francisco's literary festival, which he cofounded in 1999.

Stewart Brand dreamed up, organized, and edited the *Whole Earth Catalog*, which became the bible of the hippie movement. A few years later he coined the phrase "personal computer" in his book *Two Cybernetic Frontiers*, which helped spark a revolution of a different kind. He's still at it: leading crusades to use the tools of synthetic biology to resurrect extinct species, and the tools of engineering to erect a massive, subterranean monument to time that will ticktock for ten thousand years.

Larry Brilliant is best known for his role in hunting the smallpox virus to extinction, but his true genius may be his gift for human connection. He amassed a pretty weird Rolodex in the course of saving the world: Ram Dass, Steve Jobs, Stewart Brand, Wavy Gravy, not to mention various high government officials. And they stuck with him through all his later incarnations: hippie professor, high-tech entrepreneur, and finally advisor and confidant to Silicon Valley's most powerful.

Sergey Brin is the cofounder of Google. A hunch about the deep structure of the internet, backed up by some of Brin's PhD-level mathematical voodoo, was what turned the web from a whopping mess into the world's most important information resource.

Po Bronson wrote a novel about the financial markets and then another about start-ups before he turned to nonfiction. It was the perfect windup for his subject: the great internet gold rush of the late nineties. *The Nudist on the Late Shift* is still, in this writer's informed opinion, the best book ever written about that era.

Paul Buchheit was a very early Google employee. He coined the phrase "Don't Be Evil," which became Google's corporate motto. He built the AdSense prototype—the software that, even to this day, generates much of Google's corporate wealth. For an encore he hacked together a little experiment that he called Gmail.

Nolan Bushnell founded Atari, the company that started the computer game industry and put Silicon Valley on its modern path. Mentor and friend to Steve Jobs, he famously declined to invest in Apple Computer.

Orkut Büyükkökten had a campus-wide social network up and running at Stanford three years before Mark Zuckerberg launched The Facebook at Harvard. As an engineer at Google, he rolled out another one—named Orkut—six months before Zuckerberg moved to Silicon Valley. He's still at it, having launched Hello.com, a social network optimized for meeting people with similar interests, in 2016.

Chris Caen worked at Atari every summer as a high school student, and by the time he was eighteen was a product manager with a hard-walled office and a good salary. By the end of that summer, Atari had collapsed and Stanford seemed like an attractive alternative. After graduating he worked in VR, and then PR, and now writes about technology.

Heather Cairns was hired to do everything that founders Larry Page and Sergey Brin didn't want to do—that was the actual job description. As Google grew, so did Cairns's job title. She eventually became head of HR, but at the end of the day Cairns was Google's agony aunt, the grown-up whom people confided in.

Ezra Callahan was a Stanford student who, by chance, ended up sharing a house with Sean Parker who, post-Napster, had fallen on hard times. Then Parker discovered Facebook and Callahan got in on the ground floor. Now he owns a very chic hotel in Palm Springs.

Doug Carlson created Brøderbund Software in 1980 to publish a computer game he wrote for the TRS-80, *Galactic Empire*. The company went on to publish *Myst*, *Prince of Persia*, and *Where in the World Is Carmen Sandiego?*

Ed Catmull and Alvy Ray Smith spun their computer graphics group out of Lucasfilm in 1986 to found Pixar—which was a computer company, at first. Animated shorts were made for marketing purposes—at first. But a decade after Catmull and Smith left Lucasfilm, Pixar made *Toy Story*. And a decade after *Toy Story*, Pixar sold itself to Disney in an enormously lucrative deal.

David Cheriton is a computer science professor at Stanford, and he was the person that Google founders Larry Page and Sergey Brin turned to for advice. Cheriton told them not to try to license the search algorithm that they had developed, and to start a company instead. At first the Google guys didn't listen, but in the end they came around.

David Choe was the graffiti artist hired to paint the mural on the walls of Facebook's first real office. He slept in the office while painting it, and ended up coughing blood and spending several weeks in the hospital with severe pneumonitis. In the long run it may have been worth it, however, as Choe chose to be paid in Facebook stock that's now worth hundreds of millions of dollars.

Jim Clark grew up in West Texas, dropped out of high school, and joined the Navy. He went from being a naïve kid from a poor part of the country to a PhD in advanced computer graphics. Then he invented a special-purpose silicon graphics chip, which he parlayed into a computer company, Silicon Graphics. That got Clark the money he needed to start Netscape, the company most responsible for making the internet what it is today.

Lee Clow is a celebrated art director in the advertising industry. He's perhaps best known for the commercial announcing the launch of the Macintosh. It aired during the Super Bowl and created an instant sensation by reframing what might have been seen as a mere corporate rivalry between Apple and IBM as a struggle between good and evil, fascism and freedom.

Blaine Cook was Twitter's long-haired lead engineer in its earliest days—an exemplar of the "hippie-hacker" type that the company liked to hire.

John Couch was one of the first high-level employees that Apple hired. He was present at the famous demo that Xerox PARC gave to Apple, and he implemented some of what he saw there in the Lisa computer. After the Lisa failed to sell, Couch left Apple, only to return years later as a marketing executive.

David Crane is one of the four programmer-founders of Activision, the Atari spin-off that made cartridges for the Atari's home game console, the VCS. His Activision game *Pitfall* was one of the best-selling original VCS games of all time.

Amy Critchett was one of the first *Wired* interns—a gofer and a go-getter turned producer of everything from (some of the very first) webcasts to (now) giant pieces of public art.

Captain Crunch, aka John Draper, was a hero to the counterculturists and computer nerds who made up Silicon Valley in the seventies. Crunch was a phone phreak, a hacker that specialized in breaking into AT&T's phone network. His bravado inspired a young Steve Wozniak and Steve Jobs to go into business together selling blue boxes—illegal gizmos that could turn anyone into a phreak.

Eddy Cue is the Apple executive responsible for everything the internet touches at Apple: including iTunes, the App Store, Siri, and streaming content of all kinds.

Andy Cunningham, an employee of Apple's public relations guru Regis McKenna, worked closely with Jobs to launch the Macintosh.

Ted Dabney is the oft-forgotten other founder of Atari. He quit the company early and moved away from Silicon Valley to a small town in the Sierra foothills, and he claims to have no interest in playing video games.

Adam D'Angelo and Mark Zuckerberg were roommates in boarding school, drawn together by a shared interest in computer programming. Together they created an artificially intelligent MP3 player called Synapse that could predict which song you wanted to play next. They could have turned it into a business but decided that they'd rather go to college.

Ram Dass was Richard Alpert before an Indian holy man, Neem Karoli Baba, renamed him in a remote ashram in 1967. Ram Dass's book about his time with the guru, *Be Here Now*, is one of the classics of the era. It inspired a generation of hippies, Steve Jobs among them.

Erik Davis is a San Francisco–based writer and intellectual who specializes in California's counterculture and its modern legacy: His 1998 book *Tech-Gnosis: Myth, Magic & Mysticism in the Age of Information* is still studied and discussed in graduate schools around the country.

Fred Davis started his writing career under Stewart Brand at the *Whole Earth Software Catalog* and rose to be the editor in chief of a number of important trade magazines, including *A+* magazine, *MacUser*, *PC Magazine*, and *PC Week*. Today he works as a professional mentor to a new generation of entrepreneurs.

Michael Dhuey is a hardware engineer who spent thirty-five years working at Apple. On his first day on the job he had lunch with his heroes, Steve Wozniak and Andy Hertzfeld—and then stayed at the company until his retirement day.

John Doerr chairs one of the oldest and most storied venture capital firms in Silicon Valley, Kleiner Perkins Caufield & Byers. He got there by being one of the most successful young VCs in the Valley during the eighties and nineties, famously finding and funding Netscape, Amazon, and Google.

Jack Dorsey was the first CEO of Twitter—and perhaps the only chief executive in Silicon Valley to have sported *both* a nose ring and dreadlocks.

Douglas Edwards arrived at Google in 1999 as one of the company's first marketing hires. After leaving in 2005, he wrote an insightful memoir about his time there: *I'm Feeling Lucky: The Confessions of Google Employee Number 59.*

Larry Ellison was for many years the richest man in Silicon Valley. The fortune was made from a very dull but extremely important company called Oracle. Ellison was a very close friend to Steve Jobs and is a keen sailor—he is a two-time winner of the America's Cup.

Doug Engelbart was the first to pull together all the elements that are familiar to us today in a standard-issue computer. His oN-Line System, or NLS, had a screen, a keyboard, and a mouse. It was a revelation to all who saw it, and he sparked a computing revolution. Yet Engelbart died a brokenhearted man, convinced that he had failed himself and the world at large.

Bill English was among the first to join Doug Engelbart's team at the Stanford Research Institute. He hand-built the first mouse and stage-managed Engelbart's famous 1968 demo. In 1970 he jumped ship to Xerox's Palo Alto Research Center, where he developed a new version of Engelbart's system.

Tony Fadell is another alumnus of General Magic, the before-its-time smartphone company. After Magic failed Fadell turned to music and came up with the essential idea of the iPod, which he then took to Apple. After the iPod came the Nest—the smart thermostat company—which he sold to Google. Today he lives in Paris.

Shawn Fanning is the creator of Napster, a search engine optimized for finding and then downloading music files. Fanning created it in his dorm room as a self-taught teenage programmer. After Napster caught on, it created a firestorm of controversy over the ethics of downloading copyrighted music.

Lee Felsenstein ran the Homebrew Computer Club, the place that inspired Steve Wozniak to build his own computer and where Apple got its start. The club spawned dozens of computer companies, and Felsenstein designed the hardware for two of the most important ones: Processor Technology and the Osborne Computer Company.

Artie Fischell is a fictional character who "worked" at Atari. He had an e-mail account, an office, even a secretary. Most who heard from Mr. Fischell thought he was a real person. His quote on page 96? Not real.

Scott Fisher was an early researcher in virtual reality, first for Atari Research and then for NASA. He was the first to develop the now-familiar goggles-and-glove interface. The idea was to make space walks safer by letting astronauts remotely control a space-walking robot.

Bob Flegal was at Xerox PARC for almost twenty-five years. He worked for both Bob Taylor in the Computer Science Laboratory as well as for Alan Kay in the Learning Research Group.

Fabrice Florin was a reviewer for the *Whole Earth Catalog* and a television producer. The documentary that he made on the first Hackers Conference led to a career in technology. Florin was one of the very earliest pioneers of what was then called multimedia—video, sound, and images of all kinds served up by a computer.

Scott Forstall was one of Apple's software wizards in the later years, after Steve Jobs returned to the company as CEO. And when the iPhone started to be developed, Forstall was put in charge of a team tasked with writing its operating system.

Mark Fraser was the first customer eBay ever had. In 1995 Fraser bought a broken laser pointer that eBay founder Pierre Omidyar was trying to sell as a test of the auction site he had created. Fraser got the laser for cheap, figuring that he could fix it. He never did, but the unlikely sale convinced Omidyar that eBay was a good idea.

Nitin Ganatra spent almost twenty years at Apple and ended up as the key engineer working on Purple—the secret software project that would eventually power the iPhone.

Jerry Garcia was the Grateful Dead's lead guitarist and vocalist. The Deadhead-technology connections go deeper than one might think.

However, the quote that Garcia is credited with on page 220 of this book—"Netscape opened at what?!"—is apocryphal. It's the punch line to a joke.

Katie Geminder was Facebook's first project manager and the person responsible for turning Zuck's thoughts about the ever-evolving Facebook into detailed plans. Zuckerberg would pour his ideas, complete with diagrams, into a plain black notebook late at night and then hand it off to Geminder each morning.

John Giannandrea is a technologist's technologist: as low-key as it gets, but with a career that has always put him right at the center of the action. He was at Silicon Graphics when SGI was the most respected company in Silicon Valley, at General Magic when Magic was the company of the moment, and at Netscape during its IPO. He ran Google's search division—the entire thing. Today he's the human behind Apple's Siri.

Noah Glass was the founder of Odeo, an early podcasting company that failed—but from its ashes arose Twitter.

Adele Goldberg was a close collaborator with Alan Kay at Xerox PARC; after Kay left for Atari, she succeeded him as the manager of the Learning Research Group.

Wayne Goodrich is an event producer—and the man behind the curtain at most of Steve Jobs's highly choreographed product launches at Apple, NeXT, and even Pixar.

Jim Griffith, aka Uncle Griff, was a seller so active—and so entertaining—on the early eBay community boards that eBay ended up turning him into a company spokesman and mascot.

Andy Grignon is a straight-talking engineer whom Apple acquired via its purchase of Pixo, a company that had the tech Apple needed to build the iPod. After the iPod launched, Grignon was reassigned to the iPhone. Grignon's specialty? The phone part of the iPhone—the transceiver.

Ralph Guggenheim was Pixar's first head of animation and the producer of Pixar's breakthrough film: *Toy Story*.

Justin Hall was perhaps the first of the web's many microcelebrities: an exhibitionist college student who started an online diary that went viral on the nascent web. The notoriety earned him a job at *HotWired* and a starring role in *Home Page*, a well-regarded documentary about the birth of blogging.

Brad Handler is a Silicon Valley programmer turned intellectual property lawyer. He joined eBay in 1997, just as the company was taking off, and had a part in shaping the Digital Millennium Copyright Act, which established the rules of the road on how copyright was to be enforced online.

Young Harvill is an artist, an inventor, and a programmer. He wrote the graphics code that made Jaron Lanier's virtual reality come alive.

Scott Hassan is known to Valley insiders as the "third founder" of Google. He wrote much of the initial code and convinced Larry Page and Sergey Brin to turn their research project into a search engine.

Trip Hawkins was an Atari-obsessed Stanford MBA in the late seventies who landed a marketing position at Apple: The job was to figure out how to sell the Apple IIs to business users. In 1982 Hawkins returned to his first love—gaming—by starting Electronic Arts, one of the first computer game publishing companies.

Jim Heller was the operations person at Atari responsible for disposing of returned and defective game cartridges. There were so many returned after the 1982 Christmas season that he had to bury several million in a landfill in Texas. That event, the so-called Great Video Game Burial, marked the beginning of the end for Atari.

Pete Helme is a programmer who went from working at Apple to General Magic—the Apple spin-out that famously flamed out while trying to build a smartphone—to eBay.

Andy Hertzfeld is one of the software wizards behind the original Macintosh and that project's unofficial historian. After the Mac, he went on to cofound a number of important companies, including General Magic. There he mentored a new generation of entrepreneurial engineers at Magic, the ones that run the Valley today.

Joanna Hoffman was with the Macintosh team from the very beginning. She codified the Mac's user interface and was the marketing whiz behind the original Macintosh launch.

Stan Honey was Nolan Bushnell's navigator on a sailing race from Los Angeles to Hawaii. Mid-Pacific the two dreamed up a business that would prevent people in cars from getting lost. The result, Etak, was the first real in-car navigation system—long before GPS made such things commonplace.

Bruce Horn was only a teenager when he got a job at Xerox PARC, Silicon Valley's legendary research and development institution. Later he jumped ship to Apple in order to work on the Macintosh. There he, and others, invented a number of things that we now take for granted, including the "drag and drop" mouse gesture. Today Horn is a research fellow at Intel.

Dean Hovey, of Hovey-Kelley Design, was who Steve Jobs turned to when he needed someone to design a mouse. Jobs needed something that was far easier to use and manufacture than the mouse that he saw at Xerox PARC.

Dan Ingalls worked at Alan Kay's lab at Xerox PARC. He was the principal programmer behind Smalltalk, Kay's pioneering object-oriented computer language, which ran on the Alto computer. When Steve Jobs visited PARC to see the Alto, Ingalls famously gave the demonstration. Later, Ingalls worked for Jobs at Apple.

Steve Jarrett was a project manager for General Magic, the smartphone company that spun out of Apple in the midnineties. Later he joined Apple to help launch the original iPod. Today he's an executive at Facebook, living in London.

Steve Jobs and his friend Steve Wozniak cofounded Apple, the company most responsible for bringing the personal computer to the masses. Jobs didn't manage to gain full control of his company until twenty years later, but after he did Apple released a string of hit products—the iMac, the iPod, the iPhone, the iPad—and it became the most valuable company in the world. Jobs died of cancer in 2011, at the age of fifty-five.

Ron Johnson was the vice president of merchandising at Target, before coming to Apple in order to develop the Apple Store. Those turned out to be the most successful retail operations, in terms of sales per square foot, ever created.

Steven Johnson started the first significant digital-only online magazine, *Feed*. He's now a writer of books, having written nine of them at last count, and has emerged as one of the most trenchant and articulate interpreters of the larger meaning of the technology created in Silicon Valley and elsewhere.

Coco Jones was a very early employee of *Wired* magazine—its first national ad sales manager.

Yukari Kane is one of the most effective reporters to have ever covered Apple. For her book *Haunted Empire*, she managed to penetrate not only Apple, but also Apple's notoriously guarded subcontractor in China,

Foxconn, in order to interview the workers who actually make Apple's iPhones and iPads.

Jerry Kaplan cofounded a famous failure, the Go corporation. His book about the debacle, *Startup: A Silicon Valley Adventure*, is a classic and is still in print. Then he founded OnSale, an e-commerce auction site that predates even eBay. He is still founding companies and writing books. The latest is *Artificial Intelligence: What Everyone Needs to Know*.

Larry Kaplan, author of the games *Street Racer*, *Air-Sea Battle*, and *Super Breakout*, is one of the four VCS programmers who left Atari to found its first real competitor, Activision.

Ray Kassar succeeded Nolan Bushnell as CEO of Atari. Kassar's buttoned-up, top-down, East Coast management style could not have been more different from Bushnell's. But ultimately it was probably Kassar's failure to understand the Silicon Valley ethic of constant innovation that caused Atari to fail so spectacularly.

Alan Kay conceived and championed the creation of something he called a Dynabook: a computer so small that it can travel with you, and so easy to use that even a child can program applications for it. It's been the holy grail of computer scientists for almost fifty years.

David Kelley was one half of Hovey-Kelley Design, a product design start-up out of Stanford University, which has since evolved into the industrial design powerhouse IDEO.

Kevin Kelly is curiously old-fashioned for an editor and writer who has made a career out of roaming over the horizon and hunting down the future. His track record is nearly perfect, and his latest book, *The Inevitable*, is his best yet.

Max Kelly first came to San Francisco as part of an experimental band called Sound Traffic Control. After the 9/11 terrorist attack, he joined the FBI and worked in their Computer Forensics Lab. That led to a job with Facebook in 2005 as its first chief security officer. After five years he returned to government service—this time with the National Security Agency.

Ken Kesey, the novelist, lived on Palo Alto's bohemian Perry Lane during the fifties and early sixties. While the engineers around him were creating what would become Silicon Valley, Kesey and his sidekick, Stewart Brand, were organizing the Acid Tests—public LSD-taking parties that, more than anything else, popularized the then-obscure drug.

Howard King is the lawyer for the band Metallica. He represented the band in *Metallica v. Napster*, which was settled in Metallica's favor—making the band one of the very few to actually make money from file sharing.

Dan Kottke traveled to India with Steve Jobs in 1974. They were seeking a guru whom they learned about by reading a book called *Be Here Now*. After they returned, Kottke was among Jobs's first employees. But the two had a falling out after Jobs refused to acknowledge his out-of-wedlock daughter, Lisa. Jobs eventually reconciled with Lisa, but never with Kottke.

Sanjeev Kumar and a few other engineers, anticipating the coming digital music era, formed a company in 1999 to design the custom chips that would be needed for a Walkman-like MP3 player. Apple became their biggest client when it decided to roll out the iPod in 2001.

David Kushner is a writer specializing in the history of gaming culture. His books *Masters of Doom, Jacked: The Outlaw Story of Grand Theft Auto*, and, most recently, *Rise of the Dungeon Master: Gary Gygax and the Creation of D&D* define the genre.

Butler Lampson started designing computers while still a student at UC Berkeley in the sixties. Later, at Xerox PARC, he was a key force behind the Alto, Ethernet, and the laser printer.

Jaron Lanier was a Silicon Valley game programmer who got a modest windfall from what is considered the first true art game: *Moondust*. The money allowed him and his friends to pursue research and development into what Lanier called virtual reality. It was a term he invented.

John Lasseter was fired from the Walt Disney Company for promoting the idea of computer animation, and found refuge in the computer graphics group at Lucasfilm in Marin. That group eventually became Pixar, and Lasseter—who started out making promotional shorts for the Pixar computer—ended up making *Toy Story*. Today Lasseter is the chief creative officer for both Pixar and Disney animation.

Brenda Laurel was hired by Atari to prognosticate the future of the company. After Atari collapsed, she wrote an influential book, *Computers as Theatre*, outlining her theories.

David Levitt came to Silicon Valley from MIT in order to help set up Atari Research, and then stayed to help develop virtual reality with Jaron Lanier at VPL. At the time he was known as "the other guy with dreadlocks." He's still working on VR technology.

Jim Levy was the first CEO of Activision, a video game company made of ex–Atari programmers. Activision was the first outside games developer for Atari's home video game console, but not the last. A game glut ultimately felled Atari in 1983—a shock to the economic system that took most of Silicon Valley down with it.

Steven Levy's book *Hackers* came out in 1985: It was panned by the *New York Times* and ignored by radio and TV—but it has been in print ever since. It was the first and is still the best exploration of exactly what it means to be a computer geek.

Amy Lindburg was a chip designer and hardware engineer at Apple before she joined General Magic, a famous failure of a company that nonetheless laid the foundation for today's dominant technology: the smartphone.

Kate Losse started at Facebook in 2005 as a customer service rep—the lowest rung on the corporate totem pole. She eventually worked her way into the CEO's office as a ghostwriter before becoming disillusioned and writing a tell-all: *The Boy Kings: A Journey into the Heart of the Social Network*.

George Lucas used part of the windfall from his successful Star Wars franchise to start his own research and development group. He brought in the leading lights in computer graphics—Ed Catmull and, later, Alvy Ray Smith and others—and asked them to build him better film and sound-editing tools. Animation was an afterthought.

Chris MacAskill is another General Magic alumnus. After Magic failed, MacAskill founded and ran a couple of well-regarded dot-com companies.

Jamis MacNiven is the quintessential Silicon Valley bon vivant. As the owner of Buck's Restaurant in Woodside, California—a favored haunt of the Valley's deal-making class—MacNiven has had a ringside seat on the Valley's history: He was a fly on the wall when Jim Clark and his team dreamed up Netscape, when PayPal was funded, and when Tesla was founded. Before Buck's, MacNiven was buddies with Steve Jobs—so close that when John Lennon was shot dead, an inconsolable Jobs cried in MacNiven's arms.

Michael Malone was the world's first daily high-tech journalist. He was hired to cover Silicon Valley for the *San Jose Mercury News* in 1979—only a few years after the phrase "Silicon Valley" was coined. He is the go-to expert on that now-gone era when Silicon Valley's wealth was based on silicon chips, and his book *The Intel Trinity* is the definitive account of Silicon Valley's foundation.

Mike Markkula was the first major investor in Apple and, with reluctance, provided the "adult supervision" that young Woz and Jobs needed at the beginning. After Jobs had returned as CEO to the company that he founded in 1996, Markkula retired.

John Markoff was the *New York Times*'s Silicon Valley expert for almost thirty years. His book *What the Dormouse Said* is the definitive look at the overlap between the counterculture and Silicon Valley's engineering culture. Markoff is currently working on a biography of Stewart Brand, the man who most personifies that connection.

Scott Marlette was the programmer at Facebook most responsible for implementing Photos. The ability to upload digital photos—and then tag them—was a tipping point. With Photos, Facebook started to go viral.

Marissa Mayer graduated from Stanford in 1999 with a master's degree in computer science and, almost on a whim, accepted a Google internship and then a job as Google's first female engineer. She was employee number twenty—and eventually left Google to become the CEO of Yahoo. She stepped down in 2017, after arranging that company's sale.

Steve Mayer was part of the small team that built the prototype for Atari's at-home entertainment console: the VCS. It attached to a standard color television set and had a slot in the top that accepted plug-in game cartridges.

Ray McClure was trained as an artist but took a job at Odeo—the company that ended up creating Twitter—back when it was just a few artsy, hacker-y people working out of an apartment in San Francisco's hip Mission district.

Bob Metcalfe coinvented Ethernet—the interoffice networking protocol—at Xerox PARC and then founded 3Com to commercialize the standard. He is one of a select few in Silicon Valley to have his own "law." Metcalfe's law states that the value of a network is proportional to the square of the number of its users.

Jane Metcalfe is, with Louis Rossetto, the cofounder of *Wired* magazine. She was the publisher: the business mind that took the magazine from an idea to an institution. Today she has a new publishing project that aims to do for biology what *Wired* did for computers: *Neo.Life*, an online magazine covering the "neo-biological revolution."

Michael Mikel, aka Danger Ranger, has been working in Silicon Valley since the seventies—first for Fairchild Semiconductor, the company that first commercialized the silicon chips for which the Valley is named. In the late eighties Mikel immersed himself in San Francisco's posthippie, postpunk cyberculture: Burning Man got going because of the efforts of Mikel and a few others back in the late eighties and early nineties.

Al Miller was one of the first game programmers whom Atari hired. He wrote *Surround, Hangman,* and *Basketball*—and then jumped ship to help found Activision, an Atari competitor.

Ron Milner is responsible for placing the earliest known Easter egg in an arcade game. It's hidden in the Atari's 1977 release, *Starship 1.* Find it and the screen will flash "HI RON" and give you ten free games.

Moby, aka Richard Melville Hall, is an American musician who had his greatest commercial success—the double-platinum album *Play*—just as Napster was also peaking. To Moby, Napster just seemed like the future of music: the next thing after radio.

Lou Montulli was a student at the University of Kansas when he first saw NCSA Mosaic, which was the first web browser worthy of the name. He immediately started submitting bug reports and soon after joined Netscape in Silicon Valley. After eighteen months of all-nighters Montulli was a deca-millionaire several times over.

Gordon Moore, who in 1965 ran R&D at Fairchild Semiconductor, was bold enough to predict that the number of transistors on a chip would double every year for the next ten years. It did, and then he adjusted what is now known as Moore's law to predict a doubling once every eighteen months. Astonishingly, that prediction has held true ever since.

Eugene Mosier was a computer-literate artist and enthusiastic participant in San Francisco's early-nineties underground who became *Wired* magazine's first production director.

Dustin Moskovitz was one of the original dorm room crew that cofounded Facebook at Harvard in February 2004. With the move to Silicon Valley the following summer, Moskovitz emerged as CEO Mark Zuckerberg's right-hand man.

Mike Murray was the director of marketing for the Macintosh launch. It was an uphill battle, as IBM already had consolidated control of the personal computer market.

Michael Naimark is a Silicon Valley–based artist and inventor, with the distinction of having worked in most of the most interesting R&D labs and creative shops in the Valley: Atari Research, the Apple Multimedia Lab, Lucasfilm's interactive division, Interval Research, and Google's VR division.

Nicholas Negroponte is not a creature of Silicon Valley, but he had a big hand in making it. The Media Lab that he founded at MIT in 1985 has funneled generations of computer scientists into the Valley. And in 1992 Negroponte became the first investor in *Wired*, as well as the magazine's back-page columnist.

Ted Nelson coined the words *hypertext* and *dildonics* in his seminal 1975 work, *Computer Lib / Dream Machines*.

Pierre Omidyar was a low-level, low-key employee at General Magic with a geeky interest in libertarian philosophy. When he saw the just-invented world wide web, he realized that he could put his philosophy to the test, so he hacked up an online marketplace where ordinary people could trade ordinary things without regulation or middlemen. Astonishingly, it worked.

Larry Page is, along with Sergey Brin, the cofounder of Google. His other business idea was to build an elevator to space—an idea which has so far proved impracticable.

Sean Parker came to Silicon Valley as a teenager to build the dorm-room project of his friend Shawn Fanning into a real business. Napster didn't work out as planned, but when Parker discovered Mark Zuckerberg— another teen with a dorm-room business—he knew exactly what to do. As Facebook's first president, Parker showed Zuckerberg how to raise capital while still keeping control.

Mark Pauline is a Bay Area artist who builds giant, dangerous fire-breathing robots in order to stage gladiatorial spectacles, in which he pits the infernal machines against one another in an all-versus-all battle royal.

Bill Paxton is a computer scientist who did tours of duty at the Stanford Research Institute, at Xerox PARC, and finally at Adobe. He spends his retirement building software tools for astrophysicists.

Steve Perlman has seen it all. He was at Atari when it collapsed, at Apple when Steve Jobs was ousted, at General Magic at the very beginning. Then he started founding companies based on his inventions. The most notable was WebTV, and the latest, pCell, is built around a technology that, in theory, could provide a nearly infinite amount of wireless bandwidth.

Mark Pincus was one of the few entrepreneurs to stay the course through the dot-com crash, and in 2005 he made a very early investment in Facebook.

A few years later he bet big on Facebook again by starting Zynga, the company that brought computer games—most notably *FarmVille*—to the social network.

John Plunkett was half of the husband-and-wife team that created *Wired* magazine's distinctive look: riotous neon colors splashed across a cacophonous layout. It was meant to evoke the coming digital century—and it did.

Marc Porat coined the term "information economy" in 1976 while a graduate student at Stanford. He was subsequently recruited by Apple's Advanced Technology Group to think about the future. What he saw coming was the iPhone, and in 1990 he spun out a company, General Magic, to try to build out that future. It almost worked.

Tom Porter was a programmer and graphics researcher at Pixar who made a key early breakthrough—he discovered how to fake motion blur. In optical cameras motion blur is easy: It's a side effect of a slow shutter speed. In digital images, it requires a clever algorithm.

Rabble, aka Evan Henshaw-Plath, lived the life of an itinerant activist before moving to San Francisco to work for Odeo, which eventually morphed into Twitter. He helped set up activist media centers at the 1999 Seattle WTO protests. And in 2004 he helped build a mobile phone–based social network called TXTMob to organize protests at both the Democratic and Republican National Conventions. He continues to be active in anarchist politics.

Jef Raskin is the self-proclaimed creator of the Macintosh. It's only technically true. At Apple Raskin ran something called the Macintosh project—until Steve Jobs kicked him out. Raskin's minimalist computer designs were then rejected in favor of a computer powerful enough to drive a modern graphics display. Raskin died in 2005.

Howard Rheingold is the author of a short shelf of books about the history and culture of Silicon Valley, including 1985's *Tools for Thought*, a look at the birth of the personal computer, and 1993's *The Virtual Community*, an early assessment of social media.

Eileen Richardson was the first CEO of Napster. She was the so-called adult supervision that the company's funders—Richardson among them—thought the company needed. After Napster failed so spectacularly, Silicon Valley's venture capitalists started to sour on the whole idea of replacing a young founder with a supposedly seasoned adult.

Jordan Ritter worked on Napster from the very beginning—a founder in all but name. At one point, when the company was being formed, he even held the source code hostage until he could make sure he would be treated fairly. It was little use, as ultimately Napster itself would fall victim to an entrenched industry that would stop at nothing to destroy it.

Arthur Rock coined the term "venture capital." He helped fund the first semiconductor companies: Fairchild and then Intel. He also made a very early bet on Apple and served on its board of directors for a time.

Matt Rogers landed an internship at Apple right out of school, and by virtue of his hard work on the iPod team quickly became the protégé of Tony Fadell—the "podfather." The two eventually left Apple and started a company together.

Hilary Rosen was a longtime employee of the Recording Industry Association of America, rising to become its CEO in 1998. The RIAA, with Rosen at its helm, fought against Napster with a series of lawsuits that eventually bankrupted the company.

Ron Rosenbaum wrote "The Secrets of the Little Blue Box" for *Esquire* magazine in 1971. It was about a new kind of American outlaw, a "phone phreak" named Captain Crunch who took delight in hacking into what then was the biggest computer network in the world—AT&T's global phone system.

Louis Rossetto was the cofounder, CEO, and first editor in chief of *Wired* magazine, which quickly became a media empire and then a publicly traded company—almost. There were two botched IPOs, and then the company was sold off in pieces. *Wired*, the magazine, lives on as part of a large East Coast–based media conglomerate.

Jeff Rothschild was a Silicon Valley serial entrepreneur turned venture capitalist when he was asked by a colleague to help out this struggling little company called The Facebook for a couple weeks. That few weeks turned into a ten-year deployment. Rothschild was a fifty-year-old in a sea of twentysomethings and he worked around the clock—which, for Rothschild, was a rejuvenating return to his youthful start-up days.

Jon Rubinstein—or Ruby, as he is called—was one of Steve Jobs's most important deputies at Apple and the person who first realized that the time was ripe for a new kind of music player. The iPod, more than any other product, was responsible for transforming Apple from an also-ran into a colossus.

Adam Rugel was the director of business development at Odeo, and though he was fired when the company pivoted into Twitter, he continued working at the Twitter offices, on his own project, for several years.

Jeff Rulifson was the chief software architect of the NLS, Doug Engelbart's revolutionary computer system.

Steve Russell is the programmer most responsible for creating *Spacewar*, the first computer game worthy of the name. *Spacewar* was the game that inspired Nolan Bushnell to create Atari.

Jim Sachs at Hovey-Kelley Design was responsible for making the mouse's electronic innards cheap and reliable. At first he thought it couldn't be done.

Dom Sagolla was there—on a playground in San Francisco's South Park—when the original idea for Twitter took shape.

Ruchi Sanghvi was Facebook's first female engineer and the main programmer working on the News Feed. When the feature was first introduced in 2006, it was widely hated: One out of ten Facebook users signed a petition calling for the company to scrap the feature. But Ruchi persevered, and News Feed is fundamentally what Facebook is today.

Phil Schiller is Apple's marketing chief and has been since 1997, the year Steve Jobs returned to the company as CEO.

Eric Schmidt was Google's so-called adult supervision: a seasoned CEO, installed at the insistence of investors who believed that the company could not survive with its young cofounder, Larry Page, at the helm. Schmidt served for ten years, from 2001 to 2011, whereupon a presumably grown-up Page retook control.

Ridley Scott directed *Alien, Blade Runner, Thelma & Louise, The Martian,* and "1984"—the commercial that launched the Macintosh computer.

John Sculley was the Pepsi executive whom Jobs famously recruited to be the CEO of Apple in 1983 by asking him, "Do you really want to sell sugar water, or do you want to come with me and change the world?" Two years later Sculley ousted Jobs from the company that he had founded.

Tiffany Shlain threw the best annual party of the dot-com era: the Webbies. Today Shlain is a full-time experimental filmmaker, working to bring her craft into the twenty-first century with a concept she calls "cloud filmmaking."

Dick Shoup was a researcher at Xerox PARC. His one-of-a-kind graphics machine and the software he wrote for it, SuperPaint, could process color imagery and video. He received both an Emmy and an Academy Award for his pioneering work. As a sideline, he had a serious and lifelong interest in parapsychology. Shoup died in 2015.

Ray Sidney was one of the very first Google employees. Burned out after several years of all-nighters, Sidney left just before the Google IPO in 2004 and moved to Lake Tahoe.

Charles Simonyi defected from Hungary at age seventeen and made his way to Xerox PARC where he wrote the first modern word processor—a program that eventually became Microsoft Word. Simonyi spent more than twenty years at Microsoft and has vacationed, more than once, on the International Space Station as a space tourist.

Mona Simpson, a novelist, is the younger sister of Steve Jobs. She learned this as an adult, when Jobs—who had been given up for adoption as a child—found her after searching for his biological roots. They became close, and Simpson wrote a roman à clef about Jobs entitled *A Regular Guy*. The novel's first line: "He was a man too busy to flush toilets."

R. U. Sirius, aka Ken Goffman, edited an enormously influential magazine, *Mondo 2000*. It catalyzed something that was then called the "cyberculture"—a wild, half-imagined, half-real phantasmagoria of what the future was going to be like.

Aaron Sittig was Facebook's first graphic designer, and the guy who came up with the now-ubiquitous "like" button.

Jeff Skoll, an engineer turned Stanford MBA, wrote the business plan for eBay, which famously turned a profit every quarter of its young existence. After steering eBay through its early years, Skoll moved to Los Angeles and reinvented himself as a film producer. His socially conscious films, which include *An Inconvenient Truth* and *Lincoln*, have garnered more than fifty Academy Award nominations and eleven wins—including Best Picture, for *Spotlight*.

Mike Slade got to be good friends with Bill Gates working at Microsoft in the pre-Windows days. Then he jumped ship to NeXT and became exceptionally close friends with Steve Jobs. When Slade got married in 2005, both Gates and Jobs showed up for the wedding. It was super awkward, until Melinda Gates broke the ice by turning to Laurene Jobs and making small talk.

Alvy Ray Smith was teaching computer science at New York University when he broke his leg skiing in New Hampshire in 1973. He spent three months in a body cast, recuperating—reading books, reflecting on his life, and taking LSD. By the time he healed, he realized that he was basically working for the war machine, and that his life needed to move in a different direction. After escaping to California, he found a new calling as one of the first-ever computer animators, eventually cofounding Pixar with Ed Catmull.

Burrell Smith was to the Macintosh what Steve Wozniak was to the Apple II—the hardware genius who made the machine fast, cheap, and sexy. After the Macintosh launched, Smith founded a company, Radius, with Andy Hertzfeld and other Apple alumni. Radius made monitors and other hardware accessories for Mac products.

Megan Smith started her career at Apple but quickly moved over to General Magic, where she worked as an engineer on the smartphone technology that it was developing. She then became the CEO of PlanetOut, one of the first online media companies. After a stint at Google she joined the Obama administration to become the chief technology officer of the United States.

Mary Lou Song joined eBay in 1996, when it was still called AuctionWeb. Song was a young, idealistic Stanford graduate, and eBay was a young, idealistic company. Remarkably, that idealism proved to be the company's secret weapon, giving eBay's buyers and sellers a reason to stick with the company as it endured some remarkably acute growing pains.

Kristin Spence now goes by her given name, Blaed Spence. She started at Apple as an administrative assistant, moved to *Wired* just as it started, and now makes media full-time.

Carl Steadman, aka Webster, was the cofounder of *Suck*, a passion project devoted to showing his employers at *HotWired* how truly clueless they were about their audience and their industry-in-the-making. *Suck*, which Carl and a cabal of friends worked on in secret and at night, was arguably the first blog, and it struck a note—bratty, cynical, and deeply informed—that resonates on the web even today.

Bruce Sterling has never lived in Silicon Valley, preferring Texas, but he casts a long shadow. As a novelist and an editor he defined a new voice in science fiction: cyberpunk. It inspired—and was inspired by—real life in San Francisco's wild techno-underground during the eighties and nineties.

Michael Stern was the lawyer for General Magic. He's now a partner at Cooley, a prominent Silicon Valley law firm.

Jonathan Steuer was a nerd getting a PhD in virtual reality at Stanford and running a "cybercommune" in San Francisco when Louis Rossetto hired him to develop *Wired* magazine's online spin-off. It was a short-lived affair, as Steuer's communalist instincts collided with Rossetto's capitalist ones.

Biz Stone was an art school dropout living in his mom's basement when he discovered this weird new thing online called blogging. Stone's own blog consisted of joke-filled accounts of his totally imaginary existence designing and building a new Japanese superjet, among other things. It was the fake-it-until-you-make-it philosophy, and it succeeded. Within a year Stone was working at Google. Within five years he, along with a few others, had created Twitter.

Brad Stone is the one journalist who knows the most about Amazon, Airbnb, and Uber, because he wrote the book—two of them, in fact: *The Upstarts: How Uber, Airbnb, and the Killer Companies of the New Silicon Valley Are Changing the World* and *The Everything Store: Jeff Bezos and the Age of Amazon*. Stone's day job? Running the technology coverage for Bloomberg, natch.

Christopher Stringer was one of Apple's powerful crew of in-house industrial designers. He left the company in 2017 after twenty-five years of service.

Tony Stubblebine was the lead engineer of Odeo, a podcasting company. As it failed, its employees came up with the idea for Twitter.

Tom Suiter was the creative director whom Steve Jobs turned to in order to launch the Macintosh in 1984, and then again in 1998 for the iMac.

Bob Taylor found and funded the researchers who gave us the mouse, the internet, and the modern graphical computer interface. He died at his home in Woodside, California, in 2017.

Brad Templeton founded the first ever dot-com company. Today he is the chair for computing for Singularity University, a Silicon Valley institution that is preparing people for a coming posthuman future.

Larry Tesler invented the concept of "cut and paste" while working for Alan Kay at Xerox PARC. He was the first of many to defect to Apple, and he helped them commercialize what, at PARC, was languishing in the lab.

Chuck Thacker, a lifelong tinkerer and inventor, designed and then built the Alto for Xerox's Palo Alto–based R&D lab, PARC. The Alto is, arguably, the first real personal computer and, inarguably, the inspiration for Apple's Macintosh. In 2009 he was awarded computing's highest honor: the A. M. Turing Award—and is one of the few without a PhD to have ever received it. Thacker died in 2017.

Peter Thiel was an outspoken campus conservative at Stanford who went on to found PayPal at the height of the dot-com bubble. PayPal was quickly absorbed by eBay but the core team behind it, the so-called PayPal mafia, is still regarded with awe and even fear in the Valley. Thiel was a very early investor in Facebook, and just about the only Silicon Valley figure to voice support for Donald Trump during his presidential campaign.

Clive Thompson is a journalist specializing in science and technology. He is the author of *Smarter Than You Think: How Technology Is Changing Our Minds for the Better*. His next book will explore the distinctive thought processes of computer programmers.

Aleks Totić was on the student team that built Mosaic at the University of Illinois. Then a Silicon Valley start-up called Netscape recruited him.

Lars Ulrich is the cofounder of the American heavy metal band Metallica, for which he plays the drums. They were one of the biggest rock bands in the world when they discovered that an unfinished song they had been working on was being circulated on Napster. Ulrich's rage drove him to sue the company, and he became a symbol of the music industry that wanted to shut Napster down.

Don Valentine was one of Silicon Valley's original venture capitalists. He famously backed both Atari and Apple.

Andy van Dam literally wrote the book on computer graphics. *Computer Graphics: Principles and Practice* is so fundamental that it gets a cameo in Pixar's *Toy Story*, the first feature-length film created entirely with computer-generated animation.

Jim Warren was a regular at the Homebrew Computer Club and the editor of its de facto house organ, *Dr. Dobb's Journal of Tiny BASIC Calisthenics & Orthodontia*. Its motto? "Running light without the overbyte." A similar sensibility was at work when he founded the West Coast Computer Faire, the place where Woz and Jobs launched the Apple II.

Howard Warshaw was eccentric even by Atari's standards—but one of the best game programmers the company ever had. Two of his games for the VCS are classics: *Yar's Revenge* and *Raiders of the Lost Ark*. Yet, rightly or wrongly, he's most remembered for *E.T. the Extra-Terrestrial*, the game that destroyed the very company he worked for.

Maynard Webb is one of Silicon Valley's fixers. He saved eBay in 1999 when the then-CEO called and asked Webb to fix the site, which was constantly crashing under the strain of explosive growth. Today Webb is on the board of LinkedIn and Yahoo, and still fields calls from panicked CEOs.

Steve Westly was a professor at Stanford's Graduate School of Business when he noticed this thing called the internet and decided to give it a try. Westly's specialty became marketing metrics—tracking the numbers that e-commerce generates just by its nature. That and luck led him to a tiny start-up named eBay which, it turned out, had the best metrics of all.

Joss Whedon, a script doctor, was called after a disastrous work-in-progress screening of *Toy Story* for Disney, who was footing the bill. After two weeks of intense work at Pixar, he managed to turn the movie around.

Bob Whitehead wrote *Home Run*, *Football*, and *Video Chess* for the Atari 2600 before jumping ship to form Activision with his buddies.

Randy Wigginton was a teenaged habitué of the Homebrew Computer Club. There he apprenticed himself to Woz, the club's alpha nerd. On the way home they'd stop at a diner where Woz would tutor the young Wigginton on the finer points of programming. Later, Wigginton ended up being one of the first employees of Apple Computer.

Ev Williams, a farm boy from Nebraska, showed up in San Francisco in 1999 and taught himself to code. When the boom turned to bust, Williams stayed and eventually started a series of companies, the most notable being Twitter.

Terry Winograd is a professor of computer science at Stanford University. He was the thesis advisor to Larry Page, and he guided the very earliest incarnation of what we now know as Google.

Susan Wojcicki rented out a few spare bedrooms in her Menlo Park home to Google when they were just starting, and she quickly got sucked into the company. She is now the CEO of the company's YouTube subsidiary—the second most popular site in the world, after Google.

Gary Wolf was one of the first—and best—writers for *Wired* magazine and soon was tapped to clean up the mess that was *HotWired*, the print magazine's online spin-off. His memoir, *Wired: A Romance*, is a beautifully written document of that tumultuous era.

Robert Woodhead is the programmer behind Wizardry, a version of Dungeons and Dragons played on the Apple II computer.

Kristina Woolsey was the director of Atari Research after the first director, Alan Kay, left for Apple. Later she, too, went to Apple in order to run Apple's research project in multimedia.

Steve Wozniak, aka Woz, was the technical genius behind the Apple II, the everyman machine that launched the personal computer revolution in 1977. At the time the two Steves—Jobs and Woz—were close friends. But by the time Jobs passed away the two were so estranged that Woz skipped Jobs's memorial service.

Richard Saul Wurman founded the TED festival in 1984 after noticing a convergence in the fields of technology, entertainment, and design. The first festival had Steve Jobs demoing the Macintosh, Nicholas Negroponte talking about the future, and a 3-D graphics presentation from the proto-Pixar team at Lucasfilm. Wurman was onto something: A new culture was emerging.

Jim Yurchenco, who was helping reengineer the mouse for Apple at Hovey-Kelley Design, had a flash of inspiration when he remembered Atari's trackball design.

Jamie Zawinski was one of the young idealistic hackers who, early on, saw the potential in the Mosaic web browser. The bug reports and fixes that he submitted, for free, led to a job at Netscape. A few years later he walked away with a small fortune and a determination to leave the tech industry for good—for the nightclub industry. He's now the slightly older and wiser (but just as idealistic) owner of the DNA Lounge, a San Francisco institution.

Tom Zimmerman was Jaron Lanier's partner in VPL, the first virtual reality company. Zimmerman invented the dataglove—a glove that can sense finger positions—because he thought it would be useful for digitizing dance moves. But when Lanier saw it, he realized that the dataglove was to virtual reality what the mouse was to a screen.

Mark Zuckerberg is CEO of Facebook. Unlike virtually every other important young founder in Silicon Valley's history, he was never forced to step aside and turn his company over to someone older and, supposedly, wiser.

Acknowledgments

I'm deeply grateful to Chris Calhoun, my agent, and Sean Desmond, my editor. This book was Desmond's idea and it was Calhoun who suggested my name to him. I'm a first-time author, and the faith that those two put in me was what kept me going when the light at the end of the tunnel was remarkably dim.

A profound appreciation to all those who sat for interviews: The names are all listed in the Cast of Characters section, so I won't list them again here, but I would have never been able to pull this book together without their cooperation. Very important and extremely busy people throughout Silicon Valley not only made time for me but also spoke with remarkable candor—and it happened time and time again. That openness to new people tackling an ambitious project is, to my mind, Silicon Valley's core value, as well as the key to its success.

Thank you too to those who helped me with contact information, advice, and photo research—and who aren't quoted in this book. The full list could literally go into the hundreds, and I'm grateful to all, but there were some who showed a generosity of spirit that was humbling. In no particular order, they are: Alissa Bushnell, Clive Thompson, Steve Coast, John Markoff, Vanessa Grigoriadis, Christina Engelbart, Bruce Damer, the guys from the RetroGaming Roundup podcast (Mike Kennedy, Mike James, and Scott Schreiber), Leonard Herman, Zak Penn, Alex Soojung-Kim Pang, Ted Greenwald, John Ince, Eleanor McManus, John Tayman, Lynn Fox, Erin Trowbridge, Dan Parham, Dee Gardetti, Eric McDougall, Claudia Ceniceros, James Nestor, Joe Brown, Julian Dibbell, Hugo Lindgren, Brian Lam, Blaise Zerega, Michael Rubin, Katie Geminder, Nate Tyler, Van Burnham, Sebastien de Halleux, Richard Jenkins, Ben Mezrich, Nicholas Thompson, Gary Wolf, Steven Johnson, Meredith Arthur, Josh Abramson, Tania

Ketenjian, Michael Naimark, Maura Egan, Fred Vogelstein, Lindsey Spindle, Laurel Touby, Alexander Rose, James Chiang, Kara Swisher, Michael Caruso, Julian Guthrie, Brad Stone, and Yves Béhar. Thanks also to my transcribers, most especially David Marcus.

Every author relies most of all on the sacrifices and support of friends, family, and loved ones, and I'm no exception. My deepest and most profound love and gratitude goes to them—you know who you are.

Note on Sources

The written language is very different from the spoken word. And so I've taken the liberty of correcting slips of the tongue, dividing streams of consciousness into sentences, ordering sentences into paragraphs, and eliminating redundancies. The point is not to polish and make what was originally spoken read as if it was written, but rather to make the verbatim transcripts of what was actually said readable. That said, I've been careful to keep intact the rhythms of speech and quirks of language of everyone interviewed for this book so that what you hear in your mind's ear as you read is true in every sense of the word: true to life, true to the transcript, and true to each speaker's intended meaning.

The vast majority of the words found in this volume originate in interviews that I conducted especially for this book. Where that wasn't possible I tried, with some success, to unearth previously unpublished interviews and quote from them. And in a few cases I've resorted to quoting interviews that have been published before. A full list of these secondary sources can be found at ValleyOfGenius.com and below.

Silicon Valley, Explained

Steve Jobs died on October 5, 2011, before this book was conceived. His quotes in this chapter (and the ones that follow) come from an April 1995 interview conducted by Daniel Morrow for the Smithsonian Institution; a *Playboy* interview conducted by David Sheff in 1985; a 1994 interview for the Silicon Valley Historical Association; and a WGBH Boston interview from 1990. Gordon Moore's quote is taken from a YouTube video, "Gordon Moore About Moore's Law," posted in December 2014. Don Valentine's quote is from an interview on Sequoia Capital's website.

The Big Bang

Doug Engelbart died on July 2, 2013, before this book was conceived. His quotes have been gathered from a number of sources: "Engelbart and the Dawn of Interactive Computing," an event hosted by SRI in December 2008; "Engelbart's Unfinished Revolution," a symposium produced by Stanford University Libraries and the Institute for the Future in December 1998; a Smithsonian Institute interview by Jon Eklund, conducted in May 1994; and two interviews by Judy Adams and Henry Lowood of Stanford, the first conducted in December 1986 and the second in April 1987. Steve Jobs's quotes in this chapter and others are from *Steve Jobs: The Lost Interview* conducted by Robert Cringely, and the 1990 documentary *Memory & Imagination: New Pathways to the Library of Congress* by Michael Lawrence and Julian Krainen. Ken Kesey's "next thing after acid" quote can be found in John Markoff's excellent history, *What the Dormouse Said*.

Ready Player One

Ted Dabney's quotes are taken from the RetroGaming Roundup podcast #24, October 2010, conducted by Mike Kennedy, Mike James, and Scott Schreiber, as well as the Computer History Museum's 2012 oral history of Ted Dabney. Bob Metcalfe's quotes here and throughout the book are from the Computer History Museum's 2007 oral history. Steve Mayer's quote is from the ANTIC "Interview 65" podcast.

The Time Machine

Charles Simonyi's quotes in this chapter and others are from the Computer History Museum's 2008 oral history. Steve Russell's quote is from Dean Takahashi's January 2011 interview: "Steve Russell Talks About His Early Video Game *Spacewar!*" Larry Tesler's quotes in this chapter and others are from the Computer History Museum's 2013 oral history.

Breakout

Steve Wozniak's quotes in this chapter and others are drawn from an in-depth personal interview that he gave to me, as well as two archival sources: a December 2010 interview by Patrick Betdavid and a *Game Informer* interview from June 2013. Steve Jobs's "enlightenment" quote comes from Walter Isaacson's definitive book *Steve Jobs*. Mike Markkula's quotes are from the illuminating 2011 documentary *Something Ventured*. Arthur Rock's quote is from Stanford University's "Silicon Genesis" project, 2002.

Towel Designers

Ron Milner's quotes are from the ANTIC "Interview 74" podcast. Larry Kaplan's quotes are from an undated interview posted on Digital Press, the video game database. Ray Kassar's quotes here and elsewhere are from Tristan Donovan's April 2011 *Gamasutra* interview.

PARC Opens the Kimono

The quotes from Dean Hovey, Jim Sachs, and Jim Yurchenco come from Alex Soojung-Kim Pang's "Making the Macintosh" history project for the Stanford University Library. Bill Atkinson's quotes here and in other chapters are from Bill Moggridge's *Designing Interactions*, 2007.

Hello, I'm Macintosh

Steve Jobs's "shithead" quote can be found in Walter Isaacson's *Steve Jobs*. Jef Raskin's "king of France" quote is from Michael Moritz's *The Return to the Little Kingdom*. The quotes from Mike Murray, Joanna Hoffman, and Andy Cunningham are from a panel discussion held at the Computer History Museum in 2004: "The Macintosh Marketing Story: Fact and Fiction, 20 Years Later." Burrell Smith's quote is from the May 1985 *Whole Earth Review*. Lee Clow's quotes are from a video interview by Ann-Christine Diaz, published in 2012 by *Advertising Age*: "The Art of the Super Bowl Ad: Lee Clow on How Apple's '1984' Almost Didn't Happen—The Real Story on Why the Spot Only Aired Once." Ridley Scott's quotes are from an Apple promotional video that was distributed to Apple dealers in 1984. John Sculley's remark was made in 2011 at a celebration of the thirtieth anniversary of the IBM PC in Boca Raton and reported by the *South Florida Business Journal*.

What Information Wants

Ted Nelson's quote is in Michael Schrage's November 1984 *Washington Post* story on the first Hackers Conference. Bill Atkinson's quotes are from an August 2012 Berkeley Cybersalon event on the creation and legacy of Hypercard. Doug Carlson's, Robert Woodhead's, and Steve Wozniak's quotes from the Hackers Conference are as reported in the May 1985 *Whole Earth Review*.

The Whole Earth 'Lectronic Link

Ram Dass's quotes are from two Seva Foundation videos: "Ram Dass Talks About Neem Karoli Baba, Larry Brilliant, Service, and the Birth of Seva Foundation" and "An Evening with Ram Dass: Seva Foundation Benefit 1985."

Reality Check

Mitch Altman's quotes here and elsewhere are from the Hackertrips video blog post on YouTube: "An Interview with Mitch Altman."

From Insanely Great to Greatly Insane

John Sculley's quotes were made at the Engage 2015 conference in Prague, as reported by Jim Edwards for *Business Insider*. Nicholas Negroponte's quote is from a November 1995 interview in *Wired* magazine.

Toy Stories

Ed Catmull's quotes are from the Computer History Museum's 2013 oral history; a 2014 talk he gave at Stanford's eCorner lecture series; and a 1995 documentary, *The Making of Toy Story*. John Lasseter's quotes are taken from: an October 1996 interview by Charlie Rose; a May 2006 *Fortune* magazine profile, "Pixar's Magic Man"; a June 2011 story in *Entertainment Weekly*, "John Lasseter on Pixar's Early Days—and How *Toy Story* Couldn't Have Happened Without Tim Burton"; and a June 2009 interview by Aubrey Day in *Total Film* magazine. Joss Whedon's quote is from the book *Joss Whedon: Conversations*. George Lucas's quote is from the book *To Infinity and Beyond!: The Story of Pixar Animation Studios*.

Jerry Garcia's Last Words

Marc Andreessen's quotes are from Henry Blodget's 2009 video interview in *Business Insider* and Chris Anderson's April 2012 interview in *Wired*.

The Check Is in the Mail

Pierre Omidyar's quotes are from a 2000 interview given to the American Academy of Achievement.

The Shape of the Internet

Larry Page's and Sergey Brin's quotes are from a largely unpublished series of interviews they gave to John Ince while he was working on a story for *Upside* in January 2000. Additional quotes from Page come from an interview he gave to *Fortune* in May 2008, from the commencement speech he gave at the University of Michigan in 2009, a 2002 talk he gave at Stanford's eCorner lecture series, and from an interview with the *San Francisco Chronicle* in December 2000. Additional quotes from Brin come from an interview given to the American Academy of Achievement in 2000, a November 2000 *MIT Technology Review* interview, remarks at a December 2002 O'Reilly conference on digital infrastructure, and a September 2004 *Playboy* interview. Marissa Mayer's Burning Man quote is found in *The Google Story* by David A. Vise and Mark Malseed. Eric Schmidt's quote is from a 2014 interview given at the 92nd Street Y in New York City.

Free as in Beer

Most of Shawn Fanning's quotes, and a few of Sean Parker's, are from the 2013 documentary *Downloaded* and from an interview at the movie's premiere conducted by Reggie Ugwu of *Billboard* magazine. Additional Fanning quotes are from an *SFGate* podcast posted in May 2009. Hank Barry would like the reader to know that he is speaking as an individual and that his views do not necessarily represent the views of his firm, Sidley Austin LLP. Lars Ulrich's quotes are from a live chat hosted by Yahoo! in May 2000, and an interview on *HuffPost Live* with Mike Sacks in September 2013. David Boies's quotes are from an interview in *Wired* in October 2000.

The Dot Bomb

Peter Thiel's comments were made in the spring of 2012 in a series of lectures at Stanford and recorded in the notes of Blake Masters, who was in the audience.

The Return of the King

Larry Ellison's remarks were made on a special August 2013 episode of *CBS This Morning*, "An Hour with Larry Ellison," and in the commencement speech he gave at the University of Southern California in 2016.

I'm Feeling Lucky

Paul Buchheit's quotes are found in the book *Founders at Work* by Jessica Livingston. Susan Wojcicki's quotes can be found in Jefferson Graham's *USA Today* story, "The House That Helped Build Google," in July 2007.

I'm CEO...Bitch

Mark Zuckerberg's quotes come from a guest lecture he gave to Harvard's "Introduction to Computer Science" class in 2005, from an interview he gave to the *Harvard Crimson* in February that same year, and from Y Combinator's Startup School event in 2007. Dustin Moskovitz's quotes were taken from a keynote address given to the Alliance of Youth Movements Summit in December 2008, and from David Kirkpatrick's authoritative history, *The Facebook Effect*. David Choe's comments were made on *The Howard Stern Show* in March 2016. Steve Jobs made his remarks to his biographer, Walter Isaacson. The interview was aired on *60 Minutes* soon after Jobs died in 2011.

Purple People Eater

Phil Schiller, Scott Forstall, and Christopher Stringer testified under oath in the *Apple v. Samsung* case in August 2012. Steve Jobs's comments were made onstage at the All Things D conference in June 2010. Matt Rogers's quotes were made to Kevin Rose in episode 21 of "Foundation," Rose's series of video interviews with technology moguls. Eddy Cue's quote is borrowed from Brent Schlender and Rick Tetzeli's *Becoming Steve Jobs*, the Steve Jobs biography.

Twttr

Noah Glass's comment was made during a talk at the Berkeley School of Journalism in March 2005. Tony Stubblebine shared his memories on Quora. Mark Zuckerberg's "clown car" quote is as reported in Nick Bilton's terrific book *Hatching Twitter*.

The Endless Frontier

Larry Page's speculation about a brain implant can be found in *In the Plex*, Steven Levy's sweeping history of Google.

Photo Credits

1: SRI International. 2: Courtesy of PARC, a Xerox company. 3: Roger Ressmeyer/Corbis/VCG. 4: Courtesy of The Palo Alto Historical Association / Ken Yimm. 5: Tom Munnecke/Getty. 6: Courtesy of Lucasfilm Ltd. LLC. 7: Courtesy of Dave Staugas. 8: Courtesy of Douglas Mellor. 9, 10, 11: Courtesy of Kevin Kelly. 12: Courtesy of Mark Pauline and 6th Street Studios. 13: Courtesy of Louis Rossetto. 14: Bloomberg/Getty. 15: Scott Beale/Laughing Squid. 16: Courtesy of Ev Williams.

Index

About the Author

Adam Fisher lives a quiet life in Alameda, a small island near the center of the greater Bay Area. He has a beautiful wife and daughter and writes for many different magazines: *New York Sunday Magazine*, *MIT Technology Review*, and *Wired*, to name a few. This is his first book, and his greatest interest is in hearing from you, the reader, and staying in touch. So—switching to the first person now—please drop me a line at hello@valleyofgenius.com and I'll get back to you: I've got audio clips, rare photos, outtakes, and other neat Silicon Valley stuff to share. Or, if you prefer, see it all on this book's website: valleyofgenius.com.

Thank you for reading my book!

Mission Statement

Twelve strives to publish singular books, by authors who have unique perspectives and compelling authority. Books that explain our culture; that illuminate, inspire, provoke, and entertain. Our mission is to provide a consummate publishing experience for our authors, one truly devoted to thoughtful partnership and cutting-edge promotional sophistication that reaches as many readers as possible. For readers, we aim to spark that rare reading experience—one that opens doors, transports, and possibly changes their outlook on our ever-changing world.